Lecture Notes in Computer Science 3162

Commenced Publication in 1973
Founding and Former Series Editors:
Gerhard Goos, Juris Hartmanis, and Jan van Leeuwen

Rod Downey   Michael Fellows
Frank Dehne (Eds.)

Parameterized and Exact Computation

First International Workshop, IWPEC 2004
Bergen, Norway, September 14-17, 2004
Proceedings

 Springer

Volume Editors

Rod Downey
Victoria University, School of Mathematical and Computing Sciences
PO Box 600, Wellington, New Zealand
E-mail: Rod.Downey@mcs.vuw.ac.nz

Michael Fellows
The University of Newcastle, School of Electrical Engineering and Computer Science
Calaghan, NSW, Australia
E-mail: mfellows@cs.newcastle.edu.au

Frank Dehne
Griffith University, School of Computing and IT
Nathan, Brisbane, Qld 4111, Australia
E-mail: F.Dehne@griffith.edu.au

Library of Congress Control Number: 2004111137

CR Subject Classification (1998): F.2, F.1, E.1, G.2

ISSN 0302-9743
ISBN 3-540-23071-8 Springer Berlin Heidelberg New York

Springer is a part of Springer Science+Business Media

springeronline.com

© Springer-Verlag Berlin Heidelberg 2004
Printed in Germany

Typesetting: Camera-ready by author, data conversion by Boller Mediendesign
Printed on acid-free paper   SPIN: 11321361   06/3142   5 4 3 2 1 0

Preface

The central challenge of theoretical computer science is to deploy mathematics in ways that serve the creation of useful algorithms. In recent years there has been a growing interest in the two-dimensional framework of parameterized complexity, where, in addition to the overall input size, one also considers a parameter, with a focus on how these two dimensions interact in problem complexity.

This book presents the proceedings of the 1st International Workshop on Parameterized and Exact Computation (IWPEC 2004, http://www.iwpec.org), which took place in Bergen, Norway, on September 14–16, 2004. The workshop was organized as part of ALGO 2004. There were seven previous workshops on the theory and applications of parameterized complexity. The first was organized at the Institute for the Mathematical Sciences in Chennai, India, in September, 2000. The second was held at Dagstuhl Castle, Germany, in July, 2001. In December, 2002, a workshop on parameterized complexity was held in conjunction with the FST-TCS meeting in Kanpur, India. A second Dagstuhl workshop on parameterized complexity was held in July, 2003. Another workshop on the subject was held in Ottawa, Canada, in August, 2003, in conjunction with the WADS 2003 meeting. There have also been two Barbados workshops on applications of parameterized complexity.

In response to the IWPEC 2004 call for papers, 47 papers were submitted, and from these the program committee selected 25 for presentation at the workshop. In addition, invited lectures were accepted by the distinguished researchers Michael Langston and Gerhard Woeginger.

This first instantiation of a biennial workshop series on the theory and applications of parameterized complexity got its name in recognition of the overlap of the two research programs of *parameterized complexity* and *worst-case exponential complexity analysis*, which share the same formal framework, with an explicitly declared parameter of interest. There have been exciting synergies between these two programs, and this first workshop in the IWPEC series attempts to bring these research communities together.

The second workshop in this series is tentatively scheduled for the Gold Coast of Queensland, Australia, in July, 2006. An exact computation implementation challenge is being organized as a part of this second workshop. Details of the competition will be posted at http://www.iwpec.org.

On behalf of the program committee, we would like to express our appreciation to the invited speakers and to all authors who submitted papers. We also thank the external referees who helped with the process. We thank the program committee for excellent and thoughtful analysis of the submissions, and the organizers of ALGO 2004 in Bergen. We thank especially the tireless Frank Dehne for his efforts in almost all things relating to this conference and for co-editing these proceedings.

Rod Downey and Mike Fellows, July 2004

Organization

IWPEC Steering Committee

Jianer Chen
Frank Dehne
Rod Downey
Mike Fellows
Mike Langston
Rolf Niedermeier

IWPEC 2004 Program Committee

Rod Downey, co-chair
Michael Fellows, co-chair
Richard Beigel
Hans Bodlaender
Jianer Chen
Frank Dehne
Erik Demaine
Joerg Flum
Jens Gramm
Martin Grohe
Michael Hallett
Russell Impagliazzo
Michael Langston
Rolf Niedermeier
Mark Ragan
Venkatesh Raman
Peter Rossmanith
Jan Arne Telle
Dimitrios Thilikos
Gerhard Woeginger

Table of Contents

Parameterized Enumeration, Transversals, and Imperfect Phylogeny Reconstruction[*]

Peter Damaschke

School of Computer Science and Engineering
Chalmers University, 41296 Göteborg, Sweden
ptr@cs.chalmers.se

Abstract. We study parameterized enumeration problems where we are interested in all solutions of limited size, rather than just some minimum solution. In particular, we study the computation of the transversal hypergraph restricted to hyperedges with at most k elements. Then we apply the results and techniques to almost-perfect phylogeny reconstruction in computational biology. We also derive certain concise descriptions of all vertex covers of size at most k in a graph, within less than the trivial time bound.

1 Introduction

We suppose familiarity with the notion of fixed-parameter tractable (FPT) problems, otherwise we refer to [8]. In many combinatorial optimization problems, one wants a particular solution where the parameter k is minimized. In the present paper we deal with the generation of *all* solutions with objective values bounded by parameter k. As a concrete application we study the reconstruction of almost perfect phylogenies.

A *perfect phylogeny (PP)* is a tree with nodes labeled by bit vectors of length m, and edges with labels from $[m] = \{1, \ldots, m\}$ such that, for every $i \in [m]$, the vectors having 0 and 1, respectively, at position i are separated by exactly one edge labeled i (and hence form connected subtrees). This is a fundamental structure in computational biology, as it describes evolutionary trees where at most one mutation appeared at every position. Another application domain is linguistics [26]. Recently, PP attracted new attention as it supports haplotype inference.[1]

The bit vectors are usually represented as rows of an $n \times m$ matrix. The columns correspond to the positions, also called sites or loci. We speak of a *PP matrix* if there is a PP containing all its rows (and perhaps more bit vecors) as node labels. From a PP matrix one can uniquely reconstruct such a PP in $O(nm)$ time. (Here, uniqueness means: subject to isomorphism and to the

---

[*] This work has been supported by a grant from the Swedish Research Council (Vetenskapsrådet), file no. 621-2002-4574.

[1] It is quite impossible to cite all relevant papers here. The reader is referred to the proceedings of RECOMB 2002-2004, including satellite workshops.

R. Downey, M. Fellows, and F. Dehne (Eds.): IWPEC 2004, LNCS 3162, pp. 1–12, 2004.

ordering of edge labels on paths of degree-2 nodes.) Reconstruction can be done *incrementally*. Starting from an empty set of columns, add columns successively to the input and refine the PP. Details are not complicated, see e.g. [27, Section 14.1]. One can generalize the notion of PP to non-binary cases, complexity results are in [1,2,19,25].

However, the PP assumption is often too strict. Repeated mutations at some loci, or recent immigration into a PP population leads to deviations from PP. Sequencing errors are also common, hence corrupted data may lose the PP property even if the true data would form a PP. Thus one should allow a small number k of changes, i.e. bit flips in the matrix, or extra rows or columns, or combinations of them. This motivates a few computational problems:

PP PLUS k ROWS: Given a binary matrix, find all sets of at most k rows the deletion of which leaves a PP matrix.
PP PLUS k COLUMNS: Similarly defined.
PP WITH k ERRORS: Given a binary matrix, find all sets of k bit flips such that the resulting matrix has a PP.

Enumerating all solutions captures the applications better than the minimization problem. There is no reason to assume that the smallest number of changes is always the correct explanation of data. Rather we want an overview of all consistent solutions, for at most k changes, and we also wish to reconstruct the part of the PP (i.e. without some rows or columns) common to all these conceivable solutions, the maximum agreement structure so to speak. Another phylogeny reconstruction problem has been studied in [14] from this perspective, see also [11] for more discussion of the importance of enumeration.

More generally (but a bit vaguely perhaps) it can be said that parameterized enumeration is suitable when we want to recognize certain objects from given data which do not perfectly fit the expected structure. Then all potential solutions are required for further inspection. Applications besides phylogeny may be found e.g. in data mining.

We will use a well-known characterization of PP matrices. A pair of columns is called *complete* if each of 00, 01, 10, 11 appears as a row in the submatrix induced by these two sites. Throughout the paper we refer to 00, 01, 10, 11 as *combinations*. The following has been discovered several times, see e.g. [15,28].

Theorem 1. *A matrix is a PP matrix iff it does not contain complete pairs.* □

This connects almost-PP reconstruction to the more abstract class of *subset minimization problems*: Given a set of n elements, a property π of subsets, and some k, we want all minimal subsets of size at most k enjoying π. Note carefully that the term *minimal* refers to set inclusion, not cardinality! We say that π is closed under $\supset$ if every $Y \supset X$ has property π whenever X has. Examples are vertex covers in graphs and hitting sets of set families (hypergraphs). For such π it suffices to know the minimal solutions, as they "represent" all solutions. This motivates the following

Definition 1. *Given a subset minimization problem, a* full kernel *is a set whose size depends on k only and contains all minimal solutions of size at most k.*

We call a problem inclusion-minimally fixed parameter enumerable (IMFPE) *if, for any instance of size n, all minimal solutions with value at most k are computable in time $O(f(k)p(n))$ where p is polynomial and f any function.*

Once we have a full kernel then, trivially, we can also enumerate the minimal solutions in time depending on k only, hence a problem is IMFPE in this case. It is crucial to notice the seemingly little but important difference to the optimally/minimally fixed parameter enumerable (MFPE) problems in [11]. To avoid confusion with minimum size, we added the attribute "inclusion-".

The family of all minimal hitting sets to a given set family is known as the *transversal hypergraph*. Applications include models of boolean formulae, database design, diagnosis, and data mining. Known algorithms for generating the transversal include a pseudo-polynomial output-sensitive algorithm [12], algorithms for special hypergraph classes [3], and a practical heuristic based on simple but powerful ideas [20]. Here we are interested in the "pruned" transversal hypergraph consisting of the minimal hitting sets of size at most k. Apparently, generation of hitting sets by ascending size has not been addressed before, unlike e.g. lexicographic ordering [18].

Contributions and organization of the paper: In Section 2 we obtain IMPFE results for subset minimization probnlems. In Section 3 we apply these findings to almost-PP reconstruction. In Section 4 we give an algorithm that outputs a certain concise description of all small vertex covers of a graph within less than the trivial time bound. Due to limited space, we defer the detailed exposition of results to these sections, and we could only sketch most proofs and only convey the main ideas.

We believe that the notions of IMFPE and a full kernel are more significant than some technical results which are built very much upon known research. (In particular, our results in Section 2 are close to [11], however, the new aspect is that the bounds still hold in the more demanding IMFPE setting.) The IMPFE concept is strict and sets limits to what clever algorithms could achieve, but as argued above, it seems to reflect the goals in certain applications well. Our focus is on theoretical results. Some experiments regarding the real performance on data of reasonable size would complete the picture. A handful open problems arise from the text.

More related literature: Recently, almost-PP reconstruction has also been studied in [28] in a more general frame (destroying all occurences of a given small submatrix), however without time bounds for enumeration. Results in [10] are based on a different, distance-based imperfection measure. The viewpoint in [26] is more similar to ours, but the focus was different, and exhaustive search is used for PP with extra columns. Various computational biology problems allow FPT results, see e.g. [9,13,14]. Closely related to error correction in PP matrices is reconstruction of PP from incomplete matrices [25,16]). It might be interesting to look at this NP-hard from the FPT point of view. Papers [23,24] contain results on directed PP reconstruction with missing entries. We mentioned maximum agreement problems (e.g. [17] gives an overview). Usually they have as input an arbitrary set of structures, rather than slight variants of one structure. In [6] we proposed a simple PP haplotyping algorithm for instances with enough genotypes, and the ideas in the present paper may lead to extensions to almost-PP populations.

2   Hitting All Small Hitting Sets

The VERTEX COVER problem is FPT [5,21]: Given a graph $G = (V, E)$ with n vertices and m edges, and a number k, find a k-vertex cover, i.e. a set of *at most* k vertices that is incident to every edge. A full kernel for VERTEX COVER is any subset of V that entirely contains all minimal k-vertex covers in G.

Lemma 1. VERTEX COVER *has a full kernel of size* $(1 + o(1))k^2$. *It can be constructed in* $O(m)$ *time.*

Proof. We show that the kernel from [4] is also a full kernel: Every k-vertex cover in G must contain the set H of vertices of degree larger than k. If we remove the vertices of H, all incident edges, and all vertices that had neighbors in H only, the remaining subgraph R has at most k^2 edges (or there is no solution at all), and hence less than $2k^2$ vertices. Every minimal k-vertex cover is the union of H and some minimal vertex cover of R. Thus, $H \cup R$ is a full kernel. Factor 2 can be improved to $1 + o(1)$ by more careful counting. (Omitted due to lack of space.)   □

Remarks:

(1) For the optimization version of VERTEX COVER there exist kernels of size $2k$ [5], but $\Theta(k^2)$ is already the optimal worst-case bound for *full* kernels: In the disjoint union of m stars $K_{1,m}$ (one central vertex, joined to m leaves), the leaves of any star and the centers of all other stars build a k-vertex cover, $k = 2m - 1$. Hence the full kernel has size about $k^2/4$. The optimal constant in $\Theta(k^2)$ remains open.

(2) It was crucial to restrict the full kernel to *minimal* vertex covers. If we dropped the minmality condition, the size would not even be bounded by any function of k. A simple example is the star $K_{1,n-1}$ and $k = 2$: The center plus any leaf pair is a solution, and their union has size n. But the full kernel (merely the center) has size 1.

In order to enumerate all k-vertex covers we may construct the full kernel as and then apply the bounded search tree technique. Note that we distinguish *nodes* of the search tree from *vertices* of the graph.

Theorem 2. VERTEX COVER *is IMFPE. All minimal solutions of size at most* k *can be enumerated in* $O(m + k^2 2^k)$ *time.*

Proof. List all edges in the full kernel. Put a vertex from the first edge uv in the solution and branch for every choice (u or v). Label every new node by the vertex just selected. At any node proceed as follows: If some vertex in the edge listed next has already been chosen (i.e. it appears on the path from the root to the current node), then skip this edge. Repeat this step until the condition is false. Else, select a vertex from the next edge and branch.

Since this adds a new vertex to the solution on the considered tree path, but at most k vertices can be selected, the search tree has depth at most k, and at most 2^k leaves. Since every inner node has at least two children, the total size is $O(2^k)$. Finally we prune the tree, that is, successively remove all leaves where the edge list has not been scanned completely. From the search tree we can read off all k-vertex covers, as they are the label sets of paths from the root to the leaves. At every node we checked for every edge whether some of its vertices is already on the path. This gives immediately the time bound $O(k^2 2^k)$. Pruning costs $O(2^k)$ time.

One easiliy verifies that any minimal vertex cover X appears, in fact, as some path in the search tree.

Finally we also cut away leaves with non-minimal solutions X as follows. For every vertex in X, check whether all its neighbors are in X as well. Due to the degree bound in the kernel, this needs $O(k^2 2^k)$ time.   □

HITTING SET: Given a hypergraph G with n vertices and h hyperedges (subsets of vertices), and a number k, find a set of at most k vertices that hits every hyperedge.

In c-HITTING SET, the cardinality of hyperedges is bounded by c, hence $c = 2$ is VERTEX COVER. For recent results on $c \geq 3$ see [22]. Next we study the enumeration version of an even more general problem. By a *multiedge* we mean a family of at most c disjoint sets. We omit c if it is clear from context. The following problem statement needed in 3.1 is quite natural as such and may be of independent interest, however we are not aware of earlier mention of it.

BOUNDED UNION: Given h multiedges, i.e. families of at most c disjoint sets, each with at most d vertices, find a subset U of at most k vertices, that *entirely includes* at least one set from each multiedge. In other words, find a union of sets, one from each multiedge, with size bounded by k.

We say that U *settles* a multiedge $\{S_1, \ldots, S_c\}$ if $S_i \subseteq U$ for some i. Thus, a solution to BOUNDED UNION must settle all multiedges. Note that HITTING SET is the special case when $d = 1$. On the other hand, BOUNDED UNION is trivially reducible to HITTING SET: Replace every multiedge $\{S_1, \ldots, S_c\}$ with the collection of all $|S_1| \times \ldots \times |S_c|$ hyperedges $\{s_1, \ldots, s_c\}$ such that $s_i \in S_i$ for $i = 1, \ldots, c$. Now, a set U hits all these hyperedges iff U settles the multiedge. It follows that this reduction also preserves all solutions. However, it blows up the input size by factor $O(d^c)$. Thus, one better works directly on instances of BOUNDED UNION, without the detour via this reduction.

Theorem 3. BOUNDED UNION *is IMFPE. All minimal solutions can be found in* $O(dc^{k+1}h + \min\{kc^{2k}, hkc^k\})$ *time.*

Proof. Again, we construct a bounded search tree, but now on the whole instance. List the given multiedges. Select a set from the first multiedge and branch for every choice. At any node proceed as follows: If the multiedge listed next is already settled by the union of previously selected sets on the tree path, then skip it. Repeat this step until the condition is false. Else, select a set from the next multiedge and branch. Since this adds at least one new element to the union, the search tree has depth at most k, at most c^k leaves, and $O(c^k)$ nodes in total. From the search tree we can read off all unions: In any path from the root to a leaf, collect the sets specified by the path. Completeness of the solution space can be easily established. As for the time bound, note that on each path, every multiedge is processed only once in $O(cd)$ time.

A naive method for filtering the non-minimal solutions is pairwise comparison in $O(kc^{2k})$ time. Testing the minimality of every solution X is faster if $h < c^k$. Proceed as follows. For every multiedge e, list the vertices of X contained in e. If exactly one set S of e satisfies $S \subseteq X$, then the vertices in S are not redundant. Mark all non-redundant vertices found that way. First suppose that all multiedges are already settled by these marked vertices. In this case, X is non-minimal iff X contains further, unmarked vertices. This check needs $O(hk)$ time. The other case is that some multiedges are not yet settled by the marked vertices. But since X is a solution, we conclude two things: (1) Not all vertices in X are marked. (2) For every multiedge, either one set consists of marked vertices only, or at least two sets are completely in X. Hence, we can remove an unmarked vertex from X, and still some set of every multiedge is in X. This means, X is not minimal, and we do not need further tests. $\qquad\square$

We can show that a smaller *full* kernel exists in case $k > c$, thus generalizing a result from [22].

Theorem 4. *For any instance of* HITTING SET *or* BOUNDED UNION, *an equivalent instance with no more than* k^c *hyperedges can be obtained in time* $O(ck^{c-1}h)$. *Consequently, both problems have a full kernel of size* ck^c.

Proof. First we count how often every vertex appears in the hyperedges, in $O(cdh)$ time, going through the h hyperedges or multiedges. (For an instance of BOUNDED UNION, there is no need to perform the reduction to HITTING SET explicitly, as we know the cardinalities of sets in the multiedges.)

Suppose that each vertex appears in at most k^{c-1} hyperedges. Then, a set of size k can hit at most k^c hyperedges. If there is a solution at all, the instance contains only that many hyperedges, with a total of $k + (c-1)k^c$ vertices, and we are done. Otherwise we select a vertex and $k^{c-1} + 1$ hyperedges containing it.

Suppose by induction that we have found a set C of size i, and a family H_i of $k^{c-i} + 1$ hyperedges with C as subset. Either (1) some $C \cup \{y\}$, $y \notin C$ is in at least $k^{c-(i+1)} + 1$ hyperedges of H_i, or (2) k distinct vertices $y \notin C$ are not enough to hit all hyperedges of H_i. In case (1), the induction hypothesis holds for $i + 1$. In case (2), each hitting set of size k must also hit C. But then we can create a hyperedge C and delete supersets of C in H_i from the instance, without altering the solution space.

This case distinction can be decided in $O((c - i)k^{c-i})$ time, since it suffices to consider all y from the union of members of H_i. We find the hyperedges in H_i that are to be deleted within the same time. If case (1) still holds for $i = c$, we have two copies of the same hyperedge and can also delete one. Altogether, we reduced the number of hyperedges, in $O(ck^{c-1})$ time.

The procedure is repeated less than h times. The vertex counters can be updated in time proportional to cd times the number of deleted hyperedges, which is a total of $O(cdh)$. Finally note that $d \leq k$ can be assumed. $\square$

Combining the two results, we improve the coefficient of h from Theorem 3, provided that $k > c$:

Corollary 1. *All minimal solutions of* BOUNDED UNION *can be computed in* $O(ck^{c-1}h + dc^{k+1}k^c + c^k k^{c+1})$ *time.*

Proof. Construct an instance that has the same solutions but at most k^c (rather than h) hyperedges, as in Theorem 4, then run the algorithm from Theorem 3 on it. $\square$

3   Imperfect Phylogeny Reconstruction

3.1   Extra Rows

If an instance of PP PLUS k ROWS has a solution at all, then, in any complete pair, one of 00, 01, 10, 11 appears in at most k rows. At most 3 of these combinations appear at most k rows, unless $k \geq n/4$. In the following we implicitly assume $n > 4k$, remember that k is a fixed parameter. Destroying the complete pair means to remove all rows that contain *one of* 00, 01, 10, 11. This reduces PP PLUS k ROWS to BOUNDED UNION: The rows of the matrix are elements of the ground set, and for every complete pair of columns i, j we define a multiedge whose sets are the sets of rows containing 00, 01, 10, 11, respectively, at sites i, j. Trivially, it is enough to keep sets of at most k rows. This gives $h \leq \binom{m}{2} < m^2$, $c = 3$, and $d = k$.

Before we state our theorem, we discuss a naive application of the BOUNDED UNION results: Construct the multiedges from the matrix, then solve this instance of BOUNDED UNION in $O(k^2m^2 + 3^k k^4)$ time (Corollary 1). To this end we may check all $O(m^2)$ column pairs for completeness. But, unfortunately, for each pair we have to look at almost all rows, thus preprocessing needs $O(nm^2)$ extra time. We get $O(nm^2 + 3^k k^4)$ and lose the benefits of a small kernel. By an idea mentioned in [16], the complete pairs of an $n \times m$ matrix can be found already in $O(nm^{\omega-1})$ time, where $O(n^\omega)$ is a bound for matrix multiplication. But still, the dependency in m is not linear. We omit any details, because the following time bound is anyhow an improvement, unless $3^k > m^{\omega-2}$.

Theorem 5. PP PLUS k ROWS *is IMFPE. All minimal solutions can be computed in $O(3^k nm)$ time. We can also identify at most $3k^3$ rows such that the PP without these rows can be correctly reconstructed within this time bound.*

Proof. First find some complete pair in $O(nm)$ time as follows: Construct a PP incrementally, until a failure occurs at column j, say. Column j together with some column $i < j$ forms a complete pair. To find such i, go through the columns $i < j$ again. The complete pair i, j contains at most three of 00, 01, 10, 11 at most k times. Label the affected rows in the matrix, and branch if there are two or three such combinations. Continue on the matrix without the selected rows. In order to avoid copying the whole matrix for use on the different tree paths, we can work on the original matrix and observe the row labels. This way we spend only $O(nm)$ time at each of the $O(3^k)$ nodes. The second assertion follows from Theorem 4: All minimal solutions are in a full kernel of size $3k^3$.   □

3.2   Extra Columns

Theorem 1 immediately reduces PP PLUS k COLUMNS to VERTEX COVER: The reduction graph RG has the columns as vertices and the complete pairs as edges. This reduction costs naively $O(nm^2)$ time, or $O(nm^{\omega-1})$ time using the idea from [16] mentioned above. But we can do better, unless $k^2 > m^{\omega-2}$:.

Theorem 6. PP PLUS k COLUMNS *is IMFPE. All minimal solutions can be computed in $O(k^2 nm + k^2 2^k)$ time. We can aslo identify at most $2k^2$ columns and correctly reconstruct the PP without these sites in $O(k^2 nm)$ time.*

Proof. Build a PP incrementally, adding column by column. When we get stuck, we know a column y that is in some complete pair. Find all complete pairs with y, in the obvious way in $O(nm)$ time. If these are more than k pairs, y belongs to every solution and can be discarded. Clearly, this case can appear at most k times in the course of the algorithm. If y has degree at most k, we next consider the neighbors of y in the graph RG, we determine their neighbors as above, etc. Thus we construct some connected subgraph of RG by breadth-first-search (BFS). Recall that at most k^2 edges are incident to vertices of degree at most k, if there is a solution at all. If the BFS stops, we discard all columns reached by BFS and repeat the whole procedure, as there could exist other connected subgraphs of low-degree vertices. Thus, we eventually construct the full kernel in $O(k^2 nm)$ time. Then, use Theorem 2 to produce all minimal vertex covers of the kernel in $O(k^2 2^k)$ time.   □

Theorem 6 exploited only the fact that a few columns do not fit in a PP. These "bad" columns could look arbitrarily. It arises the question whether we can take advantage of additional assumptions, if we know the way the matrix has been produced (e.g. by random mutations).

3.3   A Full Kernel for Error Correction

Errors in data are more delicate than missing entries. Changing k bits in a
PP matrix corrupts at most k rows/colummns. We would like to find all *PP
correction sets*, i.e. combinations of at most k bits whose change gives a PP
matrix. This subset property is no longer closed under $\supset$. However we may aim
at finding all *minimal PP correction sets* of size at most k. In this formulation,
the problem has a full kernel, which is perhaps not obvious. Our bound is pretty
high, but we expect a much better "real" complexity for typical instances.

Theorem 7. PP WITH k ERRORS *has a full kernel of* $3k(k+1)^3 6^k$ *matrix entries
which can be determined in* $O(k6^k nm)$ *time.*

Proof. Rather long, omitted due to lack of space.

4   All k-Vertex Covers in $o(2^k poly(n))$ Time

Base 2 in Theorem 2 cannot be improved in general, because there are graphs
with 2^k minimal vertex covers of size k, for instance k disjoint edges. Hence $O(2^k)$
is, subject to a polynomial factor, the optimal time bound for enumerating all
minimal k-vertex covers. On the other hand, the solution space for such a trivial
graph is the Cartesian product of these edges. It can be described in just this
way and, if desired, explicitly enumerated afterwards, with small delay.

This observation motivates a modification of the problem: Compute a description of the solution space from which all solutions can be enumerated with
limited delay. (It is straightforward to give formal definitions.) Such a description would typically consist of a union of (fewer than 2^k) simple parts of the
solution space. Note that this goal is different from enumerations with e.g. polynomial delay. Also, we measure the worst-case complexity in terms of parameter
k rather than output size, i.e. we do not necessarily get output-sensitivity.

In this section we derive a result of this type for k-VERTEX COVER, by
a search tree algorithm that makes decisions for vertices rather than edges.
As usual in graph theory, $N(v)$ denotes the set of neighbors of vertex v, and
$N[v] = N(v) \cup \{v\}$. For a graph $G = (V, E)$ and a set $Y \subseteq V$, $G - Y$ is the graph
after removal of Y and all incident edges. Our algorithm needs a more general
problem statement, and (k-VERTEX COVER is the special case when $Y = \emptyset$:

MINIMAL VERTEX COVER EXTENSION (MVCE):
Given a graph $G = (V, E)$ and a set $Y \subset V$, find all sets X enjoying these two
properties: (1) $X \supseteq Y$, and (2) X is a minimal vertex cover in G. Let k-MVCE
be the same problem with the additional demand (3) $|X| \leq k$.

It is NP-hard to decide whether an instance of MVCE has a solution [3]. A
vertex in X is said to be *redundant* in X if all its neighbors are in X, too.

Our algorithm for k-MVCE sets up a search tree with the given G, k, and
Y at the root. By $t(k)$ we will denote the number of leaves of the decision tree,

where argument k is the parameter at the root, i.e. the number of vertices that may still be added to Y.[2]

At every tree node we first check if $|Y| > k$, or if some vertex in Y is redundant in Y. In these cases no solution $X \supseteq Y$ can exist, and the tree node becomes a dead end. Otherwise we choose some $v \notin Y$ and create two children of the node. In one branch we add v to Y and set $k' := k - 1$. In the other branch we decide that $v \notin Y$, which has some immediate implications: We put all neighbors of v in Y (they have to be chosen anyway) and set $k' := k - |N(v)|$. Then we diminish the graph by removing $N[v]$ and all incident edges, and we set $Y' := Y \setminus N[v]$. This routine is correct, since the solutions to k-MVCE with $v \notin Y$ correspond exactly to the solutions to k'-MVCE in $G - N[x]$ (with Y' instead of Y).

Specifically, let v in any tree node be a vertex with maximum number of neighbors outside Y. Let us run this algorithm until no more vertex $v \notin Y$ with at least *two* neighbors outside Y exists, in any tree node. For all current leaves of the search tree this means that $G - Y$ merely consists of isolated vertices and edges. Isolated vertices v in $G - Y$ cannot be added to Y anymore (otherwise v would be redundant).

If we aborted the algorithm at this moment, the number of leaves $t(k)$ would be bounded by the recurrence $t(k) \leq t(k - 1) + t(k - 2)$. Remember that we always added either v or its two (or more) neighbors to Y. The solution would be $t(k) = O(\phi^k)$, with $\phi = (\sqrt{5} + 1)/2 < 1.62$. The difficulty is the presence of yet unresolved isolated edges in $G - Y$. We have to add exactly one vertex from every isolated edge to Y, but in general not every choice gives a minimal solution to MVCE, so that a concise description is not obtained yet. To make things worse, since any v has only one neighbor in $G - Y$, also the recurrence would no longer apply to further branches. Therefore we cannot prove an $O(\phi^k)$ bound. However, a refinement of both the algorithm and the analysis will still result in a base $y < 2$.

Theorem 8. *A description of all minimal vertex covers of size at most k, using only Cartesian products of edges and set union as set operations, can be computed in $O(y^k p(n))$ time, where $y \approx 1.8$, and p is some polynomial.*

Proof. We assign a *height* to every tree node as follows. For a constant r to be specified later, the root gets height $(1+r)k$. Whenever children of a tree node are created, both inherit the height from their parent. Then, the child with $v \in Y$ reduces its height by 1, for having added v to Y. The child with $v \notin Y$ reduces its height by 1 for each neighbor just added to Y, and by r for each of the $|N(v)|$ vertices of Y removed from the graph. Since every vertex is added to Y and removed from the graph at most once along a tree path, and $|Y| \leq k$, the height never becomes negative.

In a first phase we do the search tree construction as above, until every $G - Y$ is composed of isolated edges. The height is reduced by 1 in case $v \in Y$, and by

---

[2] We will spend polynomial time at each tree node, hence the time bound is $t(k)p(n)$ for some polynomial p. Here, our analysis will focus only on the more interesting $t(k)$ factor.

at least $2 + 2r$ in case $v \notin Y$, since at least two neighbors have been put in Y and then removed from the graph.

In a second phase we continue as follows, at every leaf of the search tree. If no vertex in $G - Y$ has neighbors in Y, we can already concisely describe the minimal vertex covers $X \supseteq Y$: Every solution X is the union of Y and exactly one vertex from every isolated edge, and these vertices can be chosen independently. In this case we stop at that leaf and output the description of this part of the solution space. (But if $|Y|$ plus the number of isolated edges exceeds k, we have a dead end.)

The other case is that some vertex outside Y still has some neighbor in Y. We choose such a vertex as v. In the branch for $v \in Y$ we know that v's unique neighbor u outside Y cannot be added to Y anymore. Thus we apply the rule for $u \notin Y$ (remove $N[u]$ etc.) which reduces the height by $1 + r$, since v has just been added to Y and then removed from the graph. In the branch for $v \notin Y$ we remove at least two members of Y from the graph, due to the choice of v. This reduces the height by $1 + 2r$, including summand 1 for having added u to Y.

Recurrence $t(k) \leq t(k-1) + t(k-2-2r)$ applies to the first phase, and $t(k) \leq t(k-1-r) + t(k-1-2r)$ to the second. The characteristic equation is $x^{2+2r} = x^{1+2r} + 1$ and $x^{1+2r} = x^r + 1$, respectively. Define $y = x^{1+r}$, $s = r/(1+r)$, and $z = y^s$. Note that y is the base we obtain, because the tree size is $O(t((1+r)k))$. Since $k = 0$ leads to no branching, we can set $t(0) = 1$.

Rewrite the characteristic equations as $y^2 = yz + 1$ and $yz = z + 1$, respectively. Since $0 \leq s < 1$, our z must also satisfy $1 < z < y$. Substituting $z = 1/(y-1)$ in the first equation, we obtain $y^3 - y^2 = 2y - 1$, hence $y \approx 1.8$. $\square$

The approach can be extended far beyond this result. We briefly mention one possible refinement. Let us accept isolated paths and cycles in the subgraphs $G - Y$. (Minimal vertex covers of paths and cycles are still easy to enumerate.) Then we only have to take vertices v with at least 3 neighbors outside Y in the first phase. Following the lines of the previous proof, we can get $y \approx 1.39$.

A nice special case of the problem is bipartite graphs: If vertices v in the algorithm are restricted to one partite set, MVCE has always a solution, hence dead ends can appear only because the size bound k is exceeded. However the corresponding minimization problem is NP-hard, which extends a result from [3]. (We have to omit the proof here.)

References

1. R. Agarwala, D. Fernández-Baca: A polynomial-time algorithm for the perfect phylogeny problem when the number of character states is fixed, *SIAM J. Computing* 23 (1994), 1216-1224
2. H. Bodlaender, M. Fellows, T. Warnow: Two strikes against perfect phylogeny, *19th ICALP'92*, 273-283
3. E. Boros, K. Elbassioni, V. Gurvich, L. Khachiyan: Generating dual-bounded hypergraphs, DIMACS Tech. Report 2002-23
4. J.F. Buss, J. Goldsmith: Nondeterminism within P, *SIAM J. on Computing* 22 (1993), 560-572

5. J. Chen, I.A. Kanj, W. Jia: Vertex cover: further observations and further improvements, *J. of Algorithms* 41 (2001), 280-301
6. P. Damaschke: Incremental haplotype inference, phylogeny and almost bipartite graphs, *2nd RECOMB Satellite Workshop on Computational Methods for SNPs and Haplotypes 2004*, pre-proceedings, Carnegie Mellon Univ., 1-11
7. R.G. Downey: Parameterized complexity for the skeptic, *18th IEEE Conf. on Computational Complexity 2003*, 147-170
8. R.G. Downey, M.R. Fellows: *Parameterized Complexity*, Springer 1999
9. R.G. Downey, M.R. Fellows, U. Stege: Parameterized complexity, a framework for systematically confronting computational intractability, in: *Contemporary Trends in Discrete Mathematics: From DIMACS and DIMATIA to the Future*, AMS-DIMACS Series 49 (1999), 49-99 Springer 1999
10. D. Fernández-Baca, J. Lagergren: A polynomial-time algorithm for near-perfect phylogeny, to appear in *SIAM J. Computing*, preliminary version in *23rd ICALP'96*
11. H. Fernau: On parameterized enumeration, *COCOON'2002, LNCS* 2387, 564-573
12. M.L. Freidman, L. Khachiyan: On the complexity of dualization of monotone disjunctive normal forms, *J. of Algorithms* 21 (1996), 618-628
13. J. Gramm, R. Niedermeier: Quartet inconsistency is fixed-parameter tractable, *12th CPM'2001, LNCS* 2089, 241-256
14. J. Gramm, R. Niedermeier: Breakpoint medians and breakpoint phylogenies: a fixed-parameter approach, *1st Europ. Conf. on Comput. Biology 2002*, Supplement 2 to *Bioinformatics* 18, 128-139
15. D. Gusfield: Efficient algorithms for inferring evolutionary trees, *Networks* 21 (1991), 19-28
16. E. Halperin, R.M. Karp: Perfect phylogeny and haplotype inference, *8th RECOMB'2004*, 10-19
17. J. Jansson: Consensus algorithms for trees and strings, Dissertation 17, Dept. of Computer Science, Lund Univ. 2003
18. D.S. Johnson, M. Yannakakis, C.H. Papadimitriou: On generating all maximal independent sets, *Info. Proc. Letters* 27 (1988), 119-123
19. S. Kannan, T. Warnow: A fast algorithm for the computation and enumeration of perfect phylogenies, *SIAM J. Computing* 26 (1997), 1749-1763
20. D.J. Kavvadias, E.C. Stavropoulos: A new algorithm for the transversal hypergraph problem, *WAE'99, LNCS* 1668, 72-84
21. R. Niedermeier, P. Rossmanith: Upper bounds for vertex cover further improved, *16th STACS'99, LNCS* 1563, 561-570
22. R. Niedermeier, P. Rossmanith: An efficient fixed parameter algorithm for 3-Hitting Set, *J. of Discrete Algorithms* 1 (2003), 89-102
23. I. Pe'er, R. Shamir, R. Sharan: Incomplete directed perfect phylogeny, *11th CPM'2000, LNCS* 1848, 143-153
24. I. Pe'er, R. Shamir, R. Sharan: On the generality of phylogenies from incomplete directed characters, *8th SWAT'2002, LNCS* 2368, 358-367
25. M.A. Steel: The complexity of reconstructing trees from qualitative characters and subtrees, *J. of Classification* 9 (1992), 91-116
26. T. Warnow, D. Ringe, A. Taylor: Reconstructing the evolutionary history of natural languages, *7th SODA'96*, 314-322
27. M.S. Waterman: *Introduction to Computational Biology*, Chapman and Hall 1995
28. S. Wernicke, J. Alber, J. Gramm, J. Guo, R. Niedermeier: Avoiding forbidden submatrices by row deletions, *30th SOFSEM'2004, LNCS* 2932, 349-360

Online Problems, Pathwidth, and Persistence

Rodney G. Downey[1] and Catherine McCartin[2]

[1] Victoria University, Wellington, New Zealand
Rod.Downey@mcs.vuw.ac.nz
[2] Massey University, Palmerston North, New Zealand
C.M.McCartin@massey.ac.nz

Abstract. We explore the effects of using graph width metrics as restrictions on the input to online problems. It seems natural to suppose that, for graphs having some form of bounded width, good online algorithms may exist for a number of natural problems. In the work presented we concentrate on online graph coloring problems, where we restrict the allowed input to instances having some form of bounded pathwidth. We also consider the effects of restricting the presentation of the input to some form of bounded width decomposition or layout. A consequence of our work is the clarification of a new parameter for graphs, *persistence*, which arises naturally in the online setting, and is of interest in its own right.

1 Introduction

The last 20 years has seen a revolution in the development of graph algorithms. This revolution has been driven by the systematic use of ideas from topological graph theory, with the use of graph width metrics emerging as a fundamental paradigm in such investigations. The role of graph width metrics, such as treewidth, pathwidth, and cliquewidth, is now seen as central in both algorithm design and the delineation of what is algorithmically possible. In turn, these advances cause us to focus upon the "shape", or "the inductive nature of", much real life data. Indeed, for many real life situations, worst case, or even average case, analysis no longer seems appropriate, since the data is known to have a highly regular form, especially when considered from the parameterized point of view.

It turns out, however, that the classic algorithms generated through the use of width metrics usually rely upon dynamic programming, and so are highly unsuited to the focus of this paper, the online situation.

The usual theoretic model for online problems has the input data presented to the algorithm in small units, one unit per timestep. The algorithm produces a string of outputs: after seeing t units of input, it needs to produce the t-th unit of output. Thus, the algorithm makes a decision based only on partial information about the whole input string, namely the part that has been read so far. How good the decision of the algorithm is at any given step t may depend on the future inputs, inputs that the algorithm has not yet seen.

R. Downey, M. Fellows, and F. Dehne (Eds.): IWPEC 2004, LNCS 3162, pp. 13–24, 2004.

Real-life online situations often give rise to input patterns that seem to be "long" and "narrow", that is, pathlike. For instance, consider the online scheduling of some large collection of tasks onto a small number of processors. One might reasonably expect a pattern of precedence constraints that gives rise to only small clusters of interdependent tasks, with each cluster presented more or less contiguously. Alternatively, one might expect a pattern of precedence constraints giving rise to just a few long chains of tasks, with only a small number of dependencies between chains, where each chain is presented more or less in order. One could argue that the most compelling reason for attempting to solve a problem online is that the end of the input is "too long coming" according to some criteria that we have. Given such a situation, we attempt to do the best we can with the partial information available at each timestep.

The goal of this paper is to introduce a new program in which we plan to apply ideas from topological graph theory to online situations. In particular, we intend to study *online width metrics*. It is now commonplace, in the offline setting, to find that by restricting some width parameter for the input graphs, a particular graph problem can be solved efficiently. A number of different graph width metrics naturally arise in this context which restrict the inherent complexity of a graph in various senses. The central idea is that a useful width metric should admit efficient algorithms for many (generally) intractable problems on the class of graphs for which the width is small. One of the most successful measures in this context is the notion of *treewidth* which arose from the seminal work of Robertson and Seymour on graph minors and immersions [16]. Treewidth measures, in a precisely defined way, how "tree-like" a graph is. The idea here is that we can lift many results from trees to graphs that are "tree-like". Related to treewidth is the notion of *pathwidth* which measures, in the same way, how "path-like" a graph is.

Our long-term goals are two-fold. Firstly, we seek to understand the effects of using width metrics as restrictions on the input to online problems. As mentioned above, online situations often give rise to input patterns that seem to naturally conform to restricted width metrics, in particular to bounded pathwidth. We might expect to obtain online algorithms having good performance, for various online problems, where we restrict the allowed input to instances having some form of bounded pathwidth.

Secondly, we seek to understand the effects of restricting the *presentation* of the input to some form of bounded width decomposition or layout. The method of presentation of the input structure to an algorithm has a marked effect on performance. Indeed, this observation underpins the study of online algorithms in the first place.

To lay the foundations of the program described here, we concentrate on *online graph coloring*. There has been a considerable amount of work done on online graph coloring, much of it related to earlier work on bin packing. However, approaches similar to the one that we take are notably absent from the literature.

We remark that one consequence of our study to date has been the clarification of a parameter, *persistence*, which seems completely natural in the online situation, and is of interest in its own right. We are sure that other new parameters will come to light. We see the present paper as laying the foundations for our ideas. Future studies will be concerned both with other applications, and with making foundations for online descriptive complexity.

2   Preliminaries

2.1   Treewidth and Pathwidth

Many generally intractable problems become tractable for the class of graphs that have bounded treewidth or bounded pathwidth. Furthermore, treewidth and pathwidth subsume many graph properties that have been previously mooted, in the sense that tractability for bounded treewidth or bounded pathwidth implies tractability for many other well-studied classes of graphs. For example, planar graphs with radius k have treewidth at most $3k$, series parallel multigraphs have treewidth 2, chordal graphs (graphs having no induced cycles of length 4 or more) with maximum clique size k have treewidth at most $k - 1$, graphs with bandwidth at most k have pathwidth at most k.

A graph G has treewidth at most k if we can associate a tree T with G in which each node represents a subgraph of G having at most $k + 1$ vertices, such that all vertices and edges of G are represented in at least one of the nodes of T, and for each vertex v in G, the nodes of T where v is represented form a subtree of T. Such a tree is called a *tree decomposition* of G, of *width* k. We give a formal definition here:

Definition 1. *[Tree decomposition and Treewidth]*
Let $G = (V, E)$ be a graph. A tree decomposition of G is a pair $(T, \mathcal{X})$ where $T = (I, F)$ is a tree, and $\mathcal{X} = \{X_i \mid i \in I\}$ is a family of subsets of V, one for each node of T, such that

1. *$\bigcup_{i \in I} X_i = V$,*
2. *for every edge $\{v, w\} \in E$, there is an $i \in I$ with $v \in X_i$ and $w \in X_i$,*
3. *for all $i, j, k \in I$, if j is on the path from i to k in T, then $X_i \cap X_k \subseteq X_j$.*

The treewidth or width of a tree decomposition $((I, F), \{X_i \mid i \in I\})$ is $max_{i \in I} |X_i| - 1$. The treewidth of a graph G is the minimum width over all possible tree decompositions of G.

Definition 2. *[Path decomposition and Pathwidth]*
A path decomposition of a graph G is a tree decomposition $(P, \mathcal{X})$ of G where P is simply a path (i.e. the nodes of P have degree at most two). The pathwidth of G is the minimum width over all possible path decompositions of G.

2.2   Competitiveness

We measure the performance of an online algorithm, or gauge the difficulty of an online problem, using the concept of *competitiveness*, originally defined by Sleator and Tarjan [17].

Suppose that P is an online problem, and A is an online algorithm for P. Let $c \geq 1$ be a constant. We say that A is *c-competitive* if, for any instance I of problem P, $cost_A(I) \leq c \cdot cost_{opt}(I) + b$, where opt is an optimal offline algorithm that sees all information about the input in advance, and b is a constant independent of I. If an algorithm A is c-competitive, then we say that A has a *performance ratio* of c. We say that a given online problem P is *c-competitive* if there exists a c-competitive algorithm for P, and we say that it is *no better than c-competitive* if there exists no c'-competitive algorithm for P for any $c' \leq c$.

3   Online Presentations

The usual definition of an *online presentation* of a graph G is a structure $G^< = (V, E, <)$ where $<$ is a linear ordering of V. G is presented one vertex per timestep, v_1 at time 1, v_2 at time 2, ... and so on. At each step, the edges incident with the newly introduced vertex and those vertices already "visible" are also presented. We use the terms *online presentation* and *online graph* interchangeably.

Let $V_i = \{v_j \mid j \leq i\}$ and $G_i^< = G^<[V_i]$, the online subgraph of $G^<$ induced by V_i. An algorithm that solves some online problem on G will make a decision regarding v_i (and/or edges incident with v_i) using only information about $G_i^<$.

In this paper we introduce a different method of presenting a graphlike structure online. First, fix some arbitrary constant (parameter) k. At each timestep we present one new *active* vertex that may be incident with at most k active vertices previously presented. Once a vertex has been presented we may render some of the current set of active vertices *inactive* in preparation for the introduction of the next new vertex. At no point do we allow more than $k+1$ active vertices, and we do not allow a new vertex to be incident with any inactive vertex.

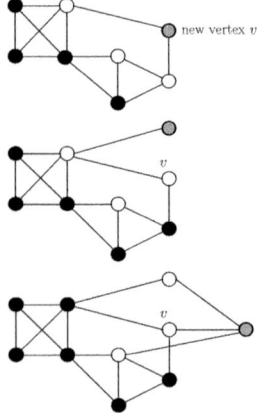

Fig. 1. Online presentation of a pathwidth 3 graph using 4 active vertices. Vertex v remains active for 3 timesteps.

These requirements mean that any graph presented in this fashion must have bounded pathwidth (pathwidth k). We are, in effect, presenting the graph as a path decomposition, one node per timestep. We denote such an online presentation of a graph G as $G^{< \text{ path } k}$.

We can add the further requirement that any vertex may *remain active* for at most l timesteps, for some arbitrary constant (parameter) l. We say that a path decomposition of width k, in which every vertex of the underlying graph belongs to at most l nodes of the path, has pathwidth k and *persistence l*, and say that a graph that admits such a decomposition has *bounded persistence pathwidth*. We explore this natural notion in Section 6, but we remark here that we believe that it truly captures the intuition behind the notion of pathwidth.

An online graph that can be presented in the form of a path decomposition with both low width and low persistence is properly pathlike, whereas graphs that have *high* persistence are, in some sense, "unnatural" or pathological. Consider the graph G presented in Figure 2. G is not really path-like, but still has a path decomposition of width only two. The reason for this is reflected in the presence of vertex a in *every* node of the path decomposition.

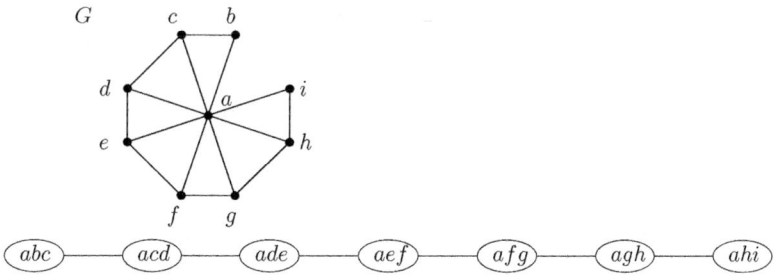

Fig. 2. A graph G having low pathwidth but high persistence.

Persistence appears to be a natural and interesting parameter in both the online setting *and* the offline setting. For many problems where the input is generated as an ordered sequence of small units, it seems natural to expect that the sphere of influence of each unit of input should be localised.

4   Online Coloring

An online algorithm A for coloring an online graph $G^<$ will determine the color of the ith vertex of $G^<$ using only information about $G_i^<$. A colors the vertices of $G^<$ one at a time in the order $v_1 < v_2, \cdots$, and at the time a color is irrevocably assigned to v_i, the algorithm can only see $G_i^<$.

A simple, but important, example of an online algorithm is *First-Fit*, which colors the vertices of $G^<$ with an initial sequence of the colors $\{1, 2, \ldots\}$ by assigning to v_i the least color that has not already been assigned to any vertex in $G_i^<$ that is adjacent to v_i.

Szegedy [18] has shown that, for any online coloring algorithm A and integer k, there is an online graph $G^<$ on at most $k(2^k - 1)$ vertices with chromatic number k on which A will use $2^k - 1$ colors. This yields a lower bound of $\Omega(\frac{n}{(\log n)^2})$ for the performance ratio of any online coloring algorithm on general graphs. Note that the worst possible performance ratio on general graphs is n. Lovasz, Saks, and Trotter [14] have given an algorithm that achieves a performance ratio $O(\frac{n}{(\log n)^*})$ on all graphs.

Online coloring of some restricted classes of graphs has been considered. In the bipartite case it can be shown that, for any online coloring algorithm A and integer k, there is an online tree $T^<$ with 2^{t-1} vertices on which A will use at least t colors. Thus, we get a lower bound of $\Omega(\log n)$ for any online algorithm on bipartite graphs. Lovasz, Saks, and Trotter [14] give an algorithm that colors any bipartite online graph using at most $1 + 2 \log n$ colors.

Kierstead and Trotter [11] have given an online coloring algorithm that achieves a performance ratio of 3 on interval graphs, which is also best possible. Kierstead [9] has shown that First-Fit has a constant performance ratio on the class of interval graphs. Gyarfas and Lehel [6] have shown that First-Fit achieves a constant performance ratio on split graphs, complements of bipartite graphs, and complements of chordal graphs.

One approach that is similar in flavour to ours is presented by Irani [7]. Irani introduces the notion of d-inductive graphs. A graph G is d-inductive if the vertices of G can be ordered in such a way that each vertex is adjacent to at most d higher numbered vertices. Such an ordering on the vertices is called an *inductive order*. As for a path or tree decomposition, an inductive order is not necessarily unique for a graph. An inductive order of a graph G defines an *inductive orientation* of G, obtained by orienting the edges from the higher numbered vertices to the lower numbered vertices. Notice that, in an inductive orientation, the indegree of each vertex is bounded by d. Hence, any d-inductive graph is $d + 1$ colorable.

In [7] it is shown that, if G is a d-inductive graph on n vertices, then First-Fit uses at most $O(d \cdot \log n)$ colors to color any online presentation $G^<$ of G. Moreover, for any online coloring algorithm A, there exists a d-inductive online graph $G^<$ such that A uses at least $\Omega(d \cdot \log n)$ colors to color $G^<$.

A connection between graphs of bounded pathwidth or bounded treewidth, and inductive graphs, is given by the following lemma (see [13], or [15]).

Lemma 1. *Any graph G of pathwidth k, or treewidth k, is k-inductive.*

5   Online Coloring of Graphs with Bounded Pathwidth

We consider two ways in which to formulate the problem of online coloring of graphs with bounded pathwidth.

We can define a parameterized "presentation" problem, where we fix a bound k on the pathwidth of any input graph G, and then proceed to present G as an implicit path decomposition, in the manner described in Section 3 above.

Alternatively, we can define a parameterized "promise" problem, where we fix a bound k on the pathwidth of any input graph G, and then proceed to present G as a structure $G^< = (V, E, <)$ where $<$ is an *arbitrary* linear ordering of V.

5.1   The Presentation Problem

If we undertake to present a graph G in the form of an implicit path decomposition, then we are effectively enforcing the presentation to be, if not best-possible, then at least "very good" for First-Fit acting on G.

Lemma 2. *If G is a graph of pathwidth k, presented in the form of an implicit path decomposition, then First-Fit will use at most $k+1$ colors to color $G^{< \text{ path } k}$.*

This is easy to see, since, at each step i, $0 \leq i \leq n$, the newly presented vertex v_i will be adjacent to at most k already-colored vertices. This result is best possible in the sense that the chromatic number (and, therefore, the online chromatic number) of the class of graphs of pathwidth k is $k + 1$. However, note that $G^{<\text{path } k}$ may not contain all of the information required to color G *optimally* online, as the following lemma shows.

Lemma 3. *For each $k \geq 0$, there is a tree T of pathwidth k presented as $T^{< \text{ path } k}$ on which First-Fit can be forced to use $k + 1$ colors.*

Proof. Suppose T_0 is a connected tree with pathwidth 0, then T must consist of a single vertex (any graph of pathwidth 0 must consist only of isolated vertices) so First-Fit will color $T_0^{< \text{ path } 0}$ with one color.

Suppose T_1 is a connected tree with pathwidth 1 that has at least two vertices. Each vertex of $T_1^{< \text{ path } 1}$ can be adjacent to at most one active vertex at the time of presentation. Since T_1 is connected, there must be vertices that are adjacent to an active vertex at the time of presentation in any $T_1^{< \text{ path } 1}$. Thus, First-Fit will need to use two colors to color any $T_1^{< \text{ path } 1}$.

Now, suppose that for any $0 \leq t < k$, there is a tree T_t of pathwidth t, and a presentation $T_t^{< \text{ path } t}$, on which First-Fit can be forced to use $t + 1$ colors. We build a connected tree T_k with pathwidth k, and a presentation $T_k^{< \text{ path } k}$, on which First-Fit will be forced to use $k + 1$ colors.

We order the trees T_t, $0 \leq t < k$, and their presentations, in descending order $[T_{k-1}, T_{k-2}, \ldots, T_0]$, and concatenate the presentations together in this order to obtain a new presentation $T_{con}^<$. Note that the subsequence $T_t^{< \text{ path } t}$ of $T_{con}^<$ will have at most $(t + 1) \leq k$ active vertices at any stage.

To obtain $T^{< \text{ path } k}$, we alter $T_{con}^<$ as follows. For each t, $0 \leq t < k$, we choose the vertex v_t from $T_t^{< \text{ path } t}$ that is colored with color $t + 1$ by First-Fit and allow it to remain active throughout the rest of $T_{con}^<$. Every other vertex from $T_t^{< \text{ path } t}$ is rendered inactive at the conclusion of $T_t^{< \text{ path } t}$ in the concatenated presentation. Thus, at any stage of $T_{con}^<$, there will be at most $k + 1$ active vertices, and at the conclusion of $T_{con}^<$ there will be k active vertices, one from each of the $T_t^{< \text{path } t}$, $0 \leq t < k$. These k active vertices will be colored with colors $1, 2, \ldots, k$ respectively. We now present one new vertex, adjacent to each of the k active vertices, which must be colored with color $k + 1$.   □

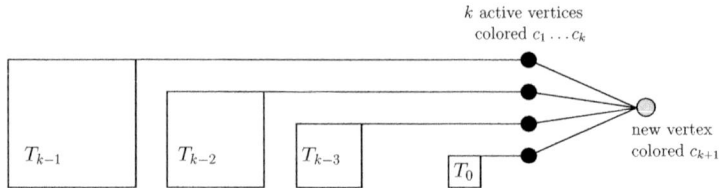

Fig. 3. Schema of $T^{< \text{path } k}$ on which First-Fit can be forced to use $k + 1$ colors.

5.2   The Promise Problem

Now, let us consider the "promise" problem for online coloring of graphs of bounded pathwidth. We first consider the case where G is a *tree* of pathwidth k.

Lemma 4. *First-Fit will use at most $3k + 1$ colors to color any $T^<$ where T is a tree of pathwidth k.*

Proof. Let $k = 0$, then T consists only of an isolated vertex, and First-Fit requires only one color to color $T^<$.

Let k be ≥ 1. Suppose that the bound holds for $k - 1$: for any tree T of pathwidth at most $k - 1$ First-Fit colors any $T^<$ with at most $3k - 2$ colors.

We rely on the fact that any tree T of pathwidth k consists of a path P and a collection of subtrees of pathwidth at most $k - 1$, each connected to a single vertex on the path P (see [4]).

Let v_i be a vertex appearing in one of the subtrees of T. When v_i is presented in $T^<$, at time i, it will be colored by First-Fit using a color chosen from the first $3k - 1$ colors of $\{1, 2, \ldots\}$.

Let T^i be the subtree in which v_i appears. Let p_i be the path vertex to which the subtree T^i is connected. Let $V_i = \{v_j \mid j \leq i\}$ and $T_i^< = T^<[V_i]$, the online subgraph of $T^<$ induced by V_i.

Suppose that p_i is not present in $T_i^<$. Then the component of $T_i^<$ containing v_i is a tree of pathwidth at most $k - 1$, disjoint from all other components of $T_i^<$. Thus, First-Fit will color v_i using a color chosen from the first $3k - 2$ colors of $\{1, 2, \ldots\}$.

Suppose that p_i is present in $T_{i-1}^<$. Then, in $T_i^<$, v_i will be adjacent to at most one vertex in the component of $T_{i-1}^<$ containing p_i. Suppose that this is the case and that this vertex has been colored with some color C_i. Consider the other components of $T_{i-1}^<$ to which v_i may become connected. Together with v_i, these form a tree of pathwidth at most $k - 1$. First-Fit will require at most $3k - 2$ colors to color this tree, but C_i cannot be used to color v_i. If

$C_i \notin \{1, 2, \ldots, 3k - 2\}$ then First-Fit will color v_i using a color chosen from $\{1, 2, \ldots, 3k - 2\}$. If $C_i \in \{1, 2, \ldots, 3k - 2\}$ then First-Fit will color v_i using a color chosen from $\{1, 2, \ldots, 3k - 1\} - C_i$. If, in $T_i^<$, v_i is not connected to the component of $T_{i-1}^<$ containing p_i, then First-Fit will color v_i using a color chosen from $\{1, 2, \ldots, 3k - 2\}$.

Let v_i be a vertex appearing in the path of T. When v_i is presented in $T^<$, at time i, it will be colored by First-Fit using a color chosen from the first $3k + 1$ colors of $\{1, 2, \ldots\}$.

Let $V_i = \{v_j \mid j \leq i\}$ and $T_i^< = T^<[V_i]$, the online subgraph of $T^<$ induced by V_i.

In $T_i^<$, v_i may be adjacent to single vertices from each of many subtrees. Note that, in $T_{i-1}^<$, each of the subtrees that becomes connected to v_i is disjoint from all other components of $T_{i-1}^<$, so any such subtree will have been colored only with colors from $\{1, 2, \ldots, 3k - 2\}$.

The path vertex v_i can also be connected to (at most) two other path vertices already colored. If v_i is not connected to any other path vertex then v_i will be colored by First-Fit using a color chosen from the first $3k - 1$ colors of $\{1, 2, \ldots\}$. If v_i is connected to only one other path vertex then v_i will be colored by First-Fit using a color chosen from the first $3k$ colors of $\{1, 2, \ldots\}$. If v_i is connected to two other path vertices then v_i will be colored by First-Fit using a color chosen from the first $3k + 1$ colors of $\{1, 2, \ldots\}$.   □

Lemma 5. *For each $k \geq 0$, there is an online tree $T^<$, of pathwidth k, such that First-Fit will use $3k + 1$ colors to color $T^<$.*

The proof given for Lemma 4 suggests a way in which to present a tree of pathwidth k that will require $3k + 1$ colors.

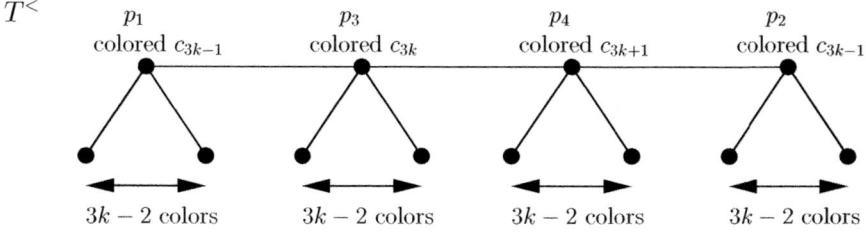

Fig. 4. Schema of online tree $T^<$ of pathwidth k on which First-Fit is forced to use $3k + 1$ colors.

We now consider the "promise" problem for online coloring in the case of general graphs of pathwidth at most k.

A graph $G = (V, E)$ is an *interval graph* if there is a function ψ which maps each vertex of V to an interval of the real line, such that for each $u, v \in V$ with $u \neq v$, $\psi(u) \cap \psi(v) \neq \emptyset \Leftrightarrow (u, v) \in E$. The function ψ is called an *interval realization* for G. The relation between interval graphs and graphs of bounded pathwidth is captured by the following lemma (see [12], or [15]).

Lemma 6. *A graph G has pathwidth at most k if and only if G is a subgraph of an interval graph G', where G' has maximum clique size at most $k + 1$.*

Kierstead and Trotter [11] have given an online algorithm that colors any online interval graph $G^<$, having maximum clique size at most $k + 1$, using $3k + 1$ colors. Thus, any graph of pathwidth k can be colored online using at most $3k + 1$ colors. Lemma 4, above, gives us a stronger result for trees – for trees of pathwidth k such a performance is possible using First-Fit.

For general graphs of pathwidth at most k, the situation for First-Fit is less well understood. First-Fit does have a constant performance ratio on graphs of pathwidth at most k. Kierstead [9] has shown that for every online interval graph $G^<$, with maximum clique size at most $k+1$, First-Fit will use at most $40(k+1)$ colors. In [10] Kierstead and Qin have improved the constant here to 25.72.

Chrobak and Slusarek [2] have used an induction argument to prove that there exists an online interval graph $G^<$, and a constant c, where c is the maximum clique size of $G^<$, such that First-Fit will require at least $4.4 \cdot c$ colors to color $G^<$. Such an interval graph $G^<$ will have pathwidth $c - 1$ and chromatic number c. Thus, the performance ratio of First-Fit on general graphs of pathwidth k must be at least 4.4. However, experimental work by Fouhy [5] indicates that we can expect a performance ratio of 3 for First-Fit on randomly generated graphs of bounded pathwidth.

6   Bounded Persistence Pathwidth

In Section 3 we introduced a quite natural online presentation scheme that gives rise to graphs having *bounded persistence pathwidth*. We present the graph as a path decomposition, one node per timestep, where every vertex of the underlying graph belongs to at most l nodes of the path. The underlying idea is that a pathwidth 2 graph should look more like a "long 2-path" than a "fuzzy ball".

A related notion is *domino treewidth* introduced by Bodlaender and Engelfriet [1]. A *domino tree decomposition* is a tree decomposition in which every vertex of the underlying graph belongs to at most two nodes of the tree. *Domino pathwidth* is a special case of bounded persistence pathwidth, where $l = 2$.

In the case of online coloring using First-Fit neither bounded persistence pathwidth, nor domino pathwidth, give us any real gain over our original pathwidth metric. However, the notion of persistence seems like it may be interesting, in the broader online setting, in its own right.

In this section we present some basic results regarding the general recognition of graphs having bounded persistence path decompositions. Note that, even though we will show that the recognition problem is *hard*, in many real life instances we may reasonably expect to "know" that the persistence is relatively low, and, indeed be given such a decomposition.

To demonstrate the hardness of basic problems associated with persistence, we will use the framework of *parameterized complexity theory*, introduced by Downey and Fellows [3]. We remind the reader that a parameterized language L is a subset of $\Sigma^* \times \Sigma^*$. If L is a parameterized language and $\langle \sigma, k \rangle \in L$ then we refer to σ as the *main part* and k as the *parameter*. The basic notion of tractability is *fixed parameter tractability* (FPT). Intuitively, we say that a parameterized problem is fixed-parameter tractable (FPT) if we can somehow confine the any "bad" complexity behaviour to some limited aspect of the problem, the parameter. Formally, we say that a parameterized language, L, is fixed-parameter tractable if there is a computable function f, an algorithm A, and a constant c such that for all k, $\langle x, k \rangle \in L$ iff $A(x, k) = 1$, and $A(x, k)$ runs in time $f(k)|x|^c$ (c is independent of k). For instance, k-VERTEX COVER is solvable in time $\mathcal{O}(|x|)$. On the other hand, for k-TURING MACHINE ACCEPTANCE, the problem of deciding if a nondeterministic Turing machine with arbitrarily large fanout has a k-step accepting path, the only known algorithm is to try all possibilities, and this takes time $\Omega(|x|^k)$. This situation, akin to NP-completeness, is described by hardness classes, and reductions. A parameterized reduction, L to L', is a transformation which takes $\langle x, k \rangle$ to $\langle x', k' \rangle$, running in time $g(k)|x|^c$, with $k \mapsto k'$ a function purely of k.

Downey and Fellows [3] observed that these reductions gave rise to a hierarchy called the W-hierarchy.

$$FPT \subset W[1] \subseteq W[2] \subseteq \ldots \subseteq W[t] \subseteq \ldots.$$

The core problem for $W[1]$ is k-TURING MACHINE ACCEPTANCE, which is equivalent to the problem WEIGHTED 3SAT. The input for WEIGHTED 3SAT is a 3CNF formula, φ and the problem is to determine whether or not φ has a satisfying assignment of Hamming weight k. $W[2]$ has the same core problem except that φ is in CNF form, with no bound on the clause size. In general, $W[t]$ has as its core problem the weighted satisfiability problem for φ of the form "products of sums of products of ..." of depth t. It is conjectured that the W-hierarchy is proper, and from $W[1]$ onwards, all parametrically intractable.

In this spirit, we define the parameterized problem BOUNDED PERSISTENCE PATHWIDTH as follows:

Instance: A graph $G = (V, E)$.
Parameter: A pair of positive integers (k, l).
Question: Is there a path decomposition of G of width at most k, and persistence at most l?

DOMINO PATHWIDTH is a special case of this problem, where $l = 2$.

The following results give strong evidence for the likely intractability of both the BOUNDED PERSISTENCE PATHWIDTH problem and the DOMINO PATHWIDTH problem. Note that these results mean that it is likely to be impossible to find FPT algorithms for either of these problems, unless an unlikely collapse occurs in the W-hierarchy. The proofs of these results can be found in [15].

Theorem 1. BOUNDED PERSISTENCE PATHWIDTH *is $W[t]$-hard, for all $t \in \mathcal{N}$.*

Theorem 2. DOMINO PATHWIDTH *is $W[2]$-hard.*

7   Conclusions

We consider the work presented in this paper to be a first step in a program to investigate the ramifications of online width metrics. Of course, there are many graph width metrics, along with many online problems, that are candidates for study. However, it does seem apparent that pathwidth, or metrics that are pathwidth-like, are a natural fit in this context.

References

1. H. L. Bodlaender, J. Engelfreit: *Domino Treewidth.* J. Alg. 24, pp 94-127, 1997.
2. M. Chrobak, M. Slusarek: *On some packing problems related to dynamic storage allocation.* RAIRO Inform. Theor. Appl. 22, pp 487-499, 1988.
3. R. G. Downey, M. R. Fellows: *Parameterized Complexity* Springer-Verlag, 1999.
4. B. de Fluiter: *Algorithms for Graphs of Small Treewidth.* ISBN 90-393-1528-0, 1997.
5. J. Fouhy: *Computational Experiments on Graph Width Metrics.* M.Sc. thesis, Victoria University, Wellington, 2003.
6. A. Gyarfas, J. Lehel: *On-line and First Fit Coloring of Graphs.* J. Graph Theory, Vol. 12, No. 2, pp 217-227, 1988.
7. S. Irani: *Coloring inductive graphs on-line.* Proceedings of the 31st Annual Symposium on Foundations of Computer Science, Vol 2, pp 470-479, 1990.
8. H. A. Kierstead: *Recursive and On-Line Graph Coloring* In Handbook of Recursive Mathematics, Volume 2, pp 1233-1269, Elsevier, 1998.
9. H. A. Kierstead: *The Linearity of First Fit Coloring of Interval Graphs.* SIAM J. on Discrete Math, Vol 1, No. 4, pp 526-530, 1988.
10. H. A. Kierstead, J. Qin: *Coloring interval graphs with First-Fit.* (Special issue: Combinatorics of Ordered Sets, papers from the 4th Oberwolfach Conf., 1991), M. Aigner and R. Wille (eds.), Discrete Math. 144, pp 47-57, 1995.
11. H. A. Kierstead, W. A. Trotter: *An Extremal Problem in Recursive Combinatorics.* Congressus Numeratium 33, pp 143-153, 1981.
12. N. G. Kinnersley: *The Vertex Separation Number of a Graph equals its Path-Width.* Information processing Letters 42(6), pp. 345-350, 1992.
13. L. M. Kirousis, D. M. Thilikos: *The Linkage of a Graph.* SIAM Journal on Computing 25(3), pp. 626-647, 1996.
14. L. Lovasz, M. E. Saks, W. A. Trotter: *An On-Line Graph Coloring Algorithm with Sublinear Performance Ratio.* Bellcore Tech Memorandum, No.TM-ARH-013-014.
15. C. M. McCartin: *Contributions to Parameterized Complexity* Ph.D. Thesis, Victoria University, Wellington, 2003.
16. N. Robertson, P. D. Seymour: *Graph minors II. Algorithmic aspects of tree-width.* Journal of Algorithms 7, pp 309-322, 1986.
17. D. D. Sleator, R. E. Tarjan: *Amortized Efficiency of List Update and Paging Rules.* Comunication of the ACM 28, pp 202-208, 1985.
18. M. Szegedy: private communication, reported in [8].

Chordless Paths Through Three Vertices

Robert Haas[1] and Michael Hoffmann[2]

[1] IBM Zurich Research Laboratory, Rüschlikon
rha@zurich.ibm.com
[2] Institute for Theoretical Computer Science, ETH Zürich
hoffmann@inf.ethz.ch

Abstract. Consider the following problem, that we call "Chordless Path through Three Vertices" or CP3V, for short: Given a simple undirected graph $G = (V, E)$, a positive integer k, and three distinct vertices s, t, and $v \in V$, is there a chordless path from s via v to t in G that consists of at most k vertices? In a chordless path, no two vertices are connected by an edge that is not in the path. Alternatively, one could say that the subgraph induced by the vertex set of the path in G is the path itself. The problem has been raised in the context of service deployment in communication networks. We resolve the parametric complexity of CP3V by proving it $W[1]$-complete with respect to its natural parameter k. Our reduction extends to a number of related problems about chordless paths. In particular, deciding on the existence of a single directed chordless (s, t)-path in a digraph is also $W[1]$-complete with respect to the length of the path.

Keywords: graph theory, induced path, parameterized complexity.

1 Introduction

The number of specialized functions such as support for quality of service and protection against denial-of-service attacks, that is being built into network nodes is growing continuously. Thus it is becoming increasingly difficult for network administrators to use such sophisticated capabilities fully, especially when new services must be deployed in a timely manner. The advent of reprogrammable network nodes, made possible by cost-efficient network processors, has aggravated this issue as new capabilities may be introduced into the network dynamically in order to provision a particular service.

An automated method to perform service deployment was presented in [20], and several categories of services with similar deployment needs were introduced. This paper focuses on chordless paths: these are paths for which no two nodes are directly connected except along the path. Chordless paths are particularly relevant to the category of path-based services, i.e., services that require each node of a path from a source to a destination to be enabled with a common function. During the deployment of path-based services, nodes are queried for specific service requirements to determine whether they have the necessary capabilities to support a certain function. Such requirements are specific to each service, so that it is not recommended to let nodes advertise all their capabilities by default, which may even vary over time. Instead, on-demand query of relevant nodes is the preferred alternative.

R. Downey, M. Fellows, and F. Dehne (Eds.): IWPEC 2004, LNCS 3162, pp. 25–36, 2004.

Often it is necessary to query only a subset of the network nodes to determine whether a path with the required capabilities exists. If such a path contains a chord, there is a shorter path through the same set of nodes for this particular source/destination pair. Thus, for this type of query we are interested in only chordless paths. In particular, nodes that do not belong to any chordless path for a given source/destination pair are irrelevant and do not have to be queried.

It has been shown that for typical Internet router-level topologies, a large fraction of nodes do not belong to chordless paths [20]. To avoid unnecessary queries, it would hence be desirable to decide the following problem efficiently.

Problem 1 (CP3V). *Given an undirected graph $G = (V, E)$, a positive integer k, and three distinct vertices $s, t, v \in V$, is there a chordless path from s via v to t in G that consists of at most k vertices?*

But as we will prove in Section 4, this problem is $W[1]$-hard with respect to k; that is— roughly speaking—it is unlikely that an algorithm exists whose complexity is bounded by an arbitrary (say, doubly exponential) function in k but polynomial in the size of the input graph. As another consequence [1, 8], there is probably no PTAS to compute an $(1 + \varepsilon)$-approximation for the shortest chordless (s, v, t)-path in time bounded by an arbitrary function in ε but polynomial in the size of the input graph.

After summarizing some related results, we complement the above-mentioned hardness claim in Section 3 by proving CP3V to be in $W[1]$, that is, altogether $W[1]$-complete. Finally, Section 5 discusses how to extend our results for CP3V to a number of problems concerning chordless paths in directed graphs.

1.1   Related Work

The following problem is a slight generalization of CP3V.

Problem 2 (**Many Chordless** (s, t)**-Paths**). *Given a simple undirected graph $G = (V, E)$, positive integers k and ℓ, and two distinct vertices $s, t \in V$, is there a set $U \subseteq V$ of at most k vertices such that the subgraph induced by U in G is a disjoint union of ℓ chordless (s, t)-paths?*

This was shown to be NP-complete by Fellows [15], already for $\ell = 2$, where it asks for a chordless cycle of length at most k through s and t. Let us refer to this problem as CC2V. As any chordless cycle through s has to pass through one of the neighbors of s, an instance of CC2V can be solved by less than $|V|$ calls to an algorithm that solves CP3V. In particular, this implies NP-completeness of CP3V. On the other hand, any CP3V instance can be converted to a CC2V instance in constant time: add a new vertex u to G and connect it to both s and t; now ask for a cycle through u and v of length at most $k + 1$. Hence, our $W[1]$-completeness result for CP3V also implies that CC2V is $W[1]$-complete.

The hardness results for CP3V and CC2V rely on graphs that contain many vertex-disjoint (s, t)-paths. It is not difficult to see that in planar graphs the existence of four vertex-disjoint (s, t)-paths basically[1] implies the existence of a chordless (s, v, t)-path.

---

[1] Except for some trivial cases which can be easily sorted out.

While this argument immediately gives an $\mathcal{O}(3^k n^c)$ time algorithm for CP3V (for some constant $c \in \mathbb{N}$), the more interesting question is whether Problem 2 is polynomial for planar graphs. This was answered in the affirmative for every fixed ℓ by McDiarmid et al. [23, 24].

If in Problem 2 we ask for vertex-disjoint paths only instead of requiring all paths to be jointly chordless, the problem is polynomial for general graphs and every fixed ℓ, even for arbitrary source-target pairs (s_i, t_i), $1 \leq i \leq \ell$ [27]. But it remains NP-complete if ℓ is considered part of the input [21].

Deciding whether a graph contains a chordless path of length at least k is one of the classical NP-complete problems (GT23 in [19]). Bienstock [3, 4] listed several other NP-complete problems related to chordless paths:

- Does a graph contain a chordless path of odd length between two specified vertices?
- Does a graph contain a chordless cycle of odd length (> 3) through a specified vertex?
- Does a graph contain a chordless path of odd length between any two vertices?

Note that these results do not imply the hardness of deciding whether there exists **any** path/cycle of odd length in a graph. This question is still open, see the discussion below.

Chordless cycles of length at least four are also called *holes*. They are tightly connected to Berge's strong perfect graph conjecture [2], whose proof has recently been announced by Chudnovsky et al. [9]. According to this conjecture, a graph is perfect[2] iff it is Berge, that is, if it contains neither an odd hole nor the complement of an odd hole. Hence, a polynomial time algorithm to decide whether there exists any odd hole in a given graph would immediately imply that perfect graphs can be recognized in polynomial time. Interestingly, no such an algorithm is known, although there are polynomial time algorithms [11, 10] to decide whether a graph is Berge, even independent of the strong perfect graph conjecture. Also, if the restriction to an odd number of vertices is omitted, the presence of holes can be detected in polynomial time: for holes on at least four vertices this is the well-studied recognition problem for chordal graphs [22, 28]. The problem of detecting holes on at least five vertices has recently been addressed by Nikolopoulos and Palios [26].

2 Notation

For a graph G denote by $V(G)$ the set of vertices in G, and denote by $E(G)$ the set of edges in G. For a vertex $v \in V$, denote by $N_G(v)$ the *neighborhood* of v in G, that is, the set of vertices from V that are adjacent to v in G. Similarly, for a set $W \subset V$ of vertices define $N_G(W) := \bigcup_{w \in W} N_G(w)$. The subscript is often omitted when it is clear which graph it refers to. A set $I \subset V$ of vertices is an *independent set* in G if E does not contain edges between any two vertices of I. For a set $W \subset V$ of vertices denote by $G[W]$ the *induced subgraph* of W in G, that is, the graph $(W, E \cap \binom{W}{2})$ or $(W, E \cap W^2)$ in the case of an undirected or a directed graph G, respectively.

---

[2] In a perfect graph, the maximum number of pairwise adjacent vertices (*clique number*) for each induced subgraph is equal to the minimum number of colors needed to color the vertices in such a way that any two adjacent vertices receive distinct colors (*chromatic number*).

A subgraph of G that has the form $(\{v_1, \ldots, v_k\}, \{\{v_i, v_{i+1}\} \mid 1 \leq i < k\})$ is called *path of length* k in G. Note that by the set notation we imply that $v_i \neq v_j$, for $1 \leq i, j \leq k$ and that the length of a path is defined as its number of vertices. The vertices v_1 and v_k are referred to as the path's *endpoints*, the other vertices are called *interior*. Two paths are called *vertex-disjoint* iff they do not share vertices except for possibly common endpoints. For two vertices $s, t \in V$ any path from s to t in G is called (s, t)-*path*. More generally, if vertices s, v, and t appear on path P in this order, we call P an (s, v, t)-*path*. A path P in G is called *chordless* if $V(P)$ is an independent set in $(V, E \setminus E(P))$. An alternative equivalent definition would be to call a path P in G chordless iff $G[V(P)] = P$. Hence, such paths are also known as *induced paths*.

Parameterized Complexity To cope with the apparent computational intractability of NP-hard problems, attempts were made to analyze more closely which parts or aspects of the input render a particular problem hard. A prototypical example is *Vertex Cover*, which asks for a set C of at most k vertices from a given graph G on n vertices such that for each edge at least one endpoint is in C. The trivial observation that for any edge at least one of the two incident vertices has to be in C, leads to an $O(2^k n)$ time algorithm: choose an arbitrary edge and branch on the two possibilities, in both cases removing one vertex and all incident edges from the graph. Hence, the intractability of Vertex Cover is connected to the number k of vertices in the cover rather than to the size of the graph G. One says that Vertex Cover is *fixed-parameter-tractable* (FPT) with respect to the parameter k because there is an algorithm that runs in $O(f(k)p(n))$ for an arbitrary, typically exponential, function f and a polynomial function p.

Naturally, there are also problems for which it is not known whether their complexity can be isolated into a particular parameter in this way. Moreover, similar to the classical complexity classes, there are classes of parameterized problems that are hard in the sense that if there is an FP algorithm for any of them, then all of them are FPT. The most important such class is called $W[1]$, which can be described in terms of the following "canonical" problem.

Problem 3 (Weighted q-CNF-Satisfiability). *Given positive integers q and k, and a boolean formula F in conjunctive normal form such that each clause contains at most q literals, is there a satisfying assignment for F with at most k variables set to true?*

A problem P parameterized by k is said to be *m-reducible* to a problem P' parameterized by k' iff there is a function h that maps an instance (x, k) of P to an instance (x', k') of P' such that $k' = g(k)$ and x' can be computed in time $f(k)p(x)$, for arbitrary functions f and g, and a polynomial function p. Now $W[1]$ is defined as the class of parameterized problems that can be m-reduced to Weighted q-CNF-Satisfiability for some constant q. Finally, a problem is $W[1]$-*hard* iff every problem in $W[1]$ can be m-reduced to it. A problem that is both $W[1]$-hard and in $W[1]$ is called $W[1]$-*complete*.

At this point, we refer the interested reader to the literature for more in-depth information about parameterized complexity. The book of Downey and Fellows [13] provides a thorough treatment of complexity-theoretic aspects, whereas the survey of Niedermeier [25] focuses more on algorithms.

3   Membership in W[1]

In this section we analyze the parameterized complexity of CP3V and prove the problem to be in $W[1]$ with respect to its natural parameter, the path length. As a first step, note that a chordless (s, t)-path P is already determined by its set of vertices. For example, only one neighbor x of s can be in $V(P)$ because any later visit of another neighbor would introduce a chord. Similarly, exactly one neighbor of x (other than s) can be in $V(P)$. In this manner P can be uniquely reconstructed from $V(P)$.

Proposition 4. *A subgraph P of G is a chordless (s, t)-path if and only if P is connected, s and t have degree one in $G[V(P)]$, and all vertices other than s and t have degree two in $G[V(P)]$.*   □

We do not know how to reduce CP3V to Weighted q-CNF-Satisfiability. Instead, we reduce to a different problem called SNTMC that is defined below. SNTMC is known to be $W[1]$-complete [5, 14], and reduction to SNTMC and its relatives has proven to be a useful tool to establish membership results within the W-hierarchy [6, 7].

Problem 5 (Short Nondeterministic Turing Machine Computation (SNTMC)).
Given a single-tape, single-head nondeterministic Turing machine M, a word x on the alphabet of M, and a positive integer k, is there a computation of M on input x that reaches a final accepting state in at most k steps?

Theorem 6. *CP3V is in $W[1]$ w.r.t. k.*

Proof. Consider an instance (G, s, v, t, k) of CP3V, where $G = (V, E)$ is a simple undirected graph, $s, v, t \in V$, and k is a positive integer. We will construct an instance (M, k') of SNTMC such that there is a computation for M that reaches a final accepting state in at most $k' = k^2 + 3k - 1$ steps iff there exists a chordless (s, v, t)-path of length at most k in G. (It is important that k' depends on k only and not on n.) A schematic view of the construction is shown in Fig. 1.

Let $M = (\Sigma, Q, \Delta, g_1, \{A\})$, where the alphabet Σ is defined as $\Sigma := \{\square\} \cup \{\sigma_u \mid u \in V\}$; the state set is

$$Q := \{A, R\} \cup \{g_i \mid 1 \le i \le k\} \cup \{a, b, c, d, l, r\} \cup \{p_u \mid u \in V\} \cup \{q_u \mid u \in V\} ;$$

the transition relation $\Delta : Q \times \Sigma \times Q \times \Sigma \times \{+, -, 0\}$ is defined below; the initial state is g_1; the final accepting state is A, and the final rejecting state is R. When the Turing machine starts, all tape cells contain the blank symbol ($\square$). The computation consists of three phases: first, the at most k vertices of a chordless (s, v, t)-path P in G are "guessed" by writing the sequence of corresponding symbols σ_u, $u \in V(P)$, onto the tape. The next two phases are completely deterministic and check that P visits s, v, and t in the order given, and that P is a chordless path in G.

First Phase: The Turing machine may write up to k arbitrary vertex symbols onto the tape: $(g_i, \square, g_{i+1}, \sigma_u, +) \in \Delta$, for all $u \in V$ and all $1 \le i < k$, and $(g_i, \square, a, \sigma_u, 0) \in \Delta$, for all $u \in V$ and all $1 \le i \le k$. (The transition specifies, in order, current state, symbol under the head, new state after transition, symbol to write to the tape, and movement of the head: $+$ for right, $-$ for left, and 0 for stay.) After the first phase, the Turing machine is in state a and the sequence of between one and k vertex symbols starts at the current tape cell, extending to the left.

Second Phase: Check whether the guessed sequence visits t, v, and s, in order. The rightmost symbol should be σ_t: $(a, \sigma_t, b, \sigma_t, -) \in \Delta$. Then somewhere σ_v must appear: $(b, \sigma_u, b, \sigma_u, -) \in \Delta$, for all $u \in V \setminus \{v\}$, and $(b, \sigma_v, c, \sigma_v, -) \in \Delta$. The final symbol has to be σ_s: $(c, \sigma_u, c, \sigma_u, -) \in \Delta$, for all $u \in V \setminus \{s\}$, and $(c, \sigma_s, d, \sigma_s, -) \in \Delta$. Nothing may follow after s: $(d, \Box, l, \Box, +) \in \Delta$. For all state/symbol combinations that are not explicitly mentioned (for example, $(b, \Box)$ or (a, σ_v)) there is a transition to the final rejecting state R. After the second phase, the machine is in state l and the head points towards the leftmost of the symbols that have been guessed in Phase 1. The content of the tape remains unchanged during Phase 2.

Third Phase: Scan and remove the first vertex: $(l, \sigma_u, p_u, \Box, +) \in \Delta$, for all $u \in V$. If no more vertex is left at this point, we are done: $(p_u, \Box, A, \Box, 0) \in \Delta$, for all $u \in V$. Otherwise, the next vertex should be adjacent: $(p_u, \sigma_w, q_u, \sigma_w, +) \in \Delta$, for all $u, w \in V$ for which $\{u, w\} \in E$. Whatever follows must not be adjacent: $(q_u, \sigma_w, q_u, \sigma_w, +) \in \Delta$, for all $u, w \in V$ with $u \neq w$ and $\{u, w\} \notin E$. If all vertices have been checked, return to the leftmost: $(q_u, \Box, r, \Box, -) \in \Delta$, for all $u \in V$, and $(r, \sigma_w, r, \sigma_w, -) \in \Delta$, for all $w \in V$. Finally, re-iterate: $(r, \Box, l, \Box, +) \in \Delta$. Again, all state/symbol combinations that are not explicitly mentioned lead to the final rejecting state R. Note that after the third phase, all tape cells contain the blank symbol again.

Phase 3 ensures that all vertices guessed in Phase 1 are distinct, as, otherwise the right scan in state q_u, for some $u \in V$, fails. Moreover, because of the transition from p_u to q_u, the vertices chosen form a path P in G. The right scan in state q_u also ensures that no two of the vertices are connected except along P. Finally, in Phase 2 we check that the endpoints of P are s and t, and that P visits v. Altogether, the machine reaches an accepting state iff it guesses the vertices of a chordless (s, v, t)-path of length at most k in Phase 1. An easy calculation reveals that if k symbols are written onto the tape in Phase 1, the remaining computation consists of exactly $k^2 + 2k - 1$ transitions.   □

4   Hardness for $W[1]$

In this section, we prove that CP3V is $W[1]$-hard using a reduction from Independent Set, which is one of the "classical" $W[1]$-hard problems [12].

Problem 7 (Independent Set). *Given a simple undirected graph $G = (V, E)$ and a positive integer k, is there an independent set of size at least k in G?*

Consider an instance of Independent Set, that is, a graph $G = (V, E)$ and an integer k, $1 \leq k \leq |V|$, and let $V = \{v_1, \ldots, v_n\}$. We construct a graph G' from G such that the answer to the CP3V problem on G' provides the solution to the independent-set problem on G.

The main ingredient for our construction is called *vertex choice diamond*: it consists of n vertices $v_1^i, \ldots, v_n^i$ plus two extra vertices s^i and t^i connected to each of the n vertices v_j^i, $1 \leq j \leq n$, as shown in Fig. 2. Clearly, there are exactly n chordless (s^i, t^i)-paths in such a diamond. As the naming of the vertices suggests, we associate each of these paths with a vertex from G in a bijective manner: routing a path through v_j^i, for some $1 \leq j \leq n$, is interpreted as selecting v_j to be part of the independent set

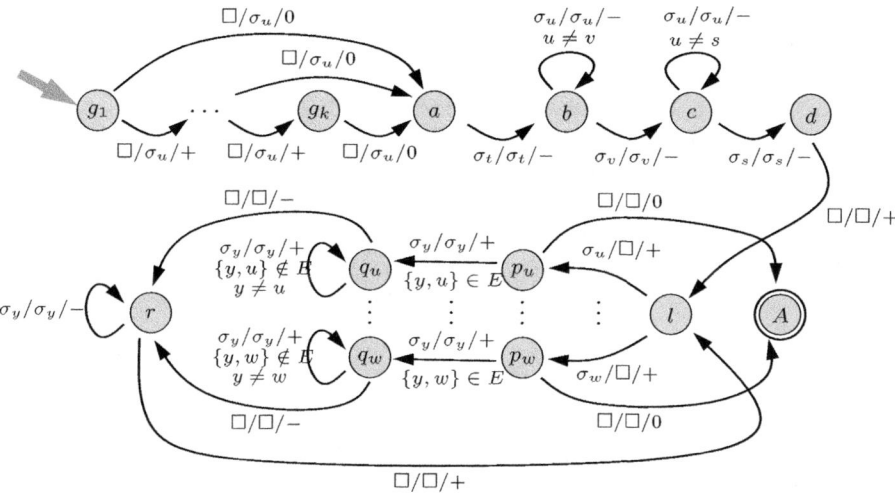

Fig. 1. A schematic description of the Turing machine defined in the proof of Theorem 6. The transition arrows are labeled by, in order, symbol under head, symbol to write, and head movement. To increase readability the final rejecting state and all transitions to it have been omitted.

I to be constructed. The construction uses k such vertex choice diamonds, which are connected by identifying s^{i+1} and t^i, for all $1 \le i < k$. Let us call the graph described so far G_{VC}, where VC stands for vertex choice.

Proposition 8. *Any chordless (s^1, t^k)-path of length ℓ in G_{VC} corresponds to a set of $\ell - k - 1$ vertices in G.* $\square$

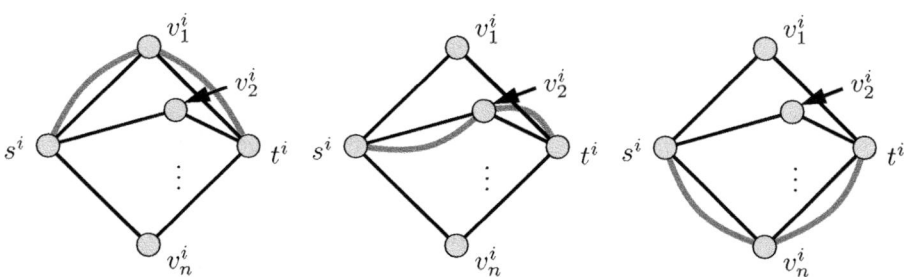

Fig. 2. A vertex choice diamond and three of its n chordless (s^i, t^i)-paths.

The next step is to ensure that the vertex sets chosen by traversing G_{VC} on a chordless path correspond to independent sets in the original graph G. To accomplish this, we construct G' from two symmetric copies of G_{VC}. Denote the vertices in the first copy C of G_{VC} by s^i, v_j^i, and t^i, whereas the vertices in the second copy Γ of G_{VC} are referred to as σ^i, φ_j^i, and τ^i, for $1 \le i \le k$ and $1 \le j \le n$. The graphs C and Γ are connected by identifying t^k and τ^k. The construction of G' is completed by adding a number of edges that encode the adjacency of G. An example is shown in Fig. 3.

- There is an edge in G' between v_j^i and φ_ℓ^i, for all $1 \leq i \leq k$ and all $1 \leq j, \ell \leq n$ with $j \neq \ell$. Such an edge is called *consistency edge*.
- For every edge $\{v_p, v_q\}$ in G, connect the vertex sets $\{v_p^i, \varphi_p^i\}$ and $\{v_q^j, \varphi_q^j\}$, for all $1 \leq i, j \leq k$ with $i \neq j$, by a complete bipartite subgraph in G'. These edges are called *independence edges*.
- The vertices v_ℓ^i and φ_ℓ^i are connected by an edge in G' to all of the vertices v_ℓ^j and φ_ℓ^j, for all $1 \leq i, j \leq k$ with $i \neq j$. These edges are called *set edges*.

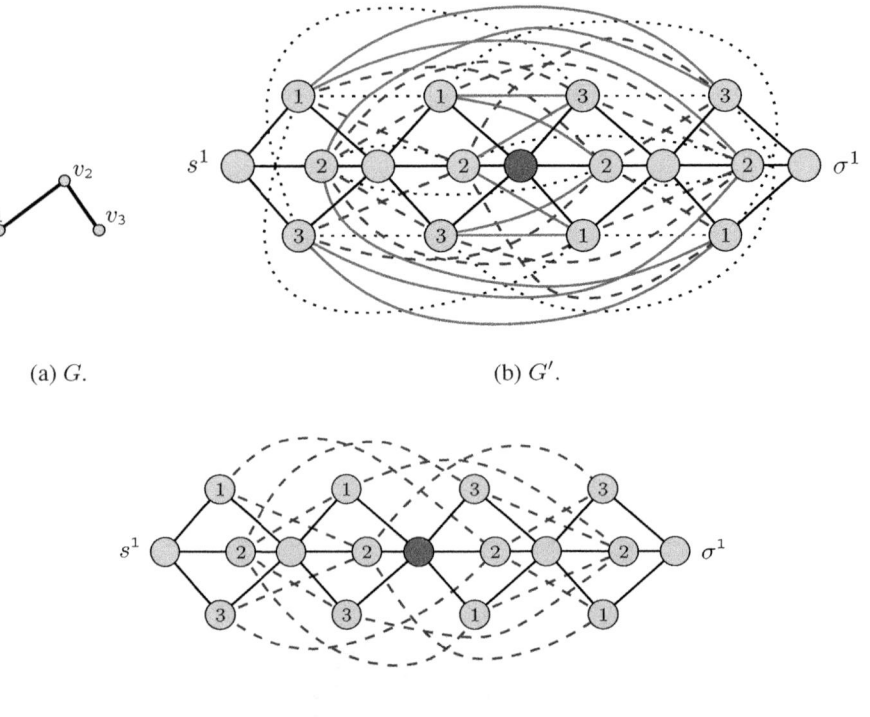

(a) G.   (b) G'.

(c) G' (independence edges only).

Fig. 3. An example illustrating the construction of G' for $k = 2$. Consistency edges are shown by solid lines, independence edges by dashed lines, and set edges by dotted lines. The vertex labels in G' indicate the correspondence to the vertices from G: for example, the vertices labeled "1" correspond to v_1. The vertex $t^k = \tau^k$ is shaded dark.

Lemma 9. *No chordless (s^1, σ^1)-path via t^k in G' uses a consistency edge or an independence edge or a set edge.*

Proof. Let P be a chordless (s^1, σ^1)-path via t^k in G'. Consider the initial part of P which traverses (part of) the first vertex choice diamond of C. By construction of G', exactly one of the vertices v_j^1, $1 \leq j \leq n$ is on P. (All of these vertices are neighbors of s^1.) Similarly, the final part of P contains exactly one of the vertices φ_ℓ^1, $1 \leq \ell \leq n$. Moreover, if $\ell \neq j$ vertices v_j^1 and φ_ℓ^1 are connected by an edge in G'. Hence, they together with s^1 and σ^1 induce an (s^1, σ^1)-path in G' that does not visit v and cannot be extended to a chordless path visiting v either. Thus, we may conclude that $\ell = j$.

Furthermore, note that by construction v_j^1 and φ_j^1 have the same neighbors along both independence and set edges. Thus, if P continues along an independence edge or set edge from either vertex, it has to do so from the other vertex as well, as otherwise the unused edge would form a chord, but, as above, P cannot reach t^k in this case.

In summary, the initial part of P goes from s^1 via v_j^1, for some $1 \leq j \leq n$, to $t^1 = s^2$, and the final part of P is completely symmetric: from τ^1 via φ_j^1 to σ^1. By induction on k, the initial part of P is an (s^1, t^k)-path Q that visits all s^i in increasing order, for $1 \leq i \leq n$, without using any consistency, independence, or set edge, and the final part of P is a (τ^k, σ^1)-path that is completely symmetric to Q.   □

Theorem 10. CP3V *is* $W[1]$-*hard w.r.t.* k.

Proof. Given an instance (G, k) of Independent Set, we construct the graph G' as described above. The graph G' contains $2k(n + 1) + 1$ vertices. To compute the number of edges in G' note that there are $4kn$ edges in the $2k$ vertex choice diamonds, plus $kn(n - 1)$ consistency edges, $4mk(k - 1)$ independence edges, and $2nk(k - 1)$ set edges, where $n := |V(G)|$ and $m := |E(G)|$. Hence, G' can be constructed from G in time and space polynomial in both n and k.

Let P be a chordless (s^1, σ^1)-path via t^k of length at most $4k + 1$ in G'. By Lemma 9, P has a very special form: in particular, its length is exactly $4k + 1$, and it visits exactly one vertex $v_{j_i}^i$ from each of the vertex choice diamonds with $1 \leq i \leq n$. Let $I := \{v_{j_i} \in V(G) \mid v_{j_i}^i \in V(P) \text{ for } 1 \leq i \leq n\}$. Suppose that $v_{j_i}^i = v_{j_\ell}^\ell$ for some $1 \leq i, \ell \leq k$ with $i \neq \ell$. As $v_{j_i}^i$ and $v_{j_\ell}^\ell$ are connected by a set edge in G' that is not in P by Lemma 9, this set edge forms a chord of P, in contradiction to our assumption that P is chordless. Therefore, the vertices $v_{j_i}^i$ visited by P correspond to mutually distinct vertices in G, that is, together with Proposition 8 it follows $|I| = 2k + 1 - k - 1 = k$.

Furthermore, we claim that I is an independent set in G. Suppose that for two vertices $v_{j_i}^i$ and $v_{j_\ell}^\ell$ on P, $1 \leq i, \ell \leq k$ and $i \neq \ell$, the corresponding vertices $v_{j_i}^i$ and v_{j_ℓ} are neighbors in G. Then by construction $v_{j_i}^i$ and $v_{j_\ell}^\ell$ are connected by an independence edge in G'. Again, this edge is not in P by Lemma 9, that is, it forms a chord of P, in contradiction to our assumption that P is chordless. Therefore, no two vertices in I are adjacent in G.

Conversely, it is easy to see that for any independent set $I = \{v_1, v_2, \ldots, v_k, \ldots\}$ (without loss of generality) of size at least k in G there is a chordless (s^1, σ^1)-path P via t^k of length $4k + 1$ in G': in the i-th vertex choice diamonds, P visits v_i^i and φ_i^i, for $1 \leq i \leq k$.

Therefore, we have a parameterized reduction from an independent set instance (G, k) to a CP3V instance $(G', 4k + 1)$, establishing $W[1]$-hardness of CP3V.   □

5   Chordless Paths in Digraphs

The notion of chordless paths generalizes in a straightforward manner to digraphs.

Problem 11 (Directed Chordless (s,t)-Path (DCP)). *Given a simple digraph $G = (V, E)$, a positive integer k, and two distinct vertices $s, t \in V$, is there a chordless directed (s, t)-path of length at most k in G?*

Fellows et al. [16] showed that DCP is NP-complete even if restricted to planar digraphs. Our constructions described above can easily be adapted to the directed setting.

Theorem 12. DCP *is $W[1]$-complete w.r.t. k.*

Proof. In the Turing machine of Theorem 6 replace all conditions that require the existence of an edge by corresponding conditions requiring the presence of a directed edge. Similarly, all conditions requiring the absence of an edge are replaced by corresponding conditions disallowing both directed edges.

The construction described in Theorem 10 is modified as follows. In the vertex choice diamonds direct all edges from s^i to v_j^i and from v_j^i to t^i, for all $1 \leq i \leq k$ and all $1 \leq j \leq n$. In the symmetric copy, direct all edges from τ^i to φ_j^i and from φ_j^i to σ^i, for all $1 \leq i \leq k$ and all $1 \leq j \leq n$. These orientations induce a linear ordering $(s^1, \ldots, t^k = \tau^k, \ldots, \sigma^1)$ on $V(G')$. The remaining edges, that is, the consistency, independence, and set edges all are oriented from the vertex that is greater with respect to this linear order to the smaller vertex. It is easy to verify that no chordless directed (s^1, σ^1)-path can use a consistency, independence, or set edge. In fact, the orientation is chosen such that any chordless directed (s^1, σ^1)-path in G' passes through $t^k = \tau^k$, although this is not required by definition, in contrast to CP3V.   □

As a consequence, also the following problem is $W[1]$-complete w.r.t. k. (Just add a single directed edge (σ^1, s^1) to the construction described in Theorem 12.)

Problem 13 (Directed Chordless Cycle). *Given a simple digraph $G = (V, E)$, a positive integer k, and a vertex $s \in V$, is there a chordless directed cycle of length at most k through s in G?*

Note that both problems are polynomial if the path or cycle is not required to be chordless: the maximum number k of vertex-disjoint directed (s, t)-paths can be computed in $O(k|E|)$ time using flow techniques [17]. However, deciding whether there exist a directed (s_1, t_1)-path and a directed (s_2, t_2)-path that are vertex-disjoint is NP-complete, even for $t_1 = s_2$ and $t_2 = s_1$ [18].

Also, if the definition of chordless is relaxed to allow "back-cutting" arcs within each path, DCP restricted to planar graphs is polynomial, even for an arbitrary but fixed number of chordless (s, t)-paths [24]. The existence of such arcs is the crucial difference between the directed and the undirected problem: in an undirected (s, t)-path P every edge joining two vertices that are non-adjacent along P can be used as a shortcut. That is, the presence of any (s, t)-path implies the existence of a chordless (s, t)-path. However, we will show below that admitting back-cutting arcs does not change the parametric complexity of the problem for general graphs.

Definition 14. *An (s,t)-path P in a graph $G = (V,E)$ is called **weakly chordless** iff P is a shortest (s,t)-path in $G[V(P)]$.*

Observe that there is no difference between chordless and weakly chordless in undirected graphs. But, in contrast to DCP, the presence of a directed weakly chordless (s,t)-path can be decided in linear time by a breadth-first search. However, the generalization to several paths defined below is again $W[1]$-complete, already for two paths.

Problem 15 (Many Weakly Chordless (s,t)-Paths). *Given a simple digraph $G = (V,E)$, positive integers k and ℓ, and two distinct vertices $s,t \in V$, is there a set $U \subseteq V$ with $|U| \leq k$ such that $G[U]$ is a disjoint union of ℓ weakly chordless (s,t)-paths?*

Theorem 16. *Two Weakly Chordless (s,t)-Paths is $W[1]$-complete w.r.t. k.*

Proof. It is clear how to adapt the Turing machine construction of Theorem 6 to establish membership in $W[1]$.

The construction described in Theorem 10 is modified as follows. First, add a directed edge from σ^1 to s^1, and let $s := t^k = \tau^k$ and $t := s^k$. In the vertex choice diamonds direct all edges from t^i to v_j^i and from v_j^i to s^i, for all $1 \leq i \leq k$ and all $1 \leq j \leq n$. Likewise, in the symmetric copy direct all edges from τ^i to φ_j^i and from φ_j^i to σ^i, for all $1 \leq i \leq k$ and all $1 \leq j \leq n$. Remove all independence and set edges within the same diamond chain, such that all remaining independence or set edges are between v_j^i and φ_q^p, for some $1 \leq i,p \leq k$ and $1 \leq j,q \leq n$. Direct those edges from v_j^i towards φ_q^p.

Consider (t^k, s^1)-paths P and Q in G' such that $G[V(P) \cup V(Q)]$ is a disjoint union of weakly chordless (t^k, s^1)-paths. The way the edges are directed, one of the paths, say, P comes via the vertex choice diamonds and visits the vertices $t^k, s^k, t^{k-1}, \ldots t^1, s^1$, in order. On the other hand, Q traverses the symmetric copy and visits the vertices $\tau^k, \sigma^k, \tau^{k-1}, \ldots \tau^1, \sigma^1$, in order, before finally reaching s^1 via the added edge. Because there must not be any edge between P and Q, we can argue as in Theorem 10 that the vertices v_j^i and φ_q^p visited by P and Q, respectively, correspond to an independent set of size at most k in G. $\square$

5.1 Acknowledgments

We thank Yoshio Okamoto and Emo Welzl for helpful discussions.

References

[1] BAZGAN, C. Schémas d'approximation et complexité paramétrée. Rapport du stage (DEA), Université Paris Sud, 1995.

[2] BERGE, C. Färbung von Graphen deren sämtliche beziehungsweise deren ungerade Kreise starr sind (Zusammenfassung). *Wiss. Z. Martin Luther Univ. Halle Wittenberg Math. Naturwiss. Reihe* (1961), 114–115.

[3] BIENSTOCK, D. On the complexity of testing for odd holes and induces odd paths. *Discrete Math. 90*, 1 (1991), 85–92.

[4] BIENSTOCK, D. Corrigendum to: On the complexity of testing for odd holes and induces odd paths. *Discrete Math. 102*, 1 (1992), 109.

[5] CAI, L., CHEN, J., DOWNEY, R. G., AND FELLOWS, M. R. On the parameterized complexity of short computation and factorization. *Arch. Math. Logic 36*, 4–5 (1997), 321–337.

[6] CESATI, M. Perfect code is W[1]-complete. *Inform. Process. Lett. 81*, 3 (2002), 163–168.

[7] CESATI, M. The Turing way to parameterized complexity. *J. Comput. Syst. Sci. 67*, 4 (2003), 654–685.

[8] CESATI, M., AND TREVISAN, L. On the efficiency of polynomial time approximation schemes. *Inform. Process. Lett. 64*, 4 (1997), 165–171.

[9] CHUDNOVSKY, M., ROBERTSON, N., SEYMOUR, P. D., AND THOMAS, R. The strong perfect graph theorem. Manuscript (2003).

[10] CHUDNOVSKY, M., AND SEYMOUR, P. D. Recognizing Berge graphs. Manuscript (2003).

[11] CORNUÉJOLS, G., LIU, X., AND VUŠKOVIĆ, K. A polynomial algorithm for recognizing perfect graphs. In *Proc. 44th Annu. IEEE Sympos. Found. Comput. Sci.* (2003), pp. 20–27.

[12] DOWNEY, R. G., AND FELLOWS, M. R. Fixed-parameter tractability and completeness II: On completeness for W[1]. *Theoret. Comput. Sci. 141* (1995), 109–131.

[13] DOWNEY, R. G., AND FELLOWS, M. R. *Parameterized Complexity*. Monographs in Computer Science. Springer-Verlag, 1999.

[14] DOWNEY, R. G., FELLOWS, M. R., KAPRON, B., HALLETT, M. T., AND WAREHAM, H. T. Parameterized complexity and some problems in logic and linguistics. In *Proc. 2nd Workshop on Structural Complexity and Recursion-theoretic methods in Logic-Programming* (1994), vol. 813 of *Lecture Notes Comput. Sci.*, Springer-Verlag, pp. 89–101.

[15] FELLOWS, M. R. The Robertson-Seymour theorems: A survey of applications. In *Proc. AMS-IMS-SIAM Joint Summer Research Conf.*, Providence, RI, 1989, pp. 1–18.

[16] FELLOWS, M. R., KRATOCHVÍL, J., MIDDENDORF, M., AND PFEIFFER, F. The complexity of induced minors and related problems. *Algorithmica 13* (1995), 266–282.

[17] FORD, L. R., AND FULKERSON, D. R. Maximal flow through a network. *Canad. J. Math. 8* (1956), 399–404.

[18] FORTUNE, S., HOPCROFT, J. E., AND WYLLIE, J. The directed subgraph homeomorphism problem. *Theoret. Comput. Sci. 10* (1980), 111–121.

[19] GAREY, M. R., AND JOHNSON, D. S. *Computers and Intractability: A Guide to the Theory of NP-Completeness*. W. H. Freeman, New York, NY, 1979.

[20] HAAS, R. *Service Deployment in Programmable Networks*. PhD thesis, ETH Zurich, Switzerland, 2003.

[21] KARP, R. M. On the complexity of combinatorial problems. *Networks 5* (1975), 45–68.

[22] LUEKER, G. S., ROSE, D. J., AND TARJAN, R. E. Algorithmic aspects of vertex elimination in graphs. *SIAM J. Comput. 5* (1976), 266–283.

[23] MCDIARMID, C., REED, B., SCHRIJVER, A., AND SHEPHERD, B. Non-interfering network flows. In *Proc. 3rd Scand. Workshop Algorithm Theory* (1992), pp. 245–257.

[24] MCDIARMID, C., REED, B., SCHRIJVER, A., AND SHEPHERD, B. Induced circuits in planar graphs. *J. Combin. Theory Ser. B 60* (1994), 169–176.

[25] NIEDERMEIER, R. *Invitation to Fixed-Parameter Algorithms*. Habilitation thesis, Wilhelm-Schickard Institut für Informatik, Universität Tübingen, Germany, 2002.

[26] NIKOLOPOULOS, S. D., AND PALIOS, L. Hole and antihole detection in graphs. In *Proc. 15th ACM-SIAM Sympos. Discrete Algorithms* (2004), pp. 843–852.

[27] ROBERTSON, N., AND SEYMOUR, P. D. Graph minors XIII. The disjoint paths problem. *J. Combin. Theory Ser. B 63* (1995), 65–110.

[28] TARJAN, R. E., AND YANNAKAKIS, M. Simple linear-time algorithms to test chordality of graphs, test acyclicity of hypergraphs, and selectively reduce acyclic hypergraphs. *SIAM J. Comput. 13* (1984), 566–579.

Computing Small Search Numbers in Linear Time[*]

Hans L. Bodlaender[1] and Dimitrios M. Thilikos[2]

[1] Institute of Information and Computing Sciences, Utrecht University,
P.O. Box 80.089, 3508 TB Utrecht, the Netherlands
hansb@cs.uu.nl
[2] Departament de Llenguatges i Sistemes Informàtics, Universitat Politècnica de
Catalunya, Campus Nord–C6. E-08034, Barcelona, Spain
sedthilk@lsi.upc.es

Abstract. Let $G = (V, E)$ be a graph. The *linear-width* of G is defined as the smallest integer k such that E can be arranged in a linear ordering $(e_1, \ldots, e_r)$ such that for every $i = 1, \ldots, r-1$, there are at most k vertices both incident to an edge that belongs to $\{e_1, \ldots, e_i\}$ and to an edge that belongs to $\{e_{i+1}, \ldots, e_r\}$. For each fixed constant k, a linear time algorithm is given, that decides for any graph $G = (V, E)$ whether the linear-width of G is at most k, and if so, finds the corresponding ordering of E. Linear-width has been proved to be related with the following graph searching parameters: mixed search number, node search number, and edge search number. A consequence of this is that we obtain for fixed k, linear time algorithms that check whether a given graph can be mixed, node, or edge searched with at most k searchers, and if so, output the corresponding search strategies.

Keywords: linear-width, pathwidth, treewidth, characteristics, graph searching.

1 Introduction

In this paper, we study algorithmic aspects of the graph parameter of *linear-width*. Apart from having interest on its own, this parameter has close relationships with some well known graph searching parameters: mixed search number, node search number, and edge search number, as well as well known parameters related or equal to these parameters, like pathwidth, vertex separation number, and proper pathwidth. The results for linear-width allow us to rederive some old results for some of these parameters, and obtain similar new results for the other parameters, basically all as consequences of one algorithm.

The *linear-width* of a graph G is defined to be the least integer k such that the edges of G can be arranged in a linear ordering $(e_1, \ldots, e_r)$ in such a way that for every $i = 1, \ldots, r - 1$, there are at most k vertices that incident to

---

[*] The second author is partially supported by the IST Program of the EU under contract number IST-2001-33116 (FLAGS) and by the Spanish CICYT project TIC-2002-04498-C05-03 (TRACER).

R. Downey, M. Fellows, and F. Dehne (Eds.): IWPEC 2004, LNCS 3162, pp. 37–48, 2004.

at least one edge that belongs to $\{e_1, \ldots, e_i\}$ and that are incident to at least one edge that belongs to $\{e_{i+1}, \ldots, e_r\}$. Linear-width was first mentioned by Thomas in [28] and is strongly connected with the notion of *crusades* introduced by Bienstock and Seymour in [2]. Linear-width can be seen as "a linear variant of branch-width", in the same way as pathwidth can be seen as "a linear variant of treewidth". In [25], it is proved that several variants of problems appearing on graph searching can be reduced to the problem of computing linear-width.

In a graph searching game a graph represents a system of tunnels where an agile, fast, and invisible fugitive is resorting. We desire to capture this fugitive by applying a search strategy while using the fewest possible searchers. Briefly said, the search number of a graph is the minimum number of searchers a searching strategy requires in order to capture the fugitive. Several variations on the way the fugitive can be captured during a search, define the the parameters of the *edge, node,* and *mixed search number* of a graph (namely, es(G), ns(G), and ms(G)). The first graph searching game was introduced by Breisch [7] and Parsons [18] and is the one of *edge searching*. *Node searching* appeared as a variant of edge searching and was introduced by Kirousis and Papadimitriou in [14]. Finally, *mixed searching* was introduced in [24] and [2] and is a natural generalisation of the two previous variants (for the formal definitions see Section 4 – for other results concerning search games on graphs see [1,8,9,15,16,23].)

The problems of computing es(G), ns(G), ms(G), or linear-width(G) is NP-complete (see [16,14,24,25]). On the other hand, since all of these parameters are closed under taking of minors, we know (see e.g. [3,20,19,22,21]) that, for any k, there exists a linear algorithm that given a graph G checks whether es(G), ns(G), ms(G), or linear-width(G) is at most k. In other words, all these parameters are fixed parameter tractable by a linear parameterized algorithm (i.e. an algorithm of time complexity $O(f(k)n)$ where f is a function not depending on n). Unfortunately, the above result is not constructive, i.e. does not provide a way to *construct* the corresponding algorithm (see [10,11]). Therefore, it is highly desired to have constructive "fixed parameter results" for the aforementioned parameters.

In this paper we carry out the above task by constructing such a linear parameterized algorithm for linear-width. This algorithm can be directly transfered to a linear parameterized algorithm for node, edge, and mixed search number thanks to their connection (see [25]) with linear-width.

So far, such a linear time algorithm has been constructed (see [3,4]) only for the parameters of treewidth and pathwidth (actually, the result in [3,4] can be directly transfered to the node search number which is known to be equal to the pathwidth plus one – see [12,13,17]). To be precise, [3,4] state that for fixed k, one can determine in linear time whether a given graph has pathwidth at most k, and if so, find a path decomposition of minimum width. This algorithm first finds a tree decomposition of width at most k, if it exists (if not, then the pathwidth is also larger than k), and then uses this tree decomposition to solve the problem, using the result in [4] that states that for fixed k and l, one can test whether the pathwidth of a graph G is at most l, and if so, find a minimum width path decomposition, assuming that G is given together with a tree decomposition of width at most k. This paper uses a similar idea: we first determine a path decomposition of bounded width (if such a path decomposition does not exist, we know that the linear-width is also not bounded), and then apply the main algorithm, presented in this paper, that, given such a path decomposition, solves our problem. It can actually be avoided to work with tree decompositions by a modification of the algorithm in [3]. The main algorithm of this paper further develops the main technique of [4] that makes use of a special type of data structures called "sets of characteristics". In general, such a structure *filters* the essential information on the existence of a bounded width layout of a the graph with respect to some of the nodes of the given path-decomposition.

We stress that the general technique of defining "sets of characteristics" has a long history of development and applications. Apart from treewidth and path-width, it has been used on the design of linear time parameterized algorithms for vertex layout parameters like cutwidth and carving-width [26]. Moreover, it was also used in [27] for a polynomial algorithm for the pathwidth of graphs with bounded treewidth and maximum degree. Moreover, its kernel ideas have been generalized so that a systematic derivation of linear parameterized algorithms is possible for a general class of graph parameters including directed cutwidth, modified cutwidth, directed modified cutwidth, and directed pathwidth. However, none of the previous results was able to cover the case of linear-width and the graph search parameters that we examine in this paper. Intuitively, the reason is that these parameters are defined using edge orderings instead of vertex orderings. Indeed it appears that the definition of a characteristic for edge orderings cannot be derived by a straightforward generalization of the previous results. In this paper we define a special (and non-trivial) type of "characteristics" for the case of linear-width which. This makes it possible to derive, in a uniform way, a linear time parameterized algorithm for the three classic search parameters.

An other parameter related to linear-width is *branch-width*. In another paper [5], we give a similar algorithm for branch-width. That algorithm uses the techniques of this paper as a building block for a more complicated algorithm.

This paper is organized as follows. In Section 2, we give some preliminary results and definitions. The main algorithm is presented in Section 3. The consequences of the main algorithm can be found in Section 4. Due to space restrictions, all the proofs are omitted in this extended abstract (see [6] for details).

2   Definitions and Preliminary Results

We first give a number of definitions and notations, dealing with sequences (i.e., linearly ordered sets) $\omega = (\omega_1, \ldots, \omega_r)$ of objects from some set $\mathcal{O}$. In this paper $\mathcal{O}$ can be a set of numbers, sequences of numbers, vertices, vertex sets, or edges. All sequences will be indexed starting by 1 except from the case were we have either sequences of vertices or sequences of sequences of numbers that will be indexed starting by 0 (we adopt this convention because it will facilitate the presentation of our results). Given a sequence ω and two of its indices i, j where $i \leq j$, we use the notation $(\omega)_{i,j}$ to denote the subsequence $(\omega_i, \ldots, \omega_j)$ of ω. Also, if "$\circ$" is an operation defined on $\mathcal{O}$, we use for any $\omega' \in \mathcal{O}$, the notation $(\omega)_{i,j} \circ \omega', i \leq j$, to denote the sequence $(\omega_i \circ \omega', \ldots, \omega_j \circ \omega')$. Given two sequences $\omega^i, i = 1, 2$, defined on $\mathcal{O}$, where $\omega^i = (\omega_s^i, \ldots, \omega_r^i), i = 1, 2$ we set $\omega^1 \circ \omega^2 = (\omega_s^1 \circ \omega_s^2, \ldots, \omega_r^1 \circ \omega_r^2)$. Also, given two sequences $\omega_i = (\omega_{s_i}^i, \ldots, \omega_{r_i}^i), i = 1, 2$ (not necessarily of the same length) we define their concatenation $\omega^1 \oplus \omega^2 = (\omega_{s_1}^1, \ldots, \omega_{r_1}^1, \omega_{s_1}^2, \ldots, \omega_{r_2}^2)$ (we stress that in the last two definitions, $s = 1$ except from the cases where we consider either sequences of sets or sequences of sequences of numbers where $s = 0$). We finally denote the length of a sequence ω by $|\omega|$.

Unless mentioned otherwise, we will assume all graphs considered in this paper to be undirected and without parallel edges or self-loops. Given a graph $G = (V, E)$ we denote its vertex set and edge set with $V(G)$ and $E(G)$ respectively. If $V' \subseteq V(G)$, we call the graph $(V', \{\{v, u\} \in E(G) : v, u \in V'\})$ *the subgraph of G induced by V'* and we denote it by $G[V']$. For any edge set $E \subseteq E(G)$ we denote by $V(E)$ the set of vertices that are incident to edges of E (i.e. $V(E) = \cup_{e \in E} e$). The degree of a vertex v in graph G is the number of edges containing it and is denoted by $d_G(v)$. We call a vertex *pendant* when it has degree 1. We call an edge of a graph *pendant* when it contains a pendant vertex.

Pathwidth: A *path decomposition* of a graph $G = (V, e)$ is a sequence $X = (X_0, \ldots, X_{|X|})$ of subsets of $V(G)$ such that **(i)** $V = \bigcup_{0 \leq i \leq |X|} X_i$, **(ii)** for any $\{v, w\} \in e$, there is an i, $1 \leq i \leq |X|$, where $\{v, w\} \subseteq X_i$, and **(iii)** for all i, i', i'', $0 \leq i \leq i' \leq i''$, $X_i \cap X_{i''} \subseteq X_{i'}$. The sets X_i are called the *nodes* of the path decomposition. The width of a path decomposition $X = (X_0, 1 \leq i \leq |X|)$ equals $\max_{0 \leq i \leq |X|} \{|X_i| - 1\}$. The *pathwidth* of a graph G is the minimum width over all path decompositions of G. If $X = (X_i, 0 \leq i \leq |X|)$ is a path decomposition of a graph G, we say that X is *nice* if $|X_0| = 1$ and $\forall_{i, 1 \leq i \leq |X|} (X_i - X_{i-1}) \cup (X_{i-1} - X_i)| = 1$. It is easy to see that for any constant k, given a path decomposition of a graph G that has width at most k and $O(|V(G)|)$ nodes, one can find a nice path decomposition of G that has width at most k and at most $2|V(G)|$ nodes in $O(|V(G)|)$ time.

Let X_i be a node of a nice path decomposition X such that $i \geq 1$. We say that X_i is an *introduce (forget)* node if $|X_i - X_{i-1}| = 1$ ($|X_{i-1} - X_i| = 1$). It is easy to observe that any node $X_i, i \geq 1$ of a nice path decomposition is either an introduce or a forget node.

Linear-width: The linear-width of a graph is defined as follows. Let G be a graph and $l = (e_1, \ldots, e_r)$ an ordering of $E(G)$. For $i = 1, \ldots, r - 1$, we define $\delta_l(e_i) = V((l)_{1,i}) \cap V((l)_{i+1,r})$ (i.e. $\delta_l(e_i)$ is the set of vertices in $V(G)$ that are endpoints of an edge in $(l)_{1,i}$ and also of an edge in $(l)_{i+1,|E(G)|}$). The linear-width of an ordering l is $\max_{1 \leq i < r} \{|\delta_l(e_i)|\}$. The linear-width of a graph is the minimum linear-width over all the orderings of $E(G)$. It is not hard to prove that for every graph G, pathwidth$(G) \leq$ linear-width(G) (see [6]).

Given an edge ordering l we define $P(l) = (\emptyset, \delta_l(e_1), \ldots, \delta_l(e_{r-1}), \emptyset)$. Moreover, we define $\mathbf{Q}(l) = ((|P(0)|), \ldots, (|P(r)|))$ Notice that $P(l)$ is a sequence of sets and $\mathbf{Q}(l)$ is a sequence of sequences of numbers each containing only one element. We use this, somewhat overloaded, definition for reasons of consistency with the terminology of a characteristic that will be introduced later. We finally define the *pendant sequence* of l as $H(l) = (\emptyset, e_1 \cap A(G), \ldots, e_r \cap A(G))$ where $A(G) = \{v \in V(G) : d_G(v) = 1\}$ (i.e. for $1 \leq i \leq r$ the element of $H(l)$ indexed by i is the set of pendant endpoints of e_i).

Given a vertex set V, a vertex $x \in V$, and a sequence of vertex sets $I = (I_0, \ldots, I_r)$ where $I_i \subseteq V, i = 0 \ldots, r$ and $x \in \cup_{i=0,\ldots,r} I_i$, we define $F_I(x) = \min\{m, 0 \leq m \leq r \mid x \in I_m\}$ and $L_I(x) = \max\{m, 0 \leq m \leq r \mid x \in I_m\}$.

3   A Decision Algorithm for Linear-Width

In this section, we give for every pair of integer constants k, l, an algorithm that, given a graph G with a path decomposition of width at most l, decides whether G has linear-width at most k.

Sequence of integers: If $A = (a_1, \ldots, a_r)$ is a sequence of integers, we define $\max(A) = \max_{1 \leq i \leq r}\{\max(a_i)\}$ and for any integer t we set $A + t = (a_1 + t, \ldots, a_r + t)$. The *typical sequence* $\tau(A)$ of a sequence of integers A is the sequence obtained after iterating the following operations, until none is possible any more.
(i) If for some $i, 1 \leq i \leq r - 1$ $a_i = a_{i+1}$, then set $A = (a_1, \ldots, a_r) \leftarrow (a_1, \ldots, a_i, a_{i+2}, \ldots, a_r)$.
(ii) If the sequence contains two elements a_i and a_j such that $i+2 \leq j$ and either $\forall_{k,i<k<j} \, a_i \leq a_k \leq a_j$ or $\forall_{k,i<k<j} \, a_i \geq a_k \geq a_j$, then set $A = (a_1, \ldots, a_r) \leftarrow (a_1, \ldots, a_i, a_j, \ldots, a_r)$.
As an example, if $A = (\mathbf{5}, 7, 6, 7, \mathbf{7}, 7, 3, 4, \mathbf{3}, 5, 4, 3, 6, \mathbf{8}, \mathbf{2}, 9, 3, 4, \mathbf{9}, 6, 7, \mathbf{2}, \mathbf{7}, 5, 4, 4, 4, \mathbf{7}, 4, \mathbf{6}, 4)$, then $\tau(A) = (5, 7, 3, 8, 2, 9, 2, 7, 4)$.

We define the set $E(A)$ of the *extensions* of A as the set containing any set A^* such that A can be obtained by A^* after applying operation **(i)** of the definition of a typical sequence until this is not possible any more. More formally:

$$E(A) = \{A^* = (a_1^*, \ldots, a_{A^*}^*) \mid \exists_{1=t_1 < \ldots < t_{|A|+1}} \, \forall_{i,1 \leq i \leq |A|} \, \forall_{k,t_i \leq k < t_{i+1}} \, a_k^* = a_i\}.$$

Let A, B be two typical sequences where $A = (a_1, \ldots, a_r)$ and $B = (b_1, \ldots, b_r)$ (notice that $|A| = |B| = r$). Then we say that $A \leq B$ if $\forall_{i,1 \leq i \leq |A|} \, a_i \leq b_i$. We also say that $A \prec B$ if there exist extensions $A^* \in E(A), B^* \in E(B)$ such that $|A^*| = |B^*|$ and $A^* \leq B^*$. For example if $A = (5, 7, 4, 8)$ and $B = (1, 7, 2, 6, 4)$ then $B \prec A$ because $A^* = (5, 7, 7, 7, 4, 8, 8, 8, 8)$ is an extension of A, $B^* = (1, 7, 2, 6, 4, 4, 4, 4, 4)$ is an extension of A, and $B^* \leq A^*$.

Suppose now that $\mathbf{A} = (A_0, \ldots, A_r)$ and $\mathbf{B} = (B_0, \ldots, B_r)$ are two equal length sequences of sequences. We say that $\mathbf{A} \prec \mathbf{B}$ if $\forall_{i,0 \leq i \leq r} \, A_i \prec B_i$. Finally, for any integer t we set $\mathbf{A} + t = (A_0 + t, \ldots, A_r + t)$ and $\max(\mathbf{A}) = \max_{0 \leq i \leq r}\{\max(A_i)\}$.

Characteristics of a typical triple: Let X be a path decomposition of a graph G and let X_i be a node in X where $0 \leq i \leq |X|$. Let $S = (S_0, \ldots, S_r)$ be a sequence of vertex sets where $S_j \subseteq X_i, 0 \leq j \leq r$, $D = (D_0, \ldots, D_r)$ a sequence of subsets of X_i where $|D_j| \leq 2, 0 \leq j \leq r$ and $\forall_{j,0 \leq j < i \leq |D|} D_i \cap D_j = \emptyset$, and $\mathbf{T} = (T_0, \ldots, T_r)$ be a sequence of typical sequences. We call such a triple $(S, D, \mathbf{T})$ a *typical triple* of X_i.

Given a typical triple $(S, D, \mathbf{T})$, of X_i we define its *characteristic* $C(S, D, \mathbf{T})$ as a typical triple $(I, K, \mathbf{A})$ defined as follows.

Let $(r_0, \ldots, r_{q+1})$ be the sequence with $0 = r_0 < \cdots < r_{q+1} = r + 1$ where

(a) $\forall_{j,0 \leq j \leq q} \ \forall_{h, r_j \leq h < r_{j+1}} \ (S_h = S_{r_j} \text{ and } D_h = D_{r_j})$,

(b) $\forall_{j,0 \leq j < q} \ (S_{r_{j+1}} \neq S_{r_j} \text{ or } D_{r_{j+1}} \neq D_{r_j})$.

We call the pair (I, K) where $I = (S_{r_0}, \ldots, S_{r_q})$ and $K = (D_0, \ldots, D_{r_q})$ the *interval model* of S and D. Notice that, as $\forall_{j,0 \leq j < i \leq r} D_i \cap D_j = \emptyset$, we can observe that if $D_{r_j} \neq \emptyset$ then $r_{j+1} = r_j + 1$ and $r_{j-1} = r_j - 1$.

For any $j, 0 \leq j \leq q$, let $A_j = (T_{r_j} \oplus \cdots \oplus T_{r_{j+1}-1})$. We also set $\mathbf{A} = (A_0, \ldots, A_q) \leftarrow (\tau(A_0), \ldots, \tau(A_q))$ and we call the typical triple $(I, K, \mathbf{A})$ the *characteristic* of the typical triple $(S, D, \mathbf{T})$. Given two typical triples $(S^1, D^1, \mathbf{T}^1)$, $j = 1, 2$ we say that $(S^1, D^1, \mathbf{T}^1) \equiv (S^2, , D^2, \mathbf{T}^2)$ if they have the same characteristic i.e. $C(S^1, D^1, \mathbf{T}^1) = C(S^2, D^2, \mathbf{T}^2)$. We also say that $(S^1, D^1, \mathbf{T}^1) \prec (S^2, D^2, \mathbf{T}^2)$ if $S^1 = S^2$, $D^1 = D^2$ (i.e. they have the same interval model), and $\mathbf{T}^1 \prec \mathbf{T}^2$. It is easy to see that relation "$\prec$" is transitive i.e. if $\mathbf{T}^1 \prec \mathbf{T}^2$ and $\mathbf{T}^2 \prec \mathbf{T}^3$ then $\mathbf{T}^1 \prec \mathbf{T}^3$. We also extend the definition of "$\oplus$" so that whenever $(S^i, D^i, \mathbf{T}^i), i = 1, 2$ are two typical triples, the typical triple $(S^1 \oplus S^2, D^1 \oplus D^2, \mathbf{T}^1 \oplus \mathbf{T}^2)$ is denoted by $(S^1, D^1, \mathbf{T}^1) \oplus (S^2, D^2, \mathbf{T}^2)$.

Characteristic of an ordering: Let X be a path decomposition of a graph G and X_i some node of X where $0 \leq i \leq |X|$. For $i = 0, \ldots, |X|$, we define $V_i = \bigcup_{0 \leq j \leq i} X_j$ and $G_i = G[V_i]$.

Let l be an edge ordering of G_i with linear-width at most k. Let also $P(l) = (P_0, \ldots, P_r)$. We define the *restriction* of l on X_i as $R(l) = (P_0 \cap X_i, \ldots, P_r \cap X_i)$.

We call the triple $C(R(l), H(l), \mathbf{Q}(l)) = (I, K, \mathbf{A})$ *characteristic* $C(l)$ of l. We say that the pair (I, K) is an *interval model at* X_i and $\mathbf{A}$ is the corresponding sequence of typical sequences. From now on, whenever we consider an edge ordering we will associate it with some node X_i of a path decomposition.

Using Lemmata 3.5 and 3.5 of [4] we can prove that the number of different characteristics of all possible edge orderings of G_i with linear-width at most k, is bounded by a function of k and l (i.e. is independent of n).

The following procedure defines function Com, that maps a typical triple $(I, K, \mathbf{A})$ to another typical triple that is a "compression" of the input triple.

PROCEDURE $\mathsf{Com}(I, K, \mathbf{A})$

Input: A typical triple $(I, K, \mathbf{A})$

Output: A typical triple $(I, K, \mathbf{A})$.

1: Set $r = |I| = |K| = |\mathbf{A}|$.

2: Apply the following operation until it is no longer possible.

• If $\exists_{0 \leq h \leq r-1} : (I_h = I_{h+1} \text{ and } K_{h+1} = \emptyset)$ then set

$I = (I_0, \ldots, I_{r-1}) \leftarrow (I_0, \ldots, I_h, I_{h+2}, \ldots, I_r)$,

$K = (K_0, \ldots, K_{r-1}) \leftarrow (K_0, \ldots, K_h, K_{h+2}, \ldots, K_r)$,

$\mathbf{A} \quad = \quad (A_0, \ldots, A_{r-1}) \quad \leftarrow \quad (A_0, \ldots, A_{h-1}, \tau(A_h \ \oplus$

$A_{h+1}), A_{h+2}, \ldots, A_r)$, and

$r \leftarrow r - 1$.

3: end.

A set $FS(i)$ of characteristics of edge orderings of a graph G_i (i is a node of the path decomposition) with width at most k is called a *full set of characteristics* at i if for each linear ordering l of G_i with linear-width at most k, there is a edge ordering l' such that $C(l') \prec C(l)$ and $C(l') \in FS(i)$, i.e. the characteristic of l' is in $FS(i)$. The following lemma can be derived directly from the definitions.

Lemma 1. *A full set of characteristics at i is non-empty if and only if the linear-width of G_i is at most k. If some full set of characteristics at i is non-empty, then every full set of characteristics at this node is non-empty.*

An important consequence of Lemma 1 is that the linear-width of G is at most k, if and only if any full set of characteristics of $G_{|X|} = G$ is non-empty. Clearly, in the cased $i = 0$, G_i is edgeless and thus $F_0 = \{((),(),())\}$. In what follows, we will show how to compute a full set of characteristics at a node X_i in $O(1)$ time, when a full set of characteristics of X_{i-1} is given. This will depend on the type of the node X_i and will be explained in the following sections.

Introducing an edge: The procedure int is an important ingredient of our algorithm. Given a typical triple $(I, K, \mathbf{A})$, it "inserts an edge": after the mth position in the jth sequence in $\mathbf{A}$, an edge is inserted; S is the set of the vertices that are endpoints of the edge and W consists of the endpoints of S that are isolated before the insertion (and therefore become pendant after it). Both I, K, and $\mathbf{A}$ are modified accordingly after the insertion.

PROCEDURE $\mathsf{Int}(I, K, \mathbf{A}, j, m, S, W)$

Input A typical triple $(I, K, \mathbf{A})$ where $I = (I_0, \ldots, I_q)$, $K = (K_0, \ldots, K_q)$, $\mathbf{A} = (A_0, \ldots, A_q)$ where $A_i = (a_1^i, \ldots, a_{q_i}^i), 1 \leq i \leq q$, two integers j, m where $0 \leq j \leq q, 1 \leq m \leq q_i$, and two vertex sets $S, W \subseteq W$ where $S - W \subseteq \bigcup_{0 \leq i \leq q} A(i) \cup K(i)$.

Output A typical triple $(I, K, \mathbf{A})$.

1: (*Splitting step*) Apply the following three steps.
- Set $I' = (I_0', \ldots, I_{q+1}') \leftarrow (I_0, \ldots, I_{j-1}, I_j, I_j, I_{j+1}, \ldots, I_q)$.
- Set $K' = (K_0', \ldots, K_{q+1}') \leftarrow (K_0 - S, \ldots, K_{j-1} - S, K_j - S, W, K_{j+1} - S, \ldots, K_q - S)$.
- Set $\mathbf{A}' = (A_0', \ldots, A_{q+1}') \leftarrow (A_0, \ldots, A_{j-1}, (a_1^j, \ldots, a_m^j), (a_m^j, \ldots, a_{q_j}^j), A_{j+1}, \ldots, A_r)$.

2: (*Insertion step*) For any $x \in S - W$, apply the following.
- For any $h, 0 \leq h \leq q+1, \min\{L_{I' \cup K'}(x), j+1\} \leq h \leq \max\{j, F_{I' \cup K'}(x)-1\}$: set $I_h' \leftarrow I_h' \cup \{x\}$ and $A_h' \leftarrow A_h' + 1$

3: Output $I', K', \mathbf{A}'$.

4: end.

The following lemmata provide important information on procedure int. The proofs is quite technical (see [6] for the details).

Lemma 2. *Let l be an ordering and assume that $(I, K, \mathbf{A}) = C(R(l), H(l), \mathbf{Q}(l))$. Then for any vertex sets S, W where $W \subseteq S$, the following hold.*

(i) *For any $j, m, 0 \leq j \leq q = |I|, 1 \leq m \leq q_j = |A_j|$, there exists an integer $\gamma, 0 \leq \gamma \leq r = |l|$ such that*
$\mathsf{Com}(\mathsf{Int}(I, K, \mathbf{A}, j, m, S, W)) = \mathsf{Com}(\mathsf{Int}(R(l), H(l), \mathbf{Q}(l), \gamma, 1, S, W))$.

(ii) *For any $\gamma, 0 \leq \gamma \leq r = |l|$, there exist two integers $j, m, 0 \leq j \leq q = |I|, 1 \leq m \leq q_j = |A_j|$ such that*
$\mathsf{Com}(\mathsf{Int}(I, K, \mathbf{A}, j, m, S, W)) \prec \mathsf{Com}(\mathsf{Int}(R(l), H(l), \mathbf{Q}(l), \gamma, 1, S, W))$.

Lemma 3. *Let $(I^i, K^i, \mathbf{A}^i), i = 1, 2$ be two typical triples such that $(I^1, K^1, \mathbf{A}^1) \succ (I^2, K^2, \mathbf{A}^2)$. Then, for any $S, W \subseteq S, j, 0 \leq j \leq q = |I^1| = |I^2|$ and $m_1, 1 \leq m_1 \leq q_j^1 = |A_j^1|$ there exists a $m_2, 1 \leq m_2 \leq q_j^2 = |A_j^2|$, such that*
$\mathsf{Com}(\mathsf{Int}(I^1, K^1, \mathbf{A}^1, j, m_1, S, W)) \succ \mathsf{Com}(\mathsf{Int}(I^2, K^2, \mathbf{A}^2, j, m_2, S, W))$.

A full set for an introduce node: We will now consider the case where X_i is an *introduce* node. Clearly $V_i = V_{i-1} \cup \{x\}$ where $x \notin V_{i-1}$. Suppose that $E_x = \{e_1, \ldots, e_t\}, 0 \leq t \leq |X_{i-1}| \leq l$ is the set of edges incident to x in G_i (notice that, $\cup_{e \in E_x} e \subseteq X_i$). If $E_x = \emptyset$, then, we simply set $FS(i) = FS(i-1)$. What remains is to examine the case where $|E_x| \geq 1$.

We define $G_i^p = (V(G_i), E(G_{i-1}) \cup \{e_1, \ldots, e_p\}), 0 \leq p \leq t$. Clearly, $FS(i-1)$ is a full set of characteristics for $G_i^0 = G_{i-1}$. Notice also that $G_i = G_i^t$. Suppose that we have a full set of characteristics $FS(i, p-1)$ for $G_i^{p-1}, 1 \leq p \leq t$ (which is the case when $p = 1$). It is sufficient to give an $O(1)$ time algorithm constructing a full set of characteristics $FS(i, p)$ for G_i^p. In what follows $W = e_p \cap A(G_i^p)$ (i.e. the vertices of the inserted edge that are pendant in G_i^p).

ALGORITHM Introduce-edge
Input: A full set of characteristics $FS(i, p-1)$ for G_i^{p-1}.
Output: A full set of characteristics $FS(i, p)$ for G_i^p.
1: Initialise $FS(i, p) = \emptyset$.
2: For each characteristic $(I, K, \mathbf{A}) \in FS(i, p-1)$ where $I = (I_0, \ldots, I_q)$ and $\mathbf{A} = (A_0, \ldots, A_q)$, apply step **3**.
3: For any set $I_j \in I$ let $A_j = (a_1^j, \ldots, a_{q_j}^j)$ and apply step **4**.
4: For any $a_m^j, 1 \leq m \leq q_j$ in A_j,
    set $(I', K', \mathbf{A}') = \mathsf{Com}(\mathsf{Int}(I', K', \mathbf{A}', j, m, e_p, W))$ and
    if $\max(\mathbf{A}') \leq k$, then set $FS(i, p) \leftarrow FS(i, p) \cup \{(I', K', \mathbf{A}')\}$.
5: end.

The proof of the following is based on Lemmata 2 and 3.

Lemma 4. *The set $FS(i, p)$ constructed by algorithm* Introduce-edge *is a full set of characteristics.*

A full set for a forget node: We will now consider the case where X_i is a *forget* node. Clearly, $G_i = G_{i-1}$ and there exists a unique vertex $v \in X_{i-1}$ with $v \notin X_i$. We call this vertex v *forgotten*. Given a full set of characteristics $F(i-1)$ for X_{i-1}, the algorithm Forget-Vertex computes a full set of characteristics $F(i)$ for X_i.

ALGORITHM Forget-Vertex
Input: A full set of characteristics $FS(i-1)$ for G_{i-1} and a forgotten vertex x.
Output: A full set of characteristics $FS(i)$ for G_i.
1: Initialise $FS(i) = \emptyset$.
2: For any $(I, K, \mathbf{A}) \in FS(i-1)$ set $FS(i) \leftarrow FS(i) \cup \{\mathrm{Com}(I - \{x\}, K - \{x\}, \mathbf{A})\}$.
3: end.

Lemma 5. *The set $FS(i)$ constructed by the algorithm* Forget-Vertex *is a full set of characteristics.*

The decision algorithm: Using the algorithms of the previous sections we can compute a full set of characteristics for $G_1, G_2, G_3, \ldots, G_{|X|} = G$ (obviously $E(G_1) = \emptyset$). Notice that if a graph consists of a single edge $e_{\mathrm{start}} = \{v_{\mathrm{start}}^1, v_{\mathrm{start}}^2\}$, its full set of characteristics is $((\{v_{\mathrm{start}}^1, v_{\mathrm{start}}^2\}), (\{v_{\mathrm{start}}^1, v_{\mathrm{start}}^2\}), ((0)))$. Using this full set of characteristics as a starting point we can use the procedures of the previous sections to compute the full sets for G_2, G_3, etc., in order. Note that the computation needs $O(1)$ time per node of the path decomposition, and thus in total, time linear on the number of vertices of G. After the full set for the last node has been computed, in $O(1)$ time one can decide whether the linear-width of G is at most k, as this holds if and only if this last full set $FS(|X|)$ is not empty. We conclude with the algorithm Check-Linear-width. It is possible to turn this algorithm to a constructive one (see [6] for the details).

Algorithm Check-Linear-width(G, X, k).
Input: A graph G, a path decomposition X of G with width l, and an integer k.
Output: Whether the linear-width of (G) is at most k.

1: Let $F_0 = \{((), (), ())\}$.
2: For any i, $1 \leq i \leq |X|$,
   compute a full set of X_i-characteristics F_i for G_i using F_{i-1} and depending on whether X_i is an introduce or a forget node.
3: Output $F_r \neq \emptyset$.
4: End.

We conclude to the following.

Theorem 1. *For all $k, l \geq 1$ there exists an algorithm that, given a graph G and a path decomposition $X = (X_i, 1 \leq i \leq |X|)$ of G with width at most l, computes whether the linear-width of G is at most k and, if so, constructs an edge ordering of G with linear-width at most k and that uses $O(V(G) + |X|)$ time.*

The results presented so far can be trivially extended to graphs with parallel edges. In such a case, we should consider the complexity of the algorithm in Theorem 1 to be $O(|E(G)| + |X|)$.

4   The Consequences of Our Algorithm

In this section we define several search game parameters and we present their relations with linear-width. Using these relations we conclude that there exist for any fixed k, linear time algorithms that check whether given graphs can be mixed, node, or edge searched with at most k searchers, and if so, output the corresponding search strategies.

A *mixed searching game* is defined in terms of a graph representing a system of tunnels where an omniscient and agile fugitive with unbounded speed is hidden (alternatively, we can formulate the same problem considering that the tunnels are contaminated by some poisonous gas). The object of the game is to *clear* all edges, using one or more *searchers*. An edge of the graph is cleared if one of the following cases occur: **A**: *both of its endpoints are occupied by a searcher*, **B**: *a searcher slides along it*, i.e., a searcher is moved from one endpoint of the edge to the other endpoint.

A search is a sequence containing some of the following moves. **a**: place a new searcher on a vertex, **b**: remove a searcher from a vertex, **c**: slide a searcher, residing on some of the endpoints of an edge e, along e and place it on the other endpoint of e. The object of a mixed search is to clear all edges using a search. The search number of a search is the maximum number of searchers on the graph during any move. The mixed search number, $\mathrm{ms}(G)$, of a graph G is the minimum search number over all the possible searches of it. A move causes *recontamination* of an edge if it causes the appearance of a path from an uncleared edge to this edge not containing any searchers on its vertices or its edges. (Recontaminated edges must be cleared again.) A search without recontamination is called *monotone*. The *node (edge) search number*, $\mathrm{ns}(G)$ $(\mathrm{es}(G))$ is defined similarly to the mixed search number with the difference that an edge can be cleared only if **A** (**B**) happens. Parts (a)–(e) of the following result were proved (or follow easily) by Bienstock and Seymour in [2] (see also [24]). The last part is proven in [25].

Theorem 2. *For any graph G the following hold:*
(a) *If $\mathrm{ms}(G) \leq k$ then there exist a monotone mixed search in G using at most k searchers.*
(b) *linear-width$(G) \leq \mathrm{ms}(G)$.*
(c) *If G^e is the graph occurring from G after subdividing each of it edges, then $\mathrm{es}(G) = \mathrm{ms}(G^e)$.*
(d) *If G^n is the graph occurring if we replace every edge in G with two edges in parallel, then $\mathrm{ns}(G) = \mathrm{ms}(G^n)$.*
(e) *If G^h is the graph occurring from G after subdividing each of its pendant edges, then $\mathrm{ms}(G) = $ linear-width(G^h).*

The result of Theorem 1 has several consequences. First, as one can find a path decomposition of a graph G with width at most l, if existing, in linear time ([3,4], but see also below) and using the fact that pathwidth$(G) \leq$ linear-width(G) we can conclude to the following result.

Theorem 3. *For all $k \geq 1$, there exists an algorithm that, given a graph G, computes whether the linear-width of G is at most k and, if so, constructs an edge ordering of G with linear-width at most k and that uses $O(V(G))$ time.*

Using now Theorem 2 , we obtain the following.

Theorem 4. *For all k, l, there exists an algorithm, that given a graph G and a path decomposition $X = (X_i, 1 \leq i \leq |X|)$ of G with width at most l, computes whether the mixed search number (edge search number; node search number) of G is at most k, and if so, constructs a mixed search (edge search; node search) that clears G with most k searchers, and that uses at most $O(|V(G)| + |X|)$ time.*

Using small modifications of techniques from [3], the result above can be used to obtain an alternative (but strongly related) proof for the result from [3,4] that for each fixed k, the problem to determine whether a given graph has pathwidth at most k, and if so, to find a path decomposition of width at most k, has a linear time parameterized algorithm. Using this fact, we obtain from Theorem 4 the following.

Theorem 5. *For all k, there exists an algorithm, that given a graph G, computes whether the mixed search number (edge search number; node search number) of G is at most k, and if so, constructs a monotone mixed search (edge search; node search) that clears G with most k searchers, and that uses at most $O(|V(G)|)$ time.*

References

1. D. Bienstock. Graph searching, path-width, tree-width and related problems (a survey). In *Reliability of computer and communication networks (New Brunswick, NJ, 1989)*, volume 5 of *DIMACS Ser. Discrete Math. Theoret. Comput. Sci.*, pages 33–49. Amer. Math. Soc., Providence, RI, 1991.
2. D. Bienstock and P. Seymour. Monotonicity in graph searching. *Journal of Algorithms*, 12(2):239–245, 1991.
3. H. L. Bodlaender. A linear-time algorithm for finding tree-decompositions of small treewidth. *SIAM J. Comput.*, 25(6):1305–1317, 1996.
4. H. L. Bodlaender and T. Kloks. Efficient and constructive algorithms for the pathwidth and treewidth of graphs. *Journal of Algorithms*, 21:358–402, 1996.
5. H. L. Bodlaender and D. M. Thilikos. Constructive linear time algorithms for branchwidth. In *Automata, languages and programming (Bologna, 1997)*, volume 1256 of *Lecture Notes in Computer Science*, pages 627–637. Springer, Berlin, 1997.
6. H. L. Bodlaender and D. M. Thilikos. Computing small search numbers in linear time. Technical Report UU-CS-1998-05, Dept. of Computer Science, Utrecht University, 1998.
7. R. Breisch. An intuitive approach to speleotopology. *A publication of the Southwestern Region of the National Speleological Society*, VI:72–78, 1967.
8. N. Dendris, L. Kirousis, and D. Thilikos. Fugitive-search games on graphs and related parameters. *Theoretical Computer Science*, 172:233–254, 1997.

9. J. A. Ellis, I. H. Sudborough, and J. S. Turner. The vertex separation and search number of a graph. *Inform. and Comput.*, 113(1):50–79, 1994.
10. M. R. Fellows and M. A. Langston. On search, decision, and the efficiency of polynomial-time algorithms. *J. Comput. System Sci.*, 49(3):769–779, 1994.
11. H. Friedman, N. Robertson, and P. Seymour. The metamathematics of the graph minor theorem. In *Logic and combinatorics (Arcata, Calif., 1985)*, volume 65 of *Contemp. Math.*, pages 229–261. Amer. Math. Soc., Providence, RI, 1987.
12. N. G. Kinnersley. The vertex separation number of a graph equals its path-width. *Inform. Process. Lett.*, 42(6):345–350, 1992.
13. L. M. Kirousis and C. H. Papadimitriou. Interval graphs and searching. *Discrete Math.*, 55(2):181–184, 1985.
14. L. M. Kirousis and C. H. Papadimitriou. Searching and pebbling. *Theoret. Comput. Sci.*, 47(2):205–218, 1986.
15. A. S. LaPaugh. Recontamination does not help to search a graph. *J. Assoc. Comput. Mach.*, 40(2):224–245, 1993.
16. N. Megiddo, S. L. Hakimi, M. R. Garey, D. S. Johnson, and C. H. Papadimitriou. The complexity of searching a graph. *J. Assoc. Comput. Mach.*, 35(1):18–44, 1988.
17. R. H. Möhring. Graph problems related to gate matrix layout and PLA folding. In *Computational graph theory*, volume 7 of *Comput. Suppl.*, pages 17–51. Springer, Vienna, 1990.
18. T. D. Parsons. Pursuit-evasion in a graph. In *Theory and applications of graphs (Proc. Internat. Conf., Western Mich. Univ., Kalamazoo, Mich., 1976)*, pages 426–441. Lecture Notes in Math., Vol. 642. Springer, Berlin, 1978.
19. N. Robertson and P. D. Seymour. Disjoint paths—a survey. *SIAM J. Algebraic Discrete Methods*, 6(2):300–305, 1985.
20. N. Robertson and P. D. Seymour. Graph minors—a survey. In *Surveys in combinatorics 1985 (Glasgow, 1985)*, volume 103 of *London Math. Soc. Lecture Note Ser.*, pages 153–171. Cambridge Univ. Press, Cambridge, 1985.
21. N. Robertson and P. D. Seymour. An outline of a disjoint paths algorithm. *Paths, Flows and VLSI Design, Algorithms and Combinatorics*, 9:267–292, 1990.
22. N. Robertson and P. D. Seymour. Graph minors. XIII. The disjoint paths problem. *Journal of Combinatorial Theory. Series B*, 63(1):65–110, 1995.
23. P. D. Seymour and R. Thomas. Graph searching and a min-max theorem for tree-width. *J. Combin. Theory Ser. B*, 58(1):22–33, 1993.
24. A. Takahashi, S. Ueno, and Y. Kajitani. Minimal forbidden minors for the family of graphs with proper-path-width at most two. *IEICE Trans. Fundamentals*, E78-A:1828–1839, 1995.
25. D. M. Thilikos. Algorithms and obstructions for linear-width and related search parameters. *Discrete Applied Mathematics*, 105:239–271, 2000.
26. D. M. Thilikos, M. J. Serna, and H. L. Bodlaender. Constructinve linear time algorithms for small cutwidth and carving-width. In D. Lee and S.-H. Teng, editors, *Proc. 11th International Conference ISAAC 2000*, volume 1969 of *Lectures Notes in Computer Science*, pages 192–203. Springer-Verlag, 2000.
27. D. M. Thilikos, M. J. Serna, and H. L. Bodlaender. A polynomial time algorithm for the cutwidth of bounded degree graphs with small treewidth. In *Algorithms— ESA 2001 (Århus)*, volume 2161 of *Lecture Notes in Comput. Sci.*, pages 380–390. Springer, Berlin, 2001.
28. R. Thomas. Tree-decompositions of graphs. Lecture notes, School of Mathematics. Georgia Institute of Technology, Atlanta, Georgia 30332, USA, 1996.

Bounded Fixed-Parameter Tractability: The Case $2^{\mathrm{poly}(k)}$

Mark Weyer

Abteilung für Mathematische Logik, Albert-Ludwigs-Universität Freiburg, Germany
mark.weyer@math.uni-freiburg.de

Abstract. We introduce a notion of fixed-parameter tractability permitting a parameter dependence of only $2^{\mathrm{poly}(k)}$. We delve into the corresponding intractability theory. In this course we define the PW-hierarchy, provide some complete problems, and characterize each of its classes in terms of model-checking problems, in terms of propositional satisfiability problems, in terms of logical definitions, and by machine models.
We also locate the complexity of some model-checking problems that are fixed-parameter tractable but not boundedly so.

1 Introduction

The aim of complexity theory can be described as a classification of combinatorial problems into tractable ones and intractable ones. The research of complexity theory consists in large parts of charting specific problems. A smaller but not less important part is the evolution of the underlying concepts of tractability and intractability. For example the notion of fixed-parameter tractability (FPT) from parameterized complexity theory can be motivated by several aspects, in which the unparameterized theory around polynomial time (P) deviates from practical reality. However, the introduction of FPT was (intentionally) simplistic, permitting arbitrary, or arbitrary computable, dependence of the running time on the parameter. In this context general techniques such as Courcelle's Theorem or the Well-Quasi-Ordering Principle spawn tractability results that cannot be transfered into remotely practical algorithms.

The remedy is, of course, to introduce bounds on the parameter dependence. There are, however, several possibilities to do so. In [1] the case $2^{O(k)}$ was explored; despite all its merits it is less robust than might be desirable. Here we will investigate the case $2^{\mathrm{poly}(k)}$. It will turn out, that this choice retains most of FPT's robustness, while new structure appears.

We will be concerned mainly with intractability exploration. We identify classes that are analogues of the W-hierarchy. However our PW-hierarchy has additional intermediate classes. After identifying some complete problems we present characterizations of all its classes in terms of propositional logic, first-order logic, and machine models. Additionally we locate the problems from [2], that initiated the examination of bounded parameterized complexity, in our new framework. Indeed they are complete for PAW [∗] and PPA; the former is our

R. Downey, M. Fellows, and F. Dehne (Eds.): IWPEC 2004, LNCS 3162, pp. 49–60, 2004.
© Springer-Verlag Berlin Heidelberg 2004

analogue of AW [*], the latter is a class the unbounded analogue of which has not yet been investigated to the author's knowledge.

Space limitations force us to defer many proofs to the full version.

2 Basic Notions

2.1 Parameterized Complexity

Let Σ be an alphabet with at least two characters. We will assume all encodings to be made in Σ. A parameterized (decision) problem is a pair $P = (Q, \kappa)$, where $Q \subseteq \Sigma^*$ is a classical decision problem and $\kappa : \Sigma^* \to \mathbb{N}$ is a parameterization. Parameterized problems will usually be stated in the following form:

> Input: A structure $\mathfrak{A}$ and a formula φ
> Question: Does $\mathfrak{A} \models \varphi$ hold?
> Parameter: $|\varphi|$

This notation assumes some encoding of structures and of formulas. Here Q is the set of encodings of pairs $(\mathfrak{A}, \varphi)$, such that $\mathfrak{A}$ is a structure, φ is a formula, and $\mathfrak{A} \models \varphi$ does hold. The parameterization κ is:

$$\kappa : x \mapsto \begin{cases} |\varphi| \text{ , if } x \text{ is of the form } (\mathfrak{A}, \varphi) \\ 0 \text{ otherwise} \end{cases}$$

The default value 0 is arbitrary.

Let $\mathcal{F}$ be a set of functions $\mathbb{N} \to \mathbb{N}$. Then the class $\mathcal{F}$-FPT consists of those parameterized problems (Q, κ), for which the question '$x \in Q$?' can be decided in deterministic time $f(\kappa(x)) \cdot |x|^c$ for some $f \in \mathcal{F}$ and $c \in \mathbb{N}$. The usual notion of fixed-parameter tractability is

$$\text{FPT} := \{ f : \mathbb{N} \to \mathbb{N} \mid f \text{ computable} \}\text{-FPT} .$$

Here we will be concerned with the stricter notion

$$\text{PPT} := \mathcal{PE}\text{-FPT} ,$$

where

$$\mathcal{PE} := \left\{ k \mapsto 2^{k^c} \ \middle| \ c \in \mathbb{N} \right\}$$

(polynomially exponential functions).

Let $P = (Q, \kappa)$ and $P' = (Q', \kappa')$ be parameterized problems. A many-one reduction R from Q to Q' is called

1. an fpt-reduction from P to P', if there are computable functions f and g and a natural number c, such that $R(x)$ is computable from x in time $f(\kappa(x)) \cdot |x|^c$, and such that $\kappa'(R(x)) \leq g(\kappa(x))$ for all but finitely many $x \in \Sigma^*$.

2. a ppt-reduction from P to P', if there are natural numbers c, d, and e, such that $R(x)$ is computable from x in time $2^{(\kappa(x))^d} \cdot |x|^c$, and such that $\kappa'(R(x)) \in (\kappa(x))^e + (\log |x|)^{o(1)}$, that is, for all natural numbers f there is a natural number g such that $\kappa'(R(x)) \leq (\kappa(x))^e + g \cdot (\log |x|)^{\frac{1}{f}}$ for all but finitely many $x \in \Sigma^*$.

3. a simple ppt-reduction from P to P', if there are natural numbers c, d, and e, such that $R(x)$ is computable from x in time $2^{(\kappa(x))^d} \cdot |x|^c$, and such that $\kappa'(R(x)) \leq (\kappa(x))^e$ for all but finitely many $x \in \Sigma^*$.

The natural properties hold: fpt-reductions preserve pertainment to FPT, ppt-reductions preserve pertainment to PPT, and simple ppt-reductions are ppt-reductions. Furthermore simple ppt-reductions are fpt-reductions. It should be noted that the only results, that require non-simple ppt-reductions, are Proposition 12 and Proposition 13. In these cases $\kappa'(R(x)) \in O((\kappa(x))^e + \log\log |x|)$. For $\alpha \in \{f, p\}$ we write $P \leq^{\alpha\text{pt}} P'$ if an αpt-reduction from P to P' exists. We write $P \equiv^{\alpha\text{pt}} P'$ if both $P \leq^{\alpha\text{pt}} P'$ and $P' \leq^{\alpha\text{pt}} P$. Finally we define the class

$$[P]^{\alpha\text{pt}} := \{ P' \mid P' \leq^{\alpha\text{pt}} P \}.$$

2.2 Propositional Logic

Propositional formulas are built from propositional variables by conjunction, disjunction, and negation. Let Prop denote the class of all propositional formulas. A propositional literal is a propositional variable (atom) or a negated propositional variable.

For natural numbers $t \geq 0$ and $d \geq 1$ we define the classes $\Gamma_{t,d}, \Delta_{t,d} \subseteq \text{Prop}$ by recursion over t:

$$\Gamma_{0,d} := \{\lambda_1 \wedge \cdots \wedge \lambda_s \mid s \leq d \text{ and all } \lambda_i \text{ are propositional literals}\}$$
$$\Delta_{0,d} := \{\lambda_1 \vee \cdots \vee \lambda_s \mid s \leq d \text{ and all } \lambda_i \text{ are propositional literals}\}$$
$$\Gamma_{t+1,d} := \{\bigwedge \Pi \mid \Pi \subseteq \Delta_{t,d}\}$$
$$\Delta_{t+1,d} := \{\bigvee \Pi \mid \Pi \subseteq \Gamma_{t,d}\}$$

For $\Gamma \subseteq \text{Prop}$ define $\Gamma^+, \Gamma^- \subseteq \Gamma$ as follows: A formula $\alpha \in \Gamma$ belongs to Γ^+, if all atoms in α occur positively and α belongs to Γ^-, if all atoms in α occur negatively.

For $\Gamma \subseteq \text{Prop}$ let $p\text{-WSAT}(\Gamma)$ (parameterized weighted satisfiability problem) be the following parameterized problem:

Input: A formula $\alpha \in \Gamma$ and $k \in \mathbb{N}$
Question: Does α have a satisfying assignment that sets exactly k variables to TRUE?
Parameter: k

Often the following problem p-PSAT(Γ) (parameterized partitioned satisfiability problem) is more convenient:

> Input: A formula $\alpha \in \Gamma$ and a partition $(\mathcal{X}_i)_{1 \leq i \leq k}$ of var(α)
> Question: Does α have a satisfying assignment that sets exactly one variable from each $\mathcal{X}_i$ to TRUE?
> Parameter: k (the number of sets in the partition)

2.3   First-Order Logic

Some aquaintance with first-order logic is assumed, we will only fix some conventions. We restrict ourselves to relational vocabularies. First-order atoms are of the form $Rx_1 \ldots x_r$, where the x_i are first-order variables and the relation symbol R has arity r. First-order formulas are built from first-order atoms by conjunction, disjunction, negation, and existential and universal quantification.

Let t be a natural number. A first-order formula φ pertains to Σ_t if

1. t is odd and φ is of the form

$$\varphi = \exists x_{1,1} \ldots x_{1,k_1} \forall x_{2,1} \ldots x_{2,k_2} \ldots \exists x_{t,1} \ldots x_{t,k_t} \psi$$

   with quantifier free ψ, or
2. t is even and φ is of the form

$$\varphi = \exists x_{1,1} \ldots x_{1,k_1} \forall x_{2,1} \ldots x_{2,k_2} \ldots \forall x_{t,1} \ldots x_{t,k_t} \psi$$

with quantifier free ψ.

For a natural number u such a φ pertains to $\Sigma_{t,u}$, if all but the first quantifier block are of length at most u, that is, if $k_i \leq u$ for all $i \geq 2$. φ furthermore pertains to $\Sigma^*_{t,u}$, if

1. t is odd and ψ is a conjunction of literals, or
2. t is even and ψ is a disjunction of literals.

φ pertains to Π_t, if the formula obtained from φ by replacing $\exists$ by $\forall$ and vice versa pertains to Σ_t. If L is any of these logics, define L^+ to be the fragment in which no atom occurs negatively and L^- to be the fragment in which no atom occurs positively.

Let $\tau = \{R_1, \ldots, R_s\}$ be a relational vocabulary, the arity of R_i being r_i. A τ-structure is a tuple $\mathfrak{A} = (A, R_1^{\mathbb{A}}, \ldots, R_s^{\mathbb{A}})$, where each $R_i^{\mathbb{A}} \subseteq A^{r_i}$. The arity of $\mathfrak{A}$ is $\max\limits_{1 \leq i \leq s} r_i$. The size $|\mathfrak{A}|$ of $\mathfrak{A}$ is defined as

$$|\mathfrak{A}| := |A| + |\tau| + \sum_{1 \leq i \leq s} r_i \cdot |R_i^{\mathbb{A}}|$$

and only polynomially deviates from the length of a reasonable encoding of $\mathfrak{A}$.

Let L be a logic and $\mathcal{C}$ be a class of structures. We present two parameterizations of the model-checking problem for L on $\mathcal{C}$, denoted $p\text{-MC}(\mathcal{C}, L)$ and $p\text{-MC}_{\mathrm{var}}(\mathcal{C}, L)$:

> Input: A structure $\mathfrak{A} \in \mathcal{C}$ and a formula $\varphi \in L$
> Question: Does $\mathfrak{A} \models \varphi$ hold?
> Parameter: $|\varphi|$

> Input: A structure $\mathfrak{A} \in \mathcal{C}$ and a formula $\varphi \in L$
> Question: Does $\mathfrak{A} \models \varphi$ hold?
> Parameter: $|\mathrm{var}(\varphi)|$

For the first case $|\varphi|$ is defined as the number of nodes in a syntax tree for φ. This is the length of φ in a rich enough alphabet. We will simply write $p\text{-MC}(L)$, if $\mathcal{C}$ is the class of all structures, $p\text{-MC}(s, L)$ for $s \in \mathbb{N}$, if $\mathcal{C}$ is the class of all structures of arity at most s, and likewise use $p\text{-MC}_{\mathrm{var}}(L)$ and $p\text{-MC}_{\mathrm{var}}(s, L)$.

3 The PW-Hierarchy

The notions $\leq^{\mathrm{fpt}}$ and $\leq^{\mathrm{ppt}}$ are incomparable (For one direction consider problems $P \in \text{FPT} \setminus \text{PPT}$ and $P' \in \text{PPT}$, with P' being non-trivial. Then $P \leq^{\mathrm{fpt}} P'$ but $P \not\leq^{\mathrm{ppt}} P'$. For the other direction consider an unparameterized problem Q that is undecidable and the parameterizations $\kappa : x \mapsto 0$ and $\kappa' : x \mapsto \log\log|x|$. Then $(Q, \kappa) \leq^{\mathrm{ppt}} (Q, \kappa')$, but $(Q, \kappa) \not\leq^{\mathrm{fpt}} (Q, \kappa')$). At first glance it would then seem, that the theory of lower bounds has to be built up from scratch. It turns out, on the other hand, that many completeness results from the FPT-theory (that is, statements of the form $P \equiv^{\mathrm{fpt}} P'$) carry over to the PPT-theory (that is, $P \equiv^{\mathrm{ppt}} P'$ also holds, often even using the same reductions). We will therefore introduce all notions for lower bounds as analogously to the ones from the FPT-theory as possible.

For natural numbers $t \geq 1$ we set

$$\mathrm{PW}[t] := \bigcup_{d \geq 1} [p\text{-WSAT}(\Gamma_{t,d})]^{\mathrm{ppt}}$$

as an analogue of $\mathrm{W}[t]$. Many normalization properties that are well-known from FPT-theory still hold. In order to save case distinctions we introduce a short notation. For even t we define $(-)^t$ to be $+$ while for odd t we define $(-)^t$ to be $-$.

Lemma 1. *For $t, d \geq 1$, such that $t \geq 2$ or $d \geq 2$, the problems $p\text{-WSAT}(\Gamma_{t,d})$, $p\text{-WSAT}\left(\Gamma_{t,d}^{(-)^t}\right)$, $p\text{-PSAT}(\Gamma_{t,d})$, and $p\text{-PSAT}\left(\Gamma_{t,d}^{(-)^t}\right)$ are $\mathrm{PW}[t]$-complete.*

The proofs are refinements of those for the W-hierarchy.

There is, however, one completeness missing. The problem p-WSAT $\left(\Gamma_{t+1,1}^{(-)^t} \right)$ is W $[t]$-complete (see [3]). The analogue does not seem to hold for ppt-reductions. We therefore introduce further classes into our hierarchy. The new classes are

$$\mathrm{PW}\left[t + \frac{1}{2} \right] := \left[p\text{-WSAT}\left(\Gamma_{t+1,1}^{(-)^t} \right) \right]^{\mathrm{ppt}}$$

for natural numbers $t \geq 1$. We retain some of the above normalizations:

Lemma 2. *For* $t, d \geq 1$ *the problems* p-WSAT $\left(\Gamma_{t+1,d}^{(-)^t} \right)$ *and* p-PSAT $\left(\Gamma_{t+1,d}^{(-)^t} \right)$ *are* PW $\left[t + \frac{1}{2} \right]$-*complete.*

The proof can be found in this paper's full version.

Since $\Gamma_{t,d}^{(-)^t} \subseteq \Gamma_{t+1,1}^{(-)^t} \subseteq \Gamma_{t+1,1}$, the following is clear now:

Corollary 3. PW $[1] \subseteq$ PW $[1.5] \subseteq$ PW $[2] \subseteq$ PW $[2.5] \subseteq \ldots$

We conjecture the strictness of these inclusions, but a proof, even for a single one, would imply P $\neq$ NP.

3.1   Some Complete Problems

We define five parameterized problems p-CLIQUE, p-ACC (short turing machine acceptance), p-HIS (hypergraph independent set), p-DS (dominating set), and p-HDS (hypergraph disconnecting set):

> Input: A graph $\mathcal{G}$ and $k \in \mathbb{N}$
> Question: Does $\mathcal{G}$ have a clique of size k?
> Parameter: k

> Input: A nondeterministic turing machine M and $k \in \mathbb{N}$
> Question: Does M have an accepting run of length k?
> Parameter: k

> Input: A hypergraph $\mathcal{H}$ and $k \in \mathbb{N}$
> Question: Does $\mathcal{H}$ have an independent set of size k?
> Parameter: k

> Input: A graph $\mathcal{G}$ and $k \in \mathbb{N}$
> Question: Does $\mathcal{G}$ have a dominating set of size k?
> Parameter: k

> Input: A hypergraph $\mathcal{H}$ and $k \in \mathbb{N}$
> Question: Does $\mathcal{H}$ have a disconnecting set of size k?
> Parameter: k

An independent set of a hypergraph (V, E) is a set $V_0 \subseteq V$, such that $e \subseteq V_0$ for no $e \in E$. A disconnecting set of (V, E) is a set $V_0 \subseteq V$, such that for all $e \in E$ there is some $e' \in E$ with $e \cap e' \subseteq V_0$ (e' is disconnected from e by removing the vertices V_0).

By variations of the proofs for the W-hierarchy one obtains:

Proposition 4. *p*-CLIQUE *and p*-ACC *are* PW [1]-*complete. p*-DS *is* PW [2]-*complete.*

The following are new:

Proposition 5. *p*-HIS *is* PW [1.5]-*complete. p*-HDS *is* PW [2.5]-*complete.*

Proof. For *p*-HIS $\leq^{\text{ppt}}$ *p*-WSAT$(\Gamma_{2,1}^-)$ let a hypergraph $\mathcal{H} = (V, E)$ be given. We introduce propositional variables X_v for all $v \in V$. Then truth assignments for these variables correspond in a natural way to subsets of V. That such a subset is an independent set is expressed by

$$\alpha := \bigwedge_{e \in E} \bigvee_{v \in e} \neg X_v \ .$$

Hence $(\mathcal{H}, k)$ pertains to *p*-HIS if and only if (α, k) pertains to *p*-WSAT.

For the converse note that the formula just constructed was generic for $\Gamma_{2,1}^-$. Indeed, let

$$\alpha = \bigwedge_{i \in I} \bigvee_{j \in J_i} \neg X_j$$

be an arbitrary $\Gamma_{2,1}^-$-formula. Then with $V := \text{var}(\alpha)$ and

$$E := \{\{X_j \mid j \in J_i\} \mid i \in I\} \ ,$$

α describes the independent set property for $\mathcal{H} := (V, E)$. Hence (α, k) pertains to *p*-WSAT if and only if $(\mathcal{H}, k)$ pertains to *p*-HIS.

For *p*-HDS $\leq^{\text{ppt}}$ *p*-WSAT$(\Gamma_{3,1}^+)$ we can proceed analogously to *p*-HIS $\leq^{\text{ppt}}$ *p*-WSAT$(\Gamma_{2,1}^-)$ above with

$$\alpha := \bigwedge_{e \in E} \bigvee_{e' \in E} \bigwedge_{v \in e \cap e'} X_v \ .$$

For *p*-WSAT$(\Gamma_{3,1}^+)$ $\leq^{\text{ppt}}$ *p*-HDS let (α, k) be given, say

$$\alpha = \bigwedge_{i \in I} \bigvee_{j \in J_i} \bigwedge X_j \ .$$

Set $\mathcal{X} := \text{var}(\alpha)$. Without loss of generality we may assume $|\mathcal{X}| \geq k$ and all J_i to be disjoint. We construct a hypergraph $\mathcal{H} = (V, E)$ as follows: V consists of vertices v_X for $X \in \mathcal{X}$, of vertices $w_{i,h}$ for $i \in I$ and $1 \leq h \leq k + 1$, and of vertices u_h for $1 \leq h \leq k + 1$. E consists of a dummy hyperedge

$$d := \{u_h \mid 1 \leq h \leq k + 1\} \ ,$$

of hyperedges

$$e_i := \left\{ v_X \ \middle| \ X \in \bigcup_{j \in J_i} \mathcal{X}_j \right\} \cup \{w_{i,h} \mid 1 \leq h \leq k+1\} \cup d$$

for $i \in I$, and of hyperedges

$$f_j := \{v_X \mid X \in \mathcal{X}_j\} \cup \{w_{i',h} \mid i' \in I \setminus \{i\}, 1 \leq h \leq k+1\}$$

for $i \in I$ and $j \in J_i$. What would a disconnecting set of size (at most) k look like? $d \cap f_j = \emptyset$, hence d and all f_j are disconnected from some other hyperedge by *any* set of vertices. $d \cap e_i \supseteq d$ and $e_i \cap e_{i'} \supseteq d$, hence a set of size at most k can disconnect e_i at most from some f_j. If $j \notin J_i$, then $e_i \cap f_j \supseteq \{w_{i,h} \mid 1 \leq h \leq k+1\}$, hence e_i can be disconnected at most from those f_j with $j \in J_i$. In this case $e_i \cap f_j = \{v_X \mid X \in \mathcal{X}_j\}$. Now $\mathcal{H}$ has a disconnecting set of exact size k, if and only if $\mathcal{H}$ has a disconnecting set of size at most k, if and only if there is some $V_0 \subseteq \{v_X \mid X \in \mathcal{X}\}$ of size at most k such that for all $i \in I$ there is a $j \in J_i$ such that $v_X \in V_0$ for all $X \in \mathcal{X}_j$, if and only if α has a satisfying assignment that sets at most k variables to TRUE, if and only if α has a satisfying assignment that sets exactly k variables to TRUE.   □

Interestingly, if the sizes of hyperedges are bounded, then p-HDS $\in$ P. This is due to the fact that every hyperedge is a disconnecting set.

3.2   Characterizations by Model-Checking

The following result carries over from the FPT-world (see [4]):

Proposition 6. *For $t, u \geq 1$ and $s \geq 2$ the parameterized problems p-MC($\Sigma_{t,u}$) and p-MC($s, \Sigma_{t,u}$) are PW$[t]$-complete.*

This can be extended to:

Proposition 7. *For $t \geq 1$ and $u, s \geq 2$ the problems p-MC$_{\mathrm{var}}(s, \Sigma^*_{t,u})$ and p-MC$_{\mathrm{var}}\left(s, \Sigma^{*(-)^t}_{t,u}\right)$ are PW$[t]$-complete.*

The new result for our intermediate classes reads:

Proposition 8. *For $t, u \geq 1$ the parameterized problems p-MC$_{\mathrm{var}}(\Sigma^*_{t,u})$ and p-MC$_{\mathrm{var}}\left(\Sigma^{*(-)^t}_{t,u}\right)$ are PW$[t + \frac{1}{2}]$-complete.*

The proofs of the latter two can be found in this paper's full version.

3.3   Characterizations by Fagin-Definitions

Let τ be a first-order vocabulary and φ an FO $[\tau]$-formula possibly containing a second-order variable Y. For simplicity of notation let Y be monadic. The parameterized problem FD_φ (Fagin-defined by φ) is the following:

> Input: A τ-structure $\mathfrak{A}$ and $k \in \mathbb{N}$
> Question: Is there some $B \subseteq A$ with $|B| = k$, such that $\mathfrak{A} \models \varphi\,[B]$?
> Parameter: k

Proposition 9. *Let $t \geq 1$ be a natural number.*

1. $\mathrm{PW}\,[t] = \bigcup\limits_{\varphi \in \Pi_t} [\mathrm{FD}_\varphi]^{\mathrm{ppt}} = \bigcup\limits_{\varphi \in \Pi_t^{(-)^t}} [\mathrm{FD}_\varphi]^{\mathrm{ppt}}.$

2. $\mathrm{PW}\,\big[t + \tfrac{1}{2}\big] = \bigcup\limits_{\varphi \in \Pi_{t+1}^{(-)^t}} [\mathrm{FD}_\varphi]^{\mathrm{ppt}}.$

The proof can be found in this paper's full version.

3.4   Characterizations by a Machine Model

The machine model AQRAM (alternating query RAM) is an extension of the RAM model. It is somewhat similar to the W-RAM-model of [5]. Where the latter introduces instructions for querying binary relations in matrix representation in one step, our AQRAM has more natural instructions that allow for querying arbitrary relations in list representation in as many steps as a normal RAM would need.

   An AQRAM has access to three sets of registers:

1. The standard registers $s(0), s(1), s(2), \dots$.
2. The data registers $d(0), d(1), d(2), \dots$.
3. The index registers $i(0), i(1), i(2), \dots$.

Data and index registers are called nonstandard registers. The AQRAM inherits the usual (deterministic) instructions from the RAM. These are performed solely on standard registers. In addition the AQRAM may perform the following instructions:

1. `D-EXISTS` $\uparrow j$: Guess a value for a data register. At this point the program branches existentially. There is one continuation for each $0 \leq n \leq s(0)$, and in this continuation the register $d(s(j))$ is set to n.
2. `D-FORALL` $\uparrow j$: Like `D-EXISTS` $\uparrow j$, but the branching is universal.
3. `I-EXISTS` $\uparrow j$ and `I-FORALL` $\uparrow j$: Like the preceeding, but these affect $i(s(j))$ instead of $d(s(j))$.
4. `IS-JEQUAL` $\uparrow\uparrow\uparrow h, \uparrow j, m$: Conditional jump to instruction m. The condition is $s(s(0) + i(s(h))) = s(j)$.

5. ID-ASSUME $\uparrow\uparrow\uparrow h, \uparrow\uparrow j$: Check the condition $s(s(0) + i(s(h))) = d(s(j))$. If it holds, continue, if it does not hold, stop and reject.
6. ID-COASSUME $\uparrow\uparrow\uparrow h, \uparrow\uparrow j$: Check the condition $s(s(0) + i(s(h))) = d(s(j))$. If it holds, continue, if it does not hold, stop and accept.

The new conditional instructions justify the name of the new registers. A standard register, which is indexed by an index register (with a base displacement of $s(0)$) is queried. It is compared with another standard register, or with a data register.

For $t \in \mathbb{N}$ and a parameterization $\kappa : \Sigma^* \to \mathbb{N}$ an AQRAM program Π is a PW $\left[t + \frac{1}{2}\right]$-program with respect to κ, if there are $c, d, u \in \mathbb{N}$, such that for all but finitely many inputs w and all paths π in the run of Π on w the following requirements hold:

1. $|\pi| \leq 2^{(\kappa(w))^d} \cdot |w|^c$.
2. At most $2^{(\kappa(w))^d} \cdot |w|^c$ registers are used.
3. No register ever contains a value greater than $2^{(\kappa(w))^d} \cdot |w|^c$.
4. There are at most $(\kappa(w))^d$ steps after the first step that is not deterministic.

5. There are at most t alternations of branching, starting existential, and only the first block may contain more than u branching instructions.
6. This first block does not contain any I-EXISTS-instructions.
7. For odd t no ID-ASSUME-instructions are used; for even t no ID-COASSUME-instructions are used.

Π is a PW $[t]$-program with respect to κ, if furthermore there is some bound h, such that no path of a run of Π contains more than h ID-ASSUME-instructions or ID-COASSUME-instructions.

Proposition 10. *Let $t \geq 1$ be such that $2t \in \mathbb{N}$. Then a parameterized problem (Q, κ) pertains to PW $[t]$ if and only if Q is decided by some AQRAM-program that is a PW $[t]$-program with respect to κ.*

The proof can be found in this paper's full version.

4   PAW $[*]$ and PPA

A well-known result from unbounded parameterized complexity theory is

$$p\text{-MC}(\text{Words}, \text{MSO}) \in \text{FPT}$$

(for example it is implicit in [6]). On the other hand [2] implies

$$p\text{-MC}(\text{Words}, \text{FO}) \notin \text{PPT}$$

(unless FPT $=$ AW $[*]$). In this section we identify the complexity of both problems with respect to PPT. The former will be an analogue of AW $[*]$.

If Γ is a class of propositional formulas, then p-AWSAT(Γ) is the following problem:

> Input: $\alpha \in \Gamma$, a partition $(\mathcal{X}_i)_{1 \leq i \leq q}$ of $\text{var}(\alpha)$, and weights $(k_i)_{1 \leq i \leq q}$
>
> Question: Is there a subset $\mathcal{Y}_1 \subseteq \mathcal{X}_1$ with $|\mathcal{Y}_1| = k_1$, such that for all subsets $\mathcal{Y}_2 \subseteq \mathcal{X}_2$ with $|\mathcal{Y}_2| = k_2$ there is a subset $\mathcal{Y}_3 \subseteq \mathcal{X}_3$ with $|\mathcal{Y}_3| = k_3$, ..., such that the truth assignment, that sets exactly those variables to TRUE, that occur in some $\mathcal{Y}_i$, satisfies α?
>
> Parameter: $\sum\limits_{1 \leq i \leq q} k_i$

We also consider the special case p-APSAT(Γ), in which $k_i = 1$ for all i:

> Input: $\alpha \in \Gamma$ and a partition $(\mathcal{X}_i)_{1 \leq i \leq k}$ of $\text{var}(\alpha)$
>
> Question: Is there some $X_1 \in \mathcal{X}_1$, such that for all $X_2 \in \mathcal{X}_2$ there is some $X_3 \in \mathcal{X}_3$, ..., such that the truth assignment, that sets exactly the variables $X_1, \ldots, X_k$ to TRUE, satisfies α?
>
> Parameter: k

By dropping all weight restrictions we obtain p-QBF(Γ):

> Input: $\alpha \in \Gamma$ and a partition $(\mathcal{X}_i)_{1 \leq i \leq k}$ of $\text{var}(\alpha)$
>
> Question: Is there a subset $\mathcal{Y}_1 \subseteq \mathcal{X}_1$, such that for all subsets $\mathcal{Y}_2 \subseteq \mathcal{X}_2$ there is a subset $\mathcal{Y}_3 \subseteq \mathcal{X}_3$, ..., such that the truth assignment, that sets exactly those variables to TRUE, that occur in some $\mathcal{Y}_i$, satisfies α?
>
> Parameter: k

Define PAW $[*]$ as $\bigcup\limits_{t,d \geq 1} [p\text{-AWSAT}(\Gamma_{t,d})]^{\text{ppt}}$ and PPA (pure alternation, or, if you prefer, parameterized by alternation) as $[p\text{-QBF(Prop)}]^{\text{ppt}}$. Note that PPA $\neq$ PPT unless P $=$ NP, because SAT is a slice of p-QBF(Prop).

Lemma 11. *Let $t, d \geq 1$.*

1. *If $t \geq 2$ or $d \geq 2$, then the parameterized problems p-AWSAT$(\Gamma_{t,d})$ and p-APSAT$(\Gamma_{t,d})$ are PAW $[*]$-complete.*
2. *If $t \geq 2$ or $d \geq 3$, then the parameterized problem p-QBF$(\Gamma_{t,d})$ is PPA-complete.*

Proposition 12. *The problems p-MC(Words, FO) and p-MC(FO) are complete for PAW $[*]$.*

Proposition 13. *The problems p-MC(Words, MSO) and p-MC(MSO) are complete for PPA.*

The proofs can be found in this paper's full version.

5 Further Remarks

The results depicted here are part of the author's doctorate research. His forthcoming dissertation [7] contains further material, especially a treatment of the PA-hierarchy, which is the PPT-analogue of the A-hierarchy.

6 Acknowledgements

The author is indebted to the anonymous reviewers for their invaluable comments.

References

[1] Flum, J., Grohe, M., Weyer, M.: Bounded fixed-parameter tractability and $\log^2 n$ nondeterministic bits. Technical Report 04, Fakultät für Mathematik und Physik, Eckerstraße 1, 79104 Freiburg, Germany (2004)

[2] Frick, M., Grohe, M.: The complexity of first-order and monadic second-order logic revisited. Preprint (2003)

[3] Downey, R., Fellows, M.: Parameterized complexity. Springer-Verlag, New York (1999)

[4] Flum, J., Grohe, M.: Model-checking problems as a basis for parameterized intractability. Technical Report 23, Fakultät für Mathematik und Physik, Eckerstraße 1, 79104 Freiburg, Germany (2003)

[5] Chen, Y., Flum, J.: Machine characterizations of classes of the W-hierarchy. In: CSL. Volume 2803 of Lecture Notes in Computer Science., Springer (2003) 114–127

[6] Thomas, W.: Languages, automata, and logic. Technical Report 9607, Institut für Informatik und Praktische Mathematik, Christian-Albrechts-Universität Kiel, Germany (1996)

[7] (Weyer, M.) Dissertation, in preparation.

Refined Memorisation for Vertex Cover

L. Sunil Chandran and Fabrizio Grandoni

Max-Planck-Institut für Informatik,
Stuhlsatzenhausweg 85, 66123 Saarbrücken, Germany
{sunil,grandoni}@mpi-sb.mpg.de

Abstract. Memorisation is a technique which allows to speed up exponential recursive algorithms at the cost of an exponential space complexity. This technique already leads to the currently fastest algorithm for fixed-parameter vertex cover, whose time complexity is $O(1.2832^k k^{1.5} + kn)$, where n is the number of nodes and k is the size of the vertex cover. Via a refined use of memorisation, we obtain a $O(1.2759^k k^{1.5} + kn)$ algorithm for the same problem. We moreover show how to further reduce the complexity to $O(1.2745^k k^4 + kn)$.

1 Introduction

A *vertex cover* of an undirected graph G is a subset C of nodes such that any edge is incident on at least one node in C. The *vertex cover problem* consists in deciding whether G admits a vertex cover of at most k nodes.

It is well known [6] that vertex cover is *fixed-parameter tractable*, that is it can be solved in $O(f(k)n^\alpha)$ steps, where n is the number of nodes, f is a function of the *parameter* k only, and α is a constant. A lot of effort was recently devoted to develop faster and faster *fixed-parameter* algorithms for vertex cover. The first fixed-parameter algorithm for vertex cover is the $O(2^k n)$ algorithm of Fellows [8] (see also [9]), which is based on a *bounded search tree* strategy. Buss and Goldsmith [2] proposed a $O(2^k k^{2k+2} + kn)$ algorithm, in which they introduced the *kernel reduction* technique. Combining their bounded search tree strategy with the kernel reduction of Buss and Goldsmith, Downey and Fellows [5] obtained a $O(2^k k^2 + kn)$ algorithm. The complexity was later reduced to $O(1.3248^k k^2 + kn)$ by Balasubramanian, Fellows and Raman [1], to $O(1.2918^k k^2 + kn)$ by Niedermeier and Rossmanith [10], and to $O(1.2906^k k + kn)$ by Downey, Fellows and Stege [7]. The current fastest polynomial-space algorithm is the $O(1.2852^k k + kn)$ algorithm of Chen, Kanj and Jia [3]. Thanks to the *interleaving technique* of Niedermeier and Rossmanith [11], it is possible to get rid of the polynomial factor in the exponential term of the complexity. For example the complexity of the algorithm of Chen et al. can be reduced to $O(1.2852^k + kn)$. It is worth to notice that, though such polynomial factor is not relevant from the asymptotic point of view (the base of the exponential factor is overestimated), it is usually indicated. The reason is that the value of k is assumed to be not "too" big.

Memorisation is a technique developed by Robson [14,15] in the context of maximum independent set, which allows to reduce the time complexity of

R. Downey, M. Fellows, and F. Dehne (Eds.): IWPEC 2004, LNCS 3162, pp. 61–70, 2004.
© Springer-Verlag Berlin Heidelberg 2004

many exponential time recursive algorithms at the cost of an exponential space complexity. The key-idea behind memorisation is that, if the same subproblem appears many times, it may be convenient to store its solution instead of recomputing such solution from scratch. With this technique Robson [15] managed to derive a $O(1.1889^n)$ exponential-space algorithm for maximum independent set from his own $O(1.2025^n)$ polynomial-space algorithm for the same problem. Memorisation cannot be applied to the algorithm of Chen et al., since their algorithm branches on subproblems involving graphs which are not induced subgraphs of the kernel. Niedermeier and Rossmanith [12,13] applied memorisation to their $O(1.2918^k k^2 + kn)$ polynomial-space algorithm for vertex cover, thus obtaining a $O(1.2832^k k^{1.5} + kn)$ exponential-space algorithm for the same problem, which is also the currently fastest algorithm for vertex cover. It is worth to notice that the polynomial factor in the exponential term cannot be removed any more with the interleaving technique of Niedermeier and Rossmanith. In fact, such technique is based on the idea that most of subproblems concern graphs with few nodes. This is not the case when memorisation is applied.

1.1   Our Results

The kind of memorisation which is currently applied to vertex cover is in some sense a weaker version of the technique originally proposed by Robson for maximum independent set. This is mainly due to the structural differences between the two problems. In this paper we present a simple technique which allows to get rid of these structural differences, thus allowing to apply memorisation to vertex cover in its full strength. By applying our refined technique to the $O(1.2918^k k^2 + kn)$ algorithm of Niedermeier and Rossmanith, we obtain a $O(1.2759^k k^{1.5} + kn)$ exponential-space algorithm for vertex cover. With a further refined technique, we reduce the complexity to $O(1.2745^k k^4 + kn)$.

2   Preliminaries

We use standard graph notation as contained for instance in [4]. An *undirected graph* G is a pair (V, E), where V is a set of n *nodes* and E is a set of m pairs of distinct nodes (*edges*). The *order* of G is n while its *size* is $n + m$. Two nodes v and w are *adjacent* if $\{v, w\} \in E$. The edge $\{v, w\}$ is *incident* on v and w. The set of nodes adjacent to a node v is denoted by $N(v)$. Given a subset W of nodes, by $G - W$ we denote the graph obtained by removing from G all the nodes in W and all the edges incident on them. The *induced subgraphs* of G are the graphs of the kind $G - W$, for any $W \subseteq V$. A *vertex cover* of an undirected graph $G = (V, E)$ is a subset C of V such that every edge in E is incident on at least one node in C. The *vertex cover problem* consists in deciding whether G admits a vertex cover of at most k nodes. A *minimum vertex cover* is a vertex cover of minimum cardinality. From now on we will consider a variant of vertex cover problem, in which the size $mvc(G)$ of the minimum vertex covers has to be returned, if it does not exceed k.

All the currently fastest algorithms for vertex cover are based on two key-ideas: *kernel reduction* and *bounded search trees*. The idea behind kernel reduction is to reduce (in polynomial time) the original problem (G, k) to an equivalent problem (G', k'), where $k' \leq k$ and the size of G', the *kernel*, is a function of k only. Buss and Goldsmith [2] showed how to obtain a kernel with $O(k^2)$ nodes and edges in $O(kn)$ time. Chen, Kanj and Jia [3] showed how to reduce the number of nodes in the kernel to at most $2k$ in $O(m\sqrt{n})$ steps. By combining the kernel reductions of Buss and Goldsmith and of Chen et al., one obtains [3] a $O(kn + k^3)$ algorithm to reduce the original problem to an equivalent problem (G', k'), where $k' \leq k$ and the order of G' is at most $2k'$.

Let us now consider the idea behind bounded search trees. For this purpose, let us define a function $fpvc(\cdot, \cdot)$ as follows:

$$fpvc(G, k) = \begin{cases} mvc(G) & \text{if } mvc(G) \leq k; \\ +\infty & \text{otherwise.} \end{cases}$$

Given a problem (F, h), the value of $fpvc(F, h)$ can be computed with a recursive algorithm of the following kind. If $h = 0$, $fpvc(F, h)$ is equal to zero if F contains no edges and it is equal to $+\infty$ otherwise. Otherwise, one solves (F, h) by *branching* on a set of subproblems $(F_1, h_1), (F_2, h_2) \ldots (F_b, h_b)$, with b upper bounded by a constant and $h_i < h$ for each $i \in \{1, 2 \ldots b\}$. These subproblems must satisfy the condition:

$$fpvc(F, h) = \min_{i \in \{1, 2 \ldots b\}} \{h - h_i + fpvc(F_i, h_i)\}. \tag{1}$$

Thus the solution of the subproblems directly leads to the solution of (F, h).

To give an intuition of how subproblems are generated, let us consider a node v of F with neighborhood $N(v) = \{w, u\}$. Every vertex cover contains v or $N(v)$. This means that F admits a vertex cover of size at most h if and only if $F - \{v\}$ admits a vertex cover of size at most $h - 1$, or $F - \{w, u\}$ admits a vertex cover of size at most $h - 2$. In other words:

$$fpvc(F, h) = \min\{1 + fpvc(F - \{v\}, h - 1), 2 + fpvc(F - \{w, u\}, h - 2)\}.$$

Thus one can branch with the subproblems $(F - \{v\}, h - 1)$ and $(F - \{w, u\}, h - 2)$.

Let $C(h)$ denote the total number of search paths in the search tree which is created to solve a problem (F, h). For $h = 0$, $C(h) = C(0) = 1$. Otherwise, let $(F_1, h_1), (F_2, h_2) \ldots (F_b, h_b)$ be the subproblems generated to solve (F, h). The following relation holds:

$$C(h) \leq \sum_{i=1}^{b} C(h_i).$$

The inequality above is satisfied by $C(h) = \hat{c}^h$, where $\hat{c}$ is the *branching factor* of the branching considered, that is the (unique [3]) positive root of the polynomial:

$$x^h - \sum_{i=1}^{b} x^{h_i}.$$

For example the branching factor associated with the branching at (F, h) on the subproblems $(F - \{v\}, h - 1)$ and $(F - \{w, u\}, h - 2)$ is the positive root $\hat{c} = 1.6180...$ of the polynomial $x^h - x^{h-1} - x^{h-2} = x^{h-2}(x^2 - x - 1)$.

The *branching factor* c of the algorithm is the maximum over all the branching factors associated to the branchings that the algorithm may execute. By induction, one obtains that $C(h)$ is $O(c^h)$. This is also a valid upper bound on the number of nodes of the search tree.

The idea behind bounded search trees is to exploit the structure of the graph such as to obtain a branching factor c as small as possible. For this purpose, a tedious case analysis is often required.

Combining the recursive algorithm above with the kernel reduction algorithm of Chen et al., one obtains a $O(c^k k^\beta + kn)$ algorithm for vertex cover, where the cost of branching at a given subproblem is $O(k^\beta)$. This complexity can be reduced to $O(c^k + kn)$ with the interleaving technique of Niedermeier and Rossmanith [11].

Let us now shortly describe how Niedermeier and Rossmanith applied memorisation to vertex cover [12,13]. Let $\mathcal{A}$ be an algorithm of the kind described above, with the extra condition that, when it branches at a problem (F, h), the corresponding subproblems (F_i, h_i) involve graphs F_i which are induced subgraph of F. In particular, let us consider the fastest algorithm which satisfies this extra condition, which is the $O(c^k k^2 + kn) = O(1.2918^k k^2 + kn)$ algorithm of Niedermeier and Rossmanith [10]. From $\mathcal{A}$ one derives a new algorithm $\mathcal{A}'$ in the following way. For any induced subgraph F of the kernel of at most $2\alpha k$ nodes, for a given $\alpha \in (0, 1)$, $\mathcal{A}'$ solves (by using $\mathcal{A}$) the problem $(F, \alpha k)$ and it stores the pair $(F, fpvc(F, \alpha k))$ in a database. Note that, since F is an induced subgraph of the kernel, one can simply store the set of nodes of F instead of F. Then $\mathcal{A}'$ works in the same way as $\mathcal{A}$, with the following difference. When the parameter in a subproblem reaches or drops below the value αk, $\mathcal{A}'$ performs a kernel reduction. This way, $\mathcal{A}'$ generates a new subproblem (F, h), where $h \leq \alpha k$ and the order of F is at most $2\alpha k$. Thus $\mathcal{A}'$ can easily derive the value of $fpvc(F, h)$ from the value of $fpvc(F, \alpha k)$ which is stored in the database.

The cost of solving each subproblem $(F, \alpha k)$ is $O(c^{\alpha k})$. The number of pairs stored in the database is at most $2\binom{2k}{2\alpha k}$, which is $O(k^{-0.5}(\alpha^\alpha(1 - \alpha)^{1-\alpha})^{-2k})$ from Stirling's approximation. Thus the database can be created in $O(c^{\alpha k} k^{-0.5}(\alpha^\alpha(1-\alpha)^{1-\alpha})^{-2k})$ time. The search tree now contains $O(c^{(1-\alpha)k})$ nodes (that is the upper bound on $C(k)$ which is obtained by assuming $C(\alpha k) = 1$). The cost associated to each node is $O(k^2)$, not considering the leaves for which the costs of the query to the database and of the kernel reduction have to be taken into account. In particular, the database can be implemented such as that the cost of each query is $O(k)$. Moreover each kernel reduction costs $O(k^{2.5})$ (since each graph contains $O(k)$ nodes and $O(k^2)$ edges). Thus the cost to create the search tree is $O(c^{(1-\alpha)k} k^{2.5})$. The value of α has to be chosen such as to balance the cost of creating the search tree and the database. The optimum is reached when $\alpha > 0.0262$ satisfies:

$$c^{1-\alpha} = c^{\alpha} \left(\frac{1}{\alpha^{\alpha}(1-\alpha)^{1-\alpha}} \right)^2.$$

Thus one obtains a $O(c^{(1-\alpha)k}k^{2.5} + kn) = O(1.2832^k k^{2.5} + kn)$ time complexity.

The complexity can be slightly reduced in the following way. All the nodes of degree greater than 6 are filtered out in a preliminary phase (in which the algorithm stores no solution in the database and does not perform kernel reductions). In particular, let (F, h) be a subproblem where F contains a node v with $|N(v)| > 6$. The algorithm branches on the subproblems $(F - \{v\}, k - 1)$ and $(F - N(v), k - |N(v)|)$. The number of subproblem generated in the preliminary phase is $O(\tilde{c}^k)$, where $\tilde{c} < 1.256 < c$. The cost of branching at these subproblems is $O(k^2)$. Thus the total cost to remove "high" degree nodes is $O(1.256^k k^2)$. All the subproblems generated after the preliminary phase, involve subgraphs with $O(k)$ nodes and edges. This means that the kernel reductions can be performed in $O(k^{1.5})$ time only. This way the complexity of the algorithm is reduced to $O(1.2832^k k^{1.5} + kn)$.

3 Refined Memorisation for Vertex Cover

In this section we present a refined way to apply memorisation to vertex cover. The complexity analysis are based on the $O(c^k k^2 + kn) = O(1.2918^k k^2 + kn)$ algorithm of Niedermeier and Rossmanith [10], which is the currently fastest algorithm for vertex cover which is compatible with memorisation (since it branches on subproblems involving only induced subgraphs of the kernel). It is worth to mention that the technique proposed can be easily adapted to other algorithms of similar structure.

The rest of this section is organized as follows. In Section (3.1), we present a variant of the exponential-space algorithm of Niedermeier and Rossmanith [12,13], of complexity $O(1.2829^k k^{1.5} + kn)$. The reduction in the complexity is achieved via a more efficient use of the database. In Section (3.2), we show how to reduce the complexity to $O(1.2759^k k^{1.5} + kn)$ by branching on subproblems involving connected induced subgraphs only. In Section (3.3), the complexity is further reduced to $O(1.2745^k k^4 + kn)$.

3.1 A More Efficient Use of the Database

Our algorithm works as follows. No database is created a priori. Then, the algorithm works as the algorithm of Niedermeier and Rossmanith (as described in Section (2)), with the following differences. There is no value $h > 0$ of the parameter for which the recursion is stopped. Let (F, h) be a problem generated after the preliminary phase. Before branching on the corresponding subproblems $(F_1, h_1), (F_2, h_2) \ldots (F_b, h_b)$, the algorithm applies a kernel reduction to each subproblem (no matter which is the value of the parameters h_i). In particular, the algorithm branches on a set of subproblems $(F'_1, h'_1), (F'_2, h'_2) \ldots (F'_b, h'_b)$, where the order of F'_i is at most $2h'_i$ (*order-condition*). The reason of enforcing the order-condition will be clearer in the analysis. Observe that this does not modify the branching factor of the algorithm. When (F, h) is solved, the triple $(F, h, fpvc(F, h))$ is stored in a database. In this case also, before solving (F, h), the algorithm checks whether the solution is already available in the database. This way one ensures that a given subproblem is solved at most once.

The cost of the preliminary phase (in which "high" degree nodes are removed from the kernel) is not modified. The cost associated to each node of the search tree, after the preliminary phase, is $O(k^{1.5})$. Let $P(h)$ denote the number of subproblems with parameter h (after the preliminary phase). From the analysis of Section (2), $P(h)$ is $O(c^{k-h})$. The subproblems considered involve induced subgraphs of the kernel of order at most $2h$. Let $N(2k, 2h)$ denote the number of such induced subgraphs. Since no subproblem is solved more than once, $P(h)$ is $O(\min\{c^{k-h}, N(2k, 2h)\})$. For $h > k/4$:

$$\min\{c^{k-h}, N(2k, 2h)\} \leq c^{k-h} \leq c^{3k/4} = O(1.2118^k).$$

For $h \leq k/4$, $N(2k, 2h) \leq 2\binom{2k}{2h}$. As h decreases from $k/4$ to 0, the function $\min\{c^{k-h}, 2\binom{2k}{2h}\}$ first increases, then reaches a peak, and eventually decreases. The peak is reached for $\tilde{h} = \alpha k + O(1)$, where $\alpha > 0.0271$ satisfies:

$$c^{1-\alpha} = \left(\frac{1}{\alpha^\alpha (1-\alpha)^{1-\alpha}}\right)^2.$$

Thus $P(h)$ is $O(c^{(1-\alpha)k}) = O(1.2829^k)$. This is also an upper bound on the number of nodes in the search tree. Then the complexity of the algorithm proposed is $O(c^{(1-\alpha)k} k^{1.5} + kn) = O(1.2829^k k^{1.5} + kn)$.

Besides the time complexity, an important difference between the algorithm proposed in this section and the exponential-space algorithm of Niedermeier and Rossmanith is the role played by the parameter α. In the algorithm of Niedermeier and Rossmanith, α influences the behavior of the algorithm, and thus it has to be fixed carefully. In our algorithm instead, α only appears in the complexity analysis.

3.2   Branching on Connected Induced Subgraphs

In previous section we described an algorithm $\mathcal{A}$ which makes use of a database in which it stores triples of the kind $(F, h, fpvc(F, h))$, where F is an induced subgraph of the kernel of order at most $2h$ (order-condition). The graphs F stored in the database may not be connected. We will now show how to derive from $\mathcal{A}$ an algorithm $\mathcal{A}_C$ with the same branching factor as $\mathcal{A}$, which branches only on subproblems involving induced subgraphs which are connected (besides satisfying the order-condition).

The difference between $\mathcal{A}_C$ and $\mathcal{A}$ is the way $\mathcal{A}_C$ branches at a given subproblem after the preliminary phase. Let (F, h) be a problem generated after the preliminary phase and (F_1, h_1), $(F_2, h_2) \ldots (F_b, h_b)$ be the corresponding subproblems generated by $\mathcal{A}$. Let moreover $F_{i,1}, F_{i,2} \ldots F_{i,p_i}$ be the connected components of F_i. A naive idea could be to branch on the subproblems $(F_{i,j}, h_i)$, for each $i \in \{1, 2 \ldots b\}$ and for each $j \in \{1, 2 \ldots p_i\}$ (the graphs $F_{i,j}$ are connected and satisfy the order-condition). In fact, it is not hard to show that:

$$f pvc(F_i, h_i) = \begin{cases} \sigma_i & \text{if } \sigma_i \leq h_i; \\ +\infty & \text{otherwise,} \end{cases}$$

where σ_i is the sum of the solutions returned by the subproblems $(F_{i,j}, h_i)$:

$$\sigma_i = \sum_{j=1}^{p_i} f pvc(F_{i,j}, h_i).$$

Once the values $f pvc(F_i, h_i)$ are available, the value of $f pvc(F, h)$ can be easily derived from Equation (1).

Though this approach is correct in principle, it may lead to a bad branching factor. The reason is that one may generate many more than b subproblems, without decreasing the parameter in each subproblem properly. To avoid this problem, one can use the following simple observation. The size of the minimum vertex covers of the connected components with less than $6\ell + 2$ nodes, for a fixed positive integer ℓ, can be computed in constant time. Thus one does not need to branch on them. The size of the minimum vertex covers of the remaining connected components is lower bounded by ℓ (since they contain at least $6\ell + 1$ edges and their degree is upper bounded by 6). This lower bound can be used to reduce the parameter in each subproblem.

In more details, for each connected component H of F_i with less than $6\ell + 2$ nodes, the algorithm computes $mvc(H)$ by brute force, and this value is added to a variable Δ_i (which is initialized to 0). Let $F_{i,1}, F_{i,2} \ldots F_{i,b_i}$ be the remaining connected components of F_i. The minimum vertex covers of each $F_{i,j}$ have cardinality at least ℓ. This implies that, if the size of the minimum vertex covers of some $F_{i,j}$ is greater than $h_i - (b_i - 1)\ell$, then $f pvc(F_i, h_i) = +\infty$. Thus one can replace the subproblem $(F_{i,j}, h_i)$ with a subproblem $(F_{i,j}, h_i - (b_i - 1)\ell)$. Note that the order-condition is satisfied by the new subproblem since each $F_{i,j}$ contains at most $2h_i - (b_i - 1)(6\ell + 2)$ nodes. Let σ'_i be the sum of the solutions returned by the subproblems $(F_{i,j}, h_i - (b_i - 1)\ell)$:

$$\sigma'_i = \sum_{j=1}^{b_i} f pvc(F_{i,j}, h_i - (b_i - 1)\ell).$$

The value of $f pvc(F_i, h_i)$ is given by:

$$f pvc(F_i, h_i) = \begin{cases} \Delta_i + \sigma'_i & \text{if } \Delta_i + \sigma'_i \leq h_i; \\ +\infty & \text{otherwise.} \end{cases}$$

In this case also one does not need to compute the solutions of the subproblems $(F_{i,j}, h_i - (b_i - 1)\ell)$ which are already available in the database.

Let $\hat{c}$ and $\hat{c}_C$ be the branching factors corresponding to the branching on (F, h) of the algorithms $\mathcal{A}$ and $\mathcal{A}_C$ respectively. We can decompose the branching of $\mathcal{A}_C$ on (F, h) in $b + 1$ branchings. First there is one branching on the subproblems $(F_1, h_1), (F_2, h_2) \ldots (F_b, h_b)$. Then, for each subproblem (F_i, h_i), $i \in \{1, 2 \ldots b\}$, there is one branching on the "big" connected components of F_i. The first branching has branching factor $\hat{c}$. The other branchings have branching factors $\hat{c}_1, \hat{c}_2 \ldots \hat{c}_b$ respectively, where:

$$\hat{c}_i = \begin{cases} 1 & \text{if } b_i = 1; \\ b_i^{\frac{1}{(b_i - 1)\ell}} \leq 2^{1/\ell} & \text{otherwise.} \end{cases}$$

The value of $\hat{c}_C$ is upper bounded by the maximum over $\hat{c}$ and $\hat{c}_1, \hat{c}_2 \ldots \hat{c}_b$. Thus, to ensure that the branching factor of $\mathcal{A}_C$ is not greater than the branching factor c of $\mathcal{A}$, $c \cong 1.2918$, it is sufficient to fix ℓ such that $2^{1/\ell} < c$. In particular, we can assume $\ell = 3$.

Let us now consider the complexity of $\mathcal{A}_C$. The cost of the preliminary phase is not modified. We can both filter out small connected components and search in the database in $O(k)$ time (with a careful implementation of the database). Thus the cost associated to each branching is the same as in $\mathcal{A}$, that is $O(k^{1.5})$. Let $R_d(m, n)$ be the number of connected induced subgraphs of order m which are contained in a graph of order n and degree upper bounded by d. Robson [14] showed that $R_d(m, n)$ is $O\left(\frac{n}{m} \left(\frac{(d-1)^{d-1}}{(d-2)^{d-2}}\right)^m\right)$. In particular, $R_6(2h, 2k)$ is $O(\frac{k}{h}(\frac{5^5}{4^4})^{2h})$. It turns out that this is also a valid upper bound for the number $N(2k, 2h)$ of the subproblems concerning graphs of order at most $2h$ which are generated after the preliminary phase. From this bound and the analysis of Section (3.1), the time complexity of $\mathcal{A}_C$ is $O(c^{(1-\alpha)k}k^{1.5} + kn) = O(1.2759^k k^{1.5} + kn)$, where

$$\alpha = \frac{\log(c)}{\log(c) + 2\log(\frac{5^5}{4^4})} > 0.04867.$$

3.3   A Further Refinement

In Section (3.2) we showed that restricting the class of graphs which are stored in the database, without increasing the branching factor, leads to a faster algorithm. In particular, the algorithm of Section (3.2) stores connected induced subgraphs only (instead of general induced subgraphs). In this section we show how it is possible to store only connected induced subgraphs of degree lower bounded by 2. This leads to a $O(1.2745^k k^4 + kn)$ algorithm for vertex cover.

Nodes of degree zero can be safely removed (since they do not belong to any minimum vertex cover). Then we just need to take care of nodes of degree one. Let v be a node of degree one, with $N(v) = \{w\}$. It is not hard to show that there exists a minimum vertex cover which contains w and does not contain v. Since we do not need to find all the minimum vertex covers (of a given size), but only one, we can simply assume that w is in the vertex cover and remove v and w from the graph. Thus nodes of degree one can be filtered out in linear time. Observe that, if one starts with a problem (F, h) which satisfies the order-condition, the graph obtained after removing nodes of degree at most one satisfies the order-condition too. Moreover this filtering out does not increase the branching factor of the algorithm.

Let $R_{2,d}(m, n)$ be the number of connected induced subgraphs of order m and degree lower bounded by 2 which are contained in a graph of order n and degree upper bounded by d. Robson [15] showed that $R_{2,d}(m, n)$ is $O(r_d^m\, n\, poly_d(m))$, where, for any fixed d, $poly_d(m)$ is a polynomial of m and r_d is a constant which comes from a maximization problem (in particular, $r_6 < 9.927405$). Following the proof of Robson, it is not hard to derive that $poly_d(m)$ is $O(m^d)$. With a more careful analysis, one obtains that $poly_d(m)$ is $O(m^{\frac{d-3}{2}})$. Thus $R_{2,6}(2h, 2k)$ is $O(r_6^{2h} poly_6(h)k) = O(9.927405^{2h} h^{1.5} k)$. By replacing $R_6(2h, 2k)$ with $R_{2,6}(2h, 2k)$ in the analysis of Section (3.2), one obtains:

$$\alpha = \frac{\log(c)}{\log(c) + 2\log(r_6)} > 0.05282,$$

and thus a $O(c^{(1-\alpha)k} k^{2.5} k^{1.5} + kn) = O(1.2745^k k^4 + kn)$ time complexity.

References

1. R. Balasubramanian, M. Fellows, and V. Raman. An improved fixed-parameter algorithm for vertex cover. *Information Processing Letters*, 65:163–168, 1998.
2. J. F. Buss and J. Goldsmith. Nondeterminism within P. *SIAM Journal on Computing*, 22(3):560–572, 1993.
3. J. Chen, I. Kanj, and W. Jia. Vertex cover: further observations and further improvements. *Journal of Algorithms*, 41:280–301, 2001.
4. T. H. Cormen, C. E. Leiserson, and R. L. Rivest. *Introduction to algorithms*. MIT Press and McGraw-Hill Book Company, 6th edition, 1992.
5. R. G. Downey and M. R. Fellows. Fixed-parameter tractability and completeness. II. On completeness for $W[1]$. *Theoretical Computer Science*, 141(1-2):109–131, 1995.
6. R. G. Downey and M. R. Fellows. *Parameterized complexity*. Monographs in Computer Science. Springer-Verlag, 1999.
7. R. G. Downey, M. R. Fellows, and U. Stege. Parameterized complexity: A framework for systematically confronting computational intractability. In J. K. F. Roberts and J. Nesetril, editors, *Contemporary Trends in Discrete Mathematics: From DIMACS and DIMATIA to the Future*, volume 49 of *DIMACS Series in Discrete Mathematics and Theoretical Computer Science*, pages 49–99, 1999.
8. M. R. Fellows. On the complexity of vertex set problems. Technical report, Computer Science Department, University of New Mexico, 1988.
9. K. Mehlhorn. *Data Structures and Algorithms 2: Graph Algorithms and NP-Completeness*. Springer-Verlag, 1984.
10. R. Niedermeier and P. Rossmanith. Upper bounds for vertex cover further improved. In *Symposium on Theoretical Aspects of Computer Science*, pages 561–570, 1999.

11. R. Niedermeier and P. Rossmanith. A general method to speed up fixed-parameter-tractable algorithms. *Information Processing Letters*, 73(3–4):125–129, 2000.
12. R. Niedermeier and P. Rossmanith. Private communication, 2003.
13. R. Niedermeier and P. Rossmanith. On efficient fixed-parameter algorithms for weighted vertex cover. *Journal of Algorithms*, 47(2):63–77, 2003.
14. J. M. Robson. Algorithms for maximum independent sets. *Journal of Algorithms*, 7(3):425–440, 1986.
15. J. M. Robson. Finding a maximum independent set in time $O(2^{n/4})$. Technical Report 1251-01, LaBRI, Université Bordeaux I, 2001.

Parameterized Graph Separation Problems[*]

Dániel Marx

Department of Computer Science and Information Theory,
Budapest University of Technology and Economics
Budapest, H-1521, Hungary
dmarx@cs.bme.hu

Abstract. We consider parameterized problems where some separation property has to be achieved by deleting as few vertices as possible. The following five problems are studied: delete k vertices such that (a) each of the given ℓ terminals is separated from the others, (b) each of the given ℓ pairs of terminals are separated, (c) exactly ℓ vertices are cut away from the graph, (d) exactly ℓ connected vertices are cut away from the graph, (e) the graph is separated into ℓ components, We show that if both k and ℓ are parameters, then (a), (b) and (d) are fixed-parameter tractable, while (c) and (e) are W[1]-hard.

1 Introduction

In this paper we study five problems where we have to delete vertices from a graph to achieve a certain goal. In all four cases, the goal is related to making the graph disconnected by deleting as few vertices as possible.

Classical flow theory gives us a way of deciding in polynomial time whether two vertices t_1 and t_2 can be disconnected by deleting at most k vertices. However, for every $\ell \geq 3$, if we have ℓ terminals t_1, t_2, ..., t_ℓ, then it is NP-hard to find k vertices such that no two terminals are in the same component after deleting these vertices [3]. In [8] a $(2 - 2/\ell)$-approximation algorithm was presented for the problem. Here we give an algorithm that is efficient if k is small: in Section 2 it is shown that the MINIMUM TERMINAL SEPARATION problem is fixed-parameter tractable with parameter k. We also consider the more general MINIMUM TERMINAL PAIR SEPARATION problem where ℓ pairs (s_1, t_1), ..., (s_ℓ, t_ℓ) are given, and it has to be decided whether there is a set of k vertices whose deletion separates each of the ℓ pairs. We show that this problem is fixed-parameter tractable if both k and ℓ are parameters. Our results can be used in the edge deletion versions of these problems as well.

In Section 3 we consider two separation problems without terminals. In the SEPARATING ℓ VERTICES problem exactly ℓ vertices have to be separated from the rest of the graph by deleting at most k vertices. In SEPARATING INTO ℓ COMPONENTS problem k vertices have to be deleted such that the remaining

---

[*] Research is supported in part by grants OTKA 44733, 42559 and 42706 of the Hungarian National Science Fund.

Problem	Parameter(s)		
	k	ℓ	k and ℓ
MINIMUM TERMINAL SEPARATION	FPT (Theorem 1)	NP-hard for $\ell \geq 3$ [3]	FPT (Theorem 1)
MINIMUM TERMINAL PAIR SEPARATION	Open	NP-hard for $\ell \geq 3$ [3]	FPT (Theorem 2)
SEPARATING ℓ VERTICES	W[1]-hard (Theorem 4)	W[1]-hard (Theorem 4)	W[1]-hard (Theorem 4)
SEPARATING ℓ CONNECTED VERTICES	W[1]-hard (Theorem 8)	W[1]-hard (Theorem 7)	FPT (Theorem 6)
SEPARATING INTO ℓ COMPONENTS	W[1]-hard (Theorem 9)	W[1]-hard (Theorem 9)	W[1]-hard (Theorem 9)

Table 1. Complexity of the problems with different parameterizations.

graph has at least ℓ connected components. The edge deletion variants of these problems were considered in [5], where it is shown that both problems are W[1]-hard with parameter ℓ. Here we show that the vertex deletion versions of both problems are W[1]-hard even if both k and ℓ are parameters. However, in the case of SEPARATING ℓ VERTICES if we restrict the problem to bounded degree graphs, then it becomes fixed-parameter tractable if both k and ℓ are parameters. Moreover, we also consider the variant SEPARATING ℓ CONNECTED VERTICES, where it is also required that the separated vertices form a connected subgraph. It turns out that this problems is fixed-parameter tractable if both k and ℓ are parameters, but W[1]-hard if only one of them is parameter.

The results of the paper are summarized on Table 1.

2   Separating Terminals

The parameterized terminal separation problem studied in this section is formally defined as follows:

MINIMUM TERMINAL SEPARATION
Input: A graph $G(V, E)$, a set of terminals $T \subseteq V$, and an integer k.
Parameter 1: k
Parameter 2: $\ell = |T|$
Question: Is there a set of vertices $S \subseteq V$ of size at most k such that no two vertices of T belong to the same connected component of $G \setminus S$?

Note that S and T do not have to be disjoint, which means that it is allowed to delete terminals. A deleted terminal is considered to be separated from all the other terminals (later we will argue that our results remain valid for the slightly different problem where the terminals cannot be deleted).

It follows from the graph minor theory of Robertson and Seymour that MINIMUM TERMINAL SEPARATION is fixed-parameter tractable. The celebrated result of Robertson and Seymour states that graphs are well-quasi ordered with respect to the minor relation. Moreover, the same holds for graphs where the edges are colored with a fixed number of colors. For every terminal $v \in T$, we add a new vertex v' and a red edge vv' (the original edges have color black). Now separating the terminals and separating the red edges are the same problem. Consider the set $\mathscr{G}_k$ that contains those red-black graphs where the red edges can be separated by deleting at most k vertices. It is easy to see that $\mathscr{G}_k$ is closed with respect to taking minors. Therefore by the Graph Minor Theorem, $\mathscr{G}_k$ has a finite set of forbidden minors. Another result of Roberson and Seymour states that for every graph H there is an $O(|V|^3)$ algorithm for finding an H-minor, therefore membership in $\mathscr{G}_k$ can be tested in $O(|V|^3)$ time. This means that for every k, MINIMUM TERMINAL SEPARATION can be solved in $O(|V|^3)$ time, thus the problem is (non-uniformly) fixed-parameter tractable. However, the constants given by this non-constructive method are incredibly large. In this section we give a direct combinatorial algorithm for the problem, which is more efficient.

The notion of important separator is the most important definition in this section:

Definition 1. *Let $G(V, E)$ be a graph. For subsets $X, S \subseteq V$, the set of vertices reachable from $X \setminus S$ in $G \setminus S$ is denoted by $R(S, X)$. For $X, Y \subseteq V$, the set S is called an (X, Y)-separator if $Y \cap R(S, X) = \emptyset$. An (X, Y)-separator is minimal if none of its proper subsets are (X, Y)-separators. An (X, Y)-separator S' dominates an (X, Y)-separator S, if $|S'| \leq |S|$ and $R(S, X) \subset R(S', X)$. A subset S is an important (X, Y)-separator if it is minimal, and there is no (X, Y)-separator S' that dominates S.*

Abusing notations, the one element set $\{v\}$ is denoted by v. We note that X and Y can have non-empty intersection, but in this case every (X, Y)-separator has to contain $X \cap Y$.

We use Figure 1 to demonstrate the notion of important separator. Let $X = \{x\}$ and $Y = \{y_1, y_2, y_3, y_4\}$, we want separate these two sets. X and Y can be separated by deleting x, this is the only separator of size 1. There are several size 2 separators, for example $\{a, f\}$, $\{b, g\}$, $\{b, j\}$, $\{c, j\}$. However, only $\{c, j\}$ is an important separator: $R(\{c, j\}, x) = \{x, a, b, f, g, h, i\}$ and the set of vertices reachable from x is smaller for the other size 2 separators. There are two size 3 important separators: $\{c, k, \ell\}$ and $\{j, d, e\}$. Separator $\{c, h, i\}$ is not important, since it is dominated both by $\{c, j\}$ and by $\{c, k, \ell\}$. Finally, there is only one important size 4 separator, Y itself.

Testing whether a given (X, Y)-separator S is important can be done as follows. First, minimality can be easily checked by testing for each vertex $s \in S$ whether $S \setminus s$ remains separating. If it is minimal, then for every vertex $s \in S$, we test whether there is an $(R(S, X) \cup s, Y)$-separator S' of size at most $|S|$. This separator can be found in $O(|V|^3)$ time using network flow techniques. If there is such a separator, then S is not important. Notice that if S is not important, then this method can be used to find an important separator that dominates S. The test can be repeated for S', and if it is not important, then we get another separator S'' that dominates S'. We repeat this as many times as necessary.

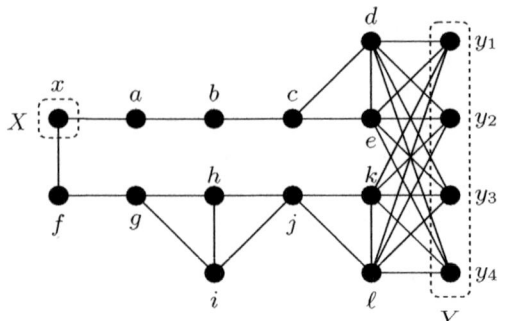

Fig. 1.

Since the set of vertices reachable from X increases at each step, eventually we arrive to an important separator.

Let X and Y be two sets of vertices, then there is at most one important (X, Y)-separator of size 1. A size 1 separator has to be a cut vertex (here we ignore the special cases where $|X| = 1$ or $|Y| = 1$). If there are multiple cut vertices that separate X and Y, then there is a unique cut vertex that is farthest from X and closest to Y. This vertex will be the only important (X, Y)-separator.

However, for larger sizes, there can be many important (X, Y)-separators of a given size. For an example, see Figure 2. To separate the two large cliques X and Y, for each $1 \leq i \leq t$, either a_i, or both b_i and c_i have to be deleted. If we choose to delete both b_i and c_i, then we have to delete two vertices instead of one, but the set of vertices reachable from X increases, it includes a_i. Therefore there are $\binom{t}{t/2}$ important (X, Y)-separators of size $3t/2$: for $t/2$ of the i's we delete a_i, and for the remaining $t/2$ we delete b_i and c_i. All these separators are important, since $R(S', X)$ and $R(S'', X)$ are pairwise incomparable for two such separators S' and S''. Thus the number of important separators of a given size k can be exponential in k. However, we show that this number is independent of the size of the graph:

Lemma 1. *For sets of vertices X, Y, there are at most 4^{k^2} important (X, Y)-separators of size k. Moreover, these separators can be enumerated in polynomial time per separator.*

Proof. The proof is by induction on k. We have seen above that the statement holds for $k = 1$. Let S be an important (X, Y)-separator of size k in G. We count how many other important separators can be in G. If H is another important (X, Y)-separator of size k, then we consider two cases depending on whether $Z = S \cap H$ is empty or not. If Z is not empty, then it is easy to see that $H \setminus Z$ is an important $(X \setminus Z, Y \setminus Z)$-separator in $G \setminus Z$. Since $|H \setminus Z| < k$, thus by the induction hypotheses the number of such separators is at most $4^{(k-1)^2}$. There are not more than 2^k possibilities for the set Z, and for each set Z there are at most $4^{(k-1)^2}$ possibilities for the set H, hence the total number of different H that intersect S is at most $2^k 4^{(k-1)^2}$.

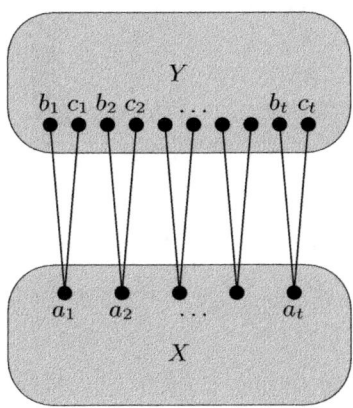

Fig. 2. A graph where there is an exponential number of important separators that separate the large cliques X and Y.

Next we count those separators that do not intersect S. Such a separator H contains ℓ vertices from $R(S, X)$ and $k-\ell$ vertices from $R(S, Y)$. It is not possible that $\ell = 0$: that would imply that $R(S, X) \cup S \subseteq R(H, X)$ and S would not be an important separator. Here we used the minimality of S: if none of $R(S, X)$ and S is deleted, then every vertex of S can be reached from X. Similarly, it is not possible that $\ell = k$ because H would not be an important separator in that case. To see this, notice that by the minimality of S, from every vertex of S a vertex of Y can be reached using only the vertices in $R(S, Y)$. Therefore no vertex of S can be reached from X in $G \setminus H$, otherwise H would not be an (X, Y)-separator. Since S is an (X, Y)-separator, thus this also means that no vertex of $R(S, Y)$ can be reached. Therefore $R(H, X)$ is contained in $R(S, X)$, and since $\ell > 0$, the containment is proper.

We divide H into two parts: let $H_1 = H \cap R(S, X)$ and $H_2 = H \cap R(S, Y)$ (see Figure 3). The separator S is also divided into two parts: $S_1 = S \cap R(H, X)$ contains those vertices that can be reached from X in $G \setminus H$, while $S_2 = S \setminus S_1$ contains those that cannot be reached. Let G_1 be the subgraph of G induced by $R(S, X) \cup S$, and G_2 be the subgraph induced by $R(S, Y) \cup S$. Now it is clear that H_1 is an $(X \cup S_1, S_2)$-separator in G_1, and H_2 is a $(S_1, Y \cup S_2)$-separator in G_2. Moreover, we claim that they are important separators. First, if H_1 is not minimal, i.e., it remains an $(X \cup S_1, S_2)$-separator without $v \in H_1$, then H would be an (X, Y)-separator without v as well. Assume therefore that an $(X \cup S_1, S_2)$-separator H_1^* in G_1 dominates H_1. In this case $H_1^* \cup H_2$ is an (X, Y)-separator in G with $R(H, X) \subset R(H_1^* \cup H_2, X)$, contradicting the assumption that H is an important separator. A similar argument shows that H_2 is an important $(S_1, Y \cup S_2)$-separator in G_2. By the induction hypotheses, we have a bound on the possible number of such separators. For a given division (S_1, S_2) and ℓ, there can be at most $4^{\ell^2} 4^{(k-\ell)^2}$ possibilities. There are at most 2^k possibilities for (S_1, S_2), and the value of ℓ is between 1 and $k - 1$. Therefore the total number

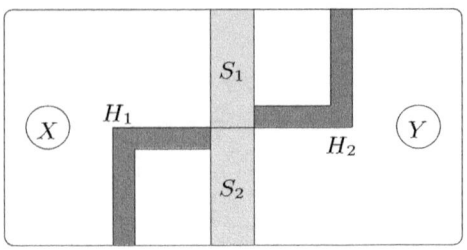

Fig. 3. Separators in the proof of Lemma 1.

of different separators (including S itself and the at most $2^k 4^{(k-1)^2}$ sets in the first case) is at most

$$1 + 2^k 4^{(k-1)^2} + \sum_{\ell=1}^{k-1} 2^k 4^{\ell^2} 4^{(k-\ell)^2} \le 1 + 2^k 4^{(k-1)^2} + (k-1)2^k 4^{(k-1)^2+1}$$

$$\le k 2^k 4^{(k-1)^2+1} \le 4^k 4^{(k-1)^2+1} = 4^{k+k^2-2k+2} \le 4^{k^2},$$

what we had to show (in the first inequality we used $\ell^2 + (k-\ell)^2 \le (k-1)^2 + 1$, which holds since $1 \le \ell \le k-1$). The proof also gives an algorithm for finding all the important separators. To handle the first case, we take every subset Z of S, and recursively find all the important size $k - |Z|$ separators in $G \setminus S$. In the second case, we consider every $1 \le \ell \le k-1$ and every division (S_1, S_2) of S. We enumerate every important $(X \cup S_1, S_2)$-separator S_1 in G_1 and every important $(S_1, Y \cup S_2)$-separator in G_2. For each S_1, S_2, it has to be checked whether $S_1 \cup S_2$ is an important (X, Y)-separator. As it was shown above, every important separator can be obtained in such a form. Our algorithm makes a constant number of recursive calls with smaller k, therefore the running time is uniformly polynomial. □

What makes important separators important is that a separator in a solution can be always replaced by an important separator:

Lemma 2. *If there is a set S of vertices that separates the terminals t_1, ..., t_r, then there is a set H with $|H| \le |S|$ that also separates the terminals and contains an important $(\{t_1\}, \{t_2, t_3, \ldots, t_r\})$-separator.*

Proof. Let $S_0 \subseteq S$ be those vertices of S that can be reached from t_1 without going through other vertices of S. Clearly, S_0 is a $(\{t_1\}, \{t_2, t_3, \ldots, t_r\})$-separator, and it contains a minimal separator S_1. If S_1 is important, then we are ready, otherwise there is an important $(\{t_1\}, \{t_2, t_3, \ldots, t_r\})$-separator S_1' that dominates S_1. We claim that $S' = (S \setminus S_1) \cup S_1'$ also separates the terminals. If this is true, then $|S_1'| \le |S_1|$ implies $|S'| \le |S|$, proving the lemma.

Since S_1' is a $(\{t_1\}, \{t_2, t_3, \ldots, t_r\})$-separator, thus S' separates t_1 from all the other vertices. Assume therefore that there is a path P in $G \setminus S'$ connecting terminals t_i and t_j. Since S separates t_i and t_j, thus this is only possible if P goes through a vertex v of S_1. Every vertex of $S_1 \subseteq S_0$ has a neighbor in $R(S, t_1)$, let w this neighbor of v. Since $R(S, t_1) \subseteq R(S', t_1)$, vertex w can be reached from t_1 in $G \setminus S'$. Therefore t_i can be reached from t_1 via w and v, which is a contradiction, since S' is a $(\{t_1\}, \{t_2, t_3, \dots, t_r\})$-separator. $\qquad\square$

Lemma 1 and Lemma 2 allows us to use the method of bounded search trees to solve the MINIMUM TERMINAL SEPARATION problem:

Theorem 1. MINIMUM TERMINAL SEPARATION *is fixed-parameter tractable with parameter k.*

Proof. We select an arbitrary terminal t that is not already separated from every other terminal. By Lemma 2, there is a solution that contains an important $(t, T \setminus t)$-separator. Using Lemma 1, we enumerate all the at most $k 4^{k^2}$ important separators of size at most k, and select a separator S from this list. We delete S from G, and recursively solve the problem for $G \setminus S$ with problem parameter $k - |S|$. At each step we can branch into at most $k 4^{k^2}$ directions, and the problem parameter is decreased by at least one, hence the search tree has height at most k and has at most $k^k 4^{k^3}$ leaves. The work to be done is polynomial at each step, hence the algorithm is uniformly polynomial. $\qquad\square$

A natural way to generalize MINIMUM TERMINAL SEPARATION is to have a more complicated restriction on which terminals should be separated. Instead of a set of terminals where every terminal has to be separated from every other terminal, in the following problem there are pairs of terminals, and every terminal has to be separated only from its pair:

MINIMUM TERMINAL PAIR SEPARATION
Input: A graph $G(V, E)$, pairs of vertices (s_1, t_1), (s_2, t_2), $\dots$, (s_ℓ, t_ℓ), and an integer k.
Parameter 1: k
Parameter 2: ℓ
Question: Is there a set of vertices $S \subseteq V$ of size at most k such that for every $1 \le i \le \ell$, vertices s_i and t_i are in different components of $G \setminus S$?

Let $T = \bigcup_{i=1}^{\ell} \{s_i, t_i\}$ be the set of terminals. We can prove an analog of Lemma 2: there is an optimal solution containing an important separator.

Lemma 3. *If there is a set S of vertices that separates every pair, then there is a set S' with $|S'| \le |S|$ that also separates the pairs and S' contains an important $(\{s_1\}, T')$-separator for some subset $T' \subseteq T$.*

Proof. We proceed similarly as in the proof of Lemma 2. Let T' be the set of those terminals that are separated from s_1 in $G \setminus S$. Let $S_0 \subseteq S$ be the vertices reachable from s_1 without going through other vertices of S. Clearly, S_0 is an (s_1, T')-separator, and it contains a minimal (s_1, T')-separator S_1. If S_1 is not important, then there is an important (s_1, T')-separator S_1' that dominates S_1. We claim that $S' = (S \setminus S_1) \cup S_1'$ also separates the pairs. Clearly, $t_1 \in T'$, hence

s_1 and t_1 are separated in S'. Assume therefore that s_i and t_i are connected by a path P in $G \setminus S'$. As in Lemma 2, path P goes through a vertex of S_1, and it follows that both s_i and t_i are connected to s_1 in $G \setminus S'$. Therefore $s_i, t_i \neq T'$. However, this implies that s_1 is connected to s_i and t_i in $G \setminus S$, hence S does not separate s_i from t_i, a contradiction. $\qquad\square$

To find k vertices that separate the pairs, we use the same method as in Theorem 1. In Lemma 3, there are 2^ℓ different possibilities for the set T', and by Lemma 1, for each T' there are at most $k4^{k^2}$ different separators of size at most k. Therefore we can generate $2^\ell \cdot k 2^{k^2}$ separators such that one of them is contained in an optimum solution. This results in a search tree with at most $2^{k\ell} \cdot k^k 4^{k^3}$ leaves.

Theorem 2. *The* Minimum Terminal Pair Separation *problem is fixed-parameter tractable with parameters k and ℓ.* $\qquad\square$

Separating the terminals in T can be expressed as separating $\binom{|T|}{2}$ pairs, hence Minimum Terminal Separation is a special case of Minimum Terminal Pair Separation. However, Theorem 2 does not imply Theorem 1. In Theorem 2 the number of pairs is a parameter, while the size of T can be unbounded in Theorem 1. We do not know the complexity of Minimum Terminal Pair Separation if only k is the parameter.

As noted above, in the separation problems we assume that any vertex can be deleted, even the terminals themselves. However, we can consider the slightly more general problem, when the input contains a set V^* of distinguished vertices, and these vertices cannot be deleted. All the results in this section hold for this variant of the problem as well. In all of the proofs, when a new separator is constructed, then it is constructed from vertices that were contained in some other separator.

We can consider the variants of Minimum Terminal Separation and Minimum Terminal Pair Separation where the terminals have to be separated by deleting at most k edges. The edge deletion problems received more attention in the literature: they were consider in e.g. [4,3,7] under the names multiway cut, multiterminal cut, and multicut. As noted in [8], it is easy to reduce the edge deletion problem to vertex deletion, therefore our algorithms can be used for these edge deletion problems as well. For completeness, we briefly describe a possible reduction. The edge deletion problem can be solved by considering the line graph (in the line graph $L(G)$ of G the vertices correspond to the edges of G, and two vertices are connected if the corresponding two edges have a common vertex.) However, we have to do some tinkering before we can define the terminals in the line graph. For each terminal v_i of G, add a new vertex v_i' and a new edge $v_i v_i'$. Let v_i' be the terminal instead of v_i. If edge $v_i v_i'$ is marked as unremovable, then this modification does not change the solvability of the instance. Now the problem can be solved by using the vertex separation algorithms (Thereom 1 and 2) on the line graph $L(G)$. The terminals in the line graph are the vertices corresponding to the edges $v_i v_i'$. These edges were marked as unremovable, hence these vertices are contained in the set V^* of distinguished vertices in the line graph.

Theorem 3. *The edge deletion versions of* Minimum Terminal Separation *(with parameter k) and* Minimum Terminal Pair Separation *(with parameters k and ℓ) are fixed-parameter tractable.*   □

3   Cutting Up a Graph

Finding a good separator that splits a graph into two parts of approximately equal size is a useful algorithmic technique (see [9,10] for classic examples). This motivates the study of the following problem, where a given number of vertices has to be separated from the rest of the graph:

> Separating ℓ Vertices
> *Input:* A graph $G(V, E)$, integers k and ℓ.
> *Parameter 1:* k
> *Parameter 2:* ℓ
> *Question:* Is there a partition $V = X \cup S \cup Y$ such that $|X| = \ell$, $|S| \leq k$ and there is no edge between X and Y?

It follows from [2] that the problem is NP-hard in general. Moreover, it is not difficult to show that the parameterized version of the problem is hard as well, even with both parameters:

Theorem 4. Separating ℓ Vertices *is* W[1]-*hard with parameters k and ℓ.*

Proof. The proof is by reduction from Maximum Clique. Let G be a graph with n vertices and m edges, it has to be determined whether G has a clique of size k. We construct G' as follows. In G' there are n vertices $v_1, \ldots, v_n$ that correspond to the vertices of G, these vertices form a clique in G'. Furthermore, G' has m vertices $e_1, \ldots, e_m$ that correspond to the edges of G. If the end points of edge e_j in G are vertices v_{j_1} and v_{j_2}, then connect vertex e_j with vertices v_{j_1} and v_{j_2} in G'. We set $\ell' = \binom{k}{2}$ and $k' = k$.

If there is a clique of size k, then we can cut ℓ' vertices by removing k' vertices. From $v_1, \ldots, v_n$ remove those k vertices that correspond to the clique. Now the $\binom{k}{2}$ vertices of G' that correspond to the edges of the clique are isolated vertices. On the other hand, assume that ℓ' vertices can be cut by deleting k' vertices. The remaining vertices of $v_1, \ldots, v_n$ form a clique of size greater than ℓ' (assuming $n > \binom{k}{2} + k$), hence the ℓ' separated vertices correspond to ℓ' edges of G. These vertices have to be isolated, since they cannot be connected to the large clique formed by the remaining v_i's. This means that the end vertices of the corresponding edges were all deleted. Therefore these $\ell' = \binom{k}{2}$ edges can have at most $k' = k$ end points, which is only possible if the end points induce a clique of size k in G.   □

If we consider only bounded degree graphs, then Separating ℓ Vertices becomes fixed-parameter tractable:

Theorem 5. Separating ℓ Vertices *is fixed-parameter tractable with parameters k, ℓ, and d, where d is the maximum degree of the graph.*

Proof. Consider a solution $V = X \cup S \cup Y$, and consider the subgraph induced by $X \cup S$. This subgraph consists of some number of connected components, let $X_i \cup S_i$ be the vertex set of the ith component. For each i, the pair (S_i, X_i) has the following two properties:

(1) in graph G the set S_i separates X_i from the rest of the graph, and
(2) $X_i \cup S_i$ induces a connected graph.

On the other hand, assume that the pairs (X_1, S_1), ..., (X_t, S_t) satisfy (1), (2), and the sets X_1, ..., X_t, S_1, ..., S_t are pairwise disjoint. In this case if $X = X_1 \cup \cdots \cup X_t$ has size exactly ℓ and $S = S_1 \cup \cdots \cup S_t$ has size at most k, then they form a solution. Therefore we generate all the pairs that satisfy these requirements, and use color coding to decide whether there are disjoint pairs with the required total size. If there is a solution, then this method will find one.

By requirement (2) a pair (X_i, S_i) induces a connected subgraph of size at most $k + \ell$. We enumerate each such connected subgraph. If a vertex v is contained in a connected subgraph of size at most $k + \ell$, then all the vertices of the subgraph are at a distance of less than $k + \ell$ from v. The maximum degree of the graph is d, thus there are at most $d^{k+\ell}$ vertices at distance less than $k + \ell$ from v. Therefore the number of connected subgraphs that contain v and have size at most $k + \ell$ is a constant, which means that there is a linear number of such subgraphs in the whole graph. We can enumerate these subgraphs in linear time. Each subgraph can be divided into a pair (X_i, S_i) in at most $2^{k+\ell}$ different ways. From these pairs we retain only those that satisfy requirement (1).

Having generated all the possible pairs (X_1, S_1), ..., (X_p, S_p), a solution can be found as follows. We consider a random coloring of the vertices with $c := k + \ell$ colors. Using dynamic programming, we try to find a solution where every vertex of $X \cup S$ has a distinct color. Subproblem (C', j, k', ℓ') asks whether it is possible to select some pairs from the first j pairs such that (a) they are pairwise disjoint, (b) they use only vertices with color C', (c) the union of the S_i's has size k', and (d) the union of the X_i's has size ℓ'. For $j = 0$, the subproblems are trivial. If the subproblems for $j - 1$ are solved, then the problem can be solved for j using the following two recurrence relations. First, if subproblem $(C', j-1, k', \ell')$ is true, then clearly (C', j, k', ℓ') is true as well. Moreover, if every vertex of $X_j \cup S_j$ has distinct color (denote by C_j these colors), and subproblem $(C' \setminus C_j, j - 1, k' - |S_j|, \ell' - |X_j|)$ is true, then a solution for this subproblem can be extended by the pair (X_j, S_j) to obtain a solution for (C', j, k', ℓ'). Using these two rules, all the subproblems can be solved.

If there is a solution $X \cup S$, then by probability at least $c!/c^c$ (where $c = k + \ell$ is the number of colors) these vertices receive distinct colors, and the algorithm described above finds a solution. Therefore if there is a solution, then on average we have to repeat the method $c^c/c!$ (constant) times to find a solution. The algorithm can be derandomized using the standard method of k-perfect hash functions, see [6, Section 8.3] and [1]. □

A variant of SEPARATING ℓ VERTICES is the SEPARATING ℓ CONNECTED VERTICES problem where we also require that X induces a connected subgraph of G. This problem is fixed-parameter tractable:

Theorem 6. *The* SEPARATING ℓ CONNECTED VERTICES *problems is fixed-parameter tractable with parameters k and ℓ.*

Proof. A vertex with degree at most $k + \ell$ will be called a *low degree* vertex, let G_0 be the subgraph induced by these vertices. A vertex v with degree more than $k + \ell$ cannot be part of X: at most k neighbors of v can be in S, hence v would have more than ℓ neighbors in X, which is impossible if $|X| = \ell$. Therefore X is a connected subgraph of G_0. As in the proof of Theorem 5, a bounded degree graph has a linear number of connected subgraphs of size ℓ. For each such subgraph, it has to be checked whether it can be separated from the rest of the graph by deleting at most k vertices. □

However, if only k is parameter, then the problem is W[1]-hard. This follows from the proof of Theorem 4. We construct the $n + m$ vertex graph as before, but instead of asking whether it is possible to separate $\binom{k}{2}$ vertices by deleting k vertices, we ask whether it is possible to separate $n + m - \binom{k}{2} - k$ connected vertices by deleting k vertices. The two questions have the same answer, thus

Theorem 7. SEPARATING ℓ CONNECTED VERTICES *is* W[1]-*hard with parameter k.* □

Similarly, the problem is W[1]-hard if only ℓ is the parameter.

Theorem 8. SEPARATING ℓ CONNECTED VERTICES *is* W[1]-*hard with parameter ℓ.*

Proof. The reduction is from MAXIMUM CLIQUE. It is not difficult to show that MAXIMUM CLIQUE remains W[1]-hard for regular graphs. Assume that we are given an r-regular graph G, and it has to be decided whether there is a clique of size k. If $r \leq k^4$, then the problem is fixed parameter tractable: for every vertex v, we select $k-1$ neighbors of v in at most $\binom{k^4}{k-1}$ possible ways, and test whether these k vertices form a clique. Thus it will be assumed that $r > k^4$.

Consider the line graph $L(G)$ of G, i.e., the vertices of $L(G)$ correspond to the edges of G. Set $\ell = \binom{k}{2}$ and $k' = k(r - k + 1)$. If G has a size k clique then the ℓ edges induced by the clique can be separated from the rest of the line graph: for each vertex of the clique, we have to delete the $r - k + 1$ edges leaving the clique. On the other hand, assume that ℓ vertices of G' can be separated by deleting k vertices. The corresponding ℓ edges in G span a set T of vertices of size $t \leq 2\ell$. We show that $t = k$, thus T is a clique of size k in G. Assume that $t > k$. Each vertex of T has at least $r - t + 1$ edges that leave T. The corresponding $t(r - t + 1)$ vertices have to be deleted from the line graph of G, hence $k' \geq t(r - t + 1)$. However, this is not possible since

$$t(r - t + 1) - k' = (t - k)r - t(t - 1) + k(k - 1) \geq (t - k)r - 4\ell^2 \geq r - k^4 > 0$$

(in the first inequality we use $4\ell^2 \geq t^2$, in the second $t > k$ and $\ell < k^2/2$). □

The vertex connectivity is the minimum number of vertices that has to be deleted to make the graph disconnected. Using network flow techniques, vertex connectivity can be determined in polynomial time. By essentially the same proof as in Theorem 4, we can show hardness for this problem as well:

SEPARATING INTO ℓ COMPONENTS
Input: A graph $G(V, E)$, integers k and ℓ
Parameter 1: k
Parameter 2: ℓ
Question: Is there a set S of k vertices such that $G \setminus S$ has at least ℓ connected components?

Theorem 9. SEPARATING INTO ℓ COMPONENTS *is* W[1]-*hard with parameters* k *and* ℓ.

Proof. The construction is the same as in Theorem 4, but this time we set $\ell' = \binom{k}{2} + 1$ and $k' = k$. By deleting the vertices corresponding to a clique of size k the graph is separated into ℓ' components. The converse is also easy to see, the argument is the same as in Theorem 4.   □

References

1. N. Alon, R. Yuster, and U. Zwick. Finding and counting given length cycles. *Algorithmica*, 17(3):209–223, 1997.
2. T. N. Bui and C. Jones. Finding good approximate vertex and edge partitions is NP-hard. *Inform. Process. Lett.*, 42(3):153–159, 1992.
3. W. H. Cunningham. The optimal multiterminal cut problem. In *Reliability of computer and communication networks (New Brunswick, NJ, 1989)*, volume 5 of *DIMACS Ser. Discrete Math. Theoret. Comput. Sci.*, pages 105–120. Amer. Math. Soc., Providence, RI, 1991.
4. E. Dahlhaus, D. S. Johnson, C. H. Papadimitriou, P. D. Seymour, and M. Yannakakis. The complexity of multiterminal cuts. *SIAM J. Comput.*, 23(4):864–894, 1994.
5. R. Downey, V. Estivill-Castro, M. Fellows, E. Prieto, and F. Rosamund. Cutting up is hard to do. In J. Harland, editor, *Electronic Notes in Theoretical Computer Science*, volume 78. Elsevier, 2003.
6. R. G. Downey and M. R. Fellows. *Parameterized complexity*. Monographs in Computer Science. Springer-Verlag, New York, 1999.
7. N. Garg, V. V. Vazirani, and M. Yannakakis. Primal-dual approximation algorithms for integral flow and multicut in trees. *Algorithmica*, 18(1):3–20, 1997.
8. N. Garg, V. V. Vazirani, and M. Yannakakis. Multiway cuts in node weighted graphs. *J. Algorithms*, 50(1):49–61, 2004.
9. R. J. Lipton and R. E. Tarjan. A separator theorem for planar graphs. *SIAM J. Appl. Math.*, 36(2):177–189, 1979.
10. R. J. Lipton and R. E. Tarjan. Applications of a planar separator theorem. *SIAM J. Comput.*, 9(3):615–627, 1980.

Parameterized Coloring Problems on Chordal Graphs*

Dániel Marx

Department of Computer Science and Information Theory,
Budapest University of Technology and Economics
Budapest, H-1521, Hungary
dmarx@cs.bme.hu

Abstract. In the precoloring extension problem (PRExT) a graph is
given with some of the vertices having a preassigned color and it has to be
decided whether this coloring can be extended to a proper coloring of the
graph with the given number of colors. Two parameterized versions of the
problem are studied in the paper: either the number of precolored vertices
or the number of colors used in the precoloring is restricted to be at most
k. We show that these problems are polynomial time solvable but W[1]-
hard in chordal graphs. For a graph class $\mathscr{F}$, let $\mathscr{F} + ke$ (resp. $\mathscr{F} + kv$)
denote those graphs that can be made to be a member of $\mathscr{F}$ by deleting
at most k edges (resp. vertices). We investigate the connection between
PRExT in $\mathscr{F}$ and the coloring of $\mathscr{F} + ke$, $\mathscr{F} + ve$ graphs. Answering
an open question of Leizhen Cai [5], we show that coloring chordal+ke
graphs is fixed-parameter tractable.

1 Introduction

In graph vertex coloring we have to assign colors to the vertices such that neigh-
boring vertices receive different colors. In the *precoloring extension* (PRExT)
problem a subset W of the vertices have a preassigned color and we have to extend this to a proper k-coloring of the whole graph. Since vertex coloring is the
special case when $W = \emptyset$, the precoloring extension problem is NP-complete in
every class of graphs where vertex coloring is NP-complete. See [2,7,8] for more
background and results on PRExT.

In this paper we study the precoloring extension problem on chordal graphs.
PRExT is NP-complete for interval graphs [2] (and for unit interval graphs [12]),
hence it is NP-complete for chordal graphs as well. On the other hand, if every
color is used only once in the precoloring (this special case is called 1-PRExT),
then the problem becomes polynomial time solvable for interval graphs [2], and
more generally, for chordal graphs [11]. Here we introduce two new restricted versions of PRExT: we investigate the complexity of the problem when either there
are only k precolored vertices, or there are only k colors used in the precoloring.

---

* Research is supported in part by grants OTKA 44733, 42559 and 42706 of the
Hungarian National Science Fund.

R. Downey, M. Fellows, and F. Dehne (Eds.): IWPEC 2004, LNCS 3162, pp. 83–95, 2004.
© Springer-Verlag Berlin Heidelberg 2004

Clearly, the former is a special case of the latter. By giving an $O(kn^{k+2})$ time algorithm, we show that for fixed k both problems are polynomial time solvable on chordal graphs. However, we cannot expect to find a uniformly polynomial time algorithm for these problems, since they are W[1]-hard even for interval graphs. To establish W[1]-hardness, we use the recent result of Slivkins [15] that the edge-disjoint paths problem is W[1]-hard.

Leizhen Cai [5] introduced a whole new family of parameterized problems. If $\mathscr{F}$ is an arbitrary class of graphs, then denote by $\mathscr{F} - kv$ (resp. $\mathscr{F} - ke$) the class of those graphs that can be obtained from a member of $\mathscr{F}$ by deleting at most k vertices (resp. k edges). Similarly, let $\mathscr{F} + kv$ (resp. $\mathscr{F} + ke$) be the class of those graphs that can be made to be a member of $\mathscr{F}$ by deleting at most k vertices (resp. k edges). For any class of graphs $\mathscr{F}$ and for any graph problem, we can ask what is the complexity of the problem restricted to these 'almost $\mathscr{F}$' graphs. This question is investigated in [5] for the vertex coloring problem. Coloring $\mathscr{F} + kv$ or $\mathscr{F} + ke$ graphs can be very different than coloring graphs in $\mathscr{F}$, and might involve significantly new approaches.

We investigate the relations between PREXT and the coloring of the modified graph classes. We show that for several reasonable graph classes, reductions are possible between PREXT for graphs in $\mathscr{F}$ and the coloring of $\mathscr{F} + kv$ or $\mathscr{F} + ke$ graphs. Based on this correspondence between the problems, we show that both chordal+ke and chordal+kv graphs can be colored in polynomial time for fixed k, but chordal+kv graph coloring is W[1]-hard. Moreover, answering an open question of Cai [5], we develop a uniformly polynomial time algorithm for coloring chordal+ke graphs.

The paper is organized as follows. Section 2 contains preliminary notions. Section 3 reviews tree decomposition, which will be our main tool when dealing with chordal graphs. In Section 4, we investigate the parameterized PREXT problems for chordal graphs. The connections between PREXT and coloring $\mathscr{F} + ke$, $\mathscr{F} + kv$ graphs are investigated in Section 5. Finally, in Section 6, we show that coloring chordal+ke graphs is fixed-parameter tractable.

2   Preliminaries

A C-coloring is a proper coloring of the vertices with color set C. We introduce two different parameterization of the precoloring extension problem. Formally, the problem is as follows:

Precoloring Extension (PrExt)
Input: A graph $G(V, E)$, a set of colors C, and a precoloring $\psi \colon W \to C$ for a set of vertices $W \subseteq V$.
Parameter 1: $|W|$, the number of precolored vertices.
Parameter 2: $|\{\psi(w) : w \in W\}| = |C_W|$, the number of colors used in the precoloring.
Question: Is there a proper C-coloring ψ' of G that extends ψ (i.e., $\psi'(w) = \psi(w)$ for every $w \in W$)?

Note that $C_W \subseteq C$ is the set of colors appearing on the precolored vertices, and can be much smaller than the set of available colors C. When we consider parameter 1, then the problem will be called PREXT with fixed number of precolored vertices, while considering parameter 2 corresponds to PREXT with fixed number of colors in the precoloring.

For every class $\mathscr{F}$ and every fixed k, one can ask what is the complexity of vertex coloring on the four classes $\mathscr{F} + ke$, $\mathscr{F} + kv$, $\mathscr{F} - ke$, $\mathscr{F} - kv$. The first question is whether the problem is NP-complete for some fixed k. If the problem is solvable in polynomial time for every fixed k, then the next question is whether the problem is fixed-parameter tractable, that is, whether there is a uniformly polynomial time algorithm for the given classes.

If $\mathscr{F}$ is hereditary with respect to taking induced subgraphs, then $\mathscr{F} - kv$ is the same as $\mathscr{F}$, hence coloring $\mathscr{F} - kv$ graphs is the same as coloring in $\mathscr{F}$. Moreover, it is shown in [5] that if $\mathscr{F}$ is closed under edge contraction and has a polynomial time algorithm for coloring, then coloring $\mathscr{F} - ke$ graphs is fixed parameter tractable. Therefore we can conclude that coloring chordal$-kv$ and chordal$-ke$ graphs are in FPT. In this paper we show that coloring chordal$+ke$ graphs is in FPT, but coloring chordal$+kv$ graphs is W[1]-hard.

The *modulator* of an $\mathscr{F} + ke$ graph G is a set of at most k edges whose removal makes G a member of $\mathscr{F}$. Similar definitions apply for the other classes. We will call the vertices and edges of the modulator *special edges and vertices*. In the case of $\mathscr{F} + e$ and $\mathscr{F} - e$ graphs, the vertices incident to the special edges are the special vertices.

When considering the complexity of coloring in a given parameterized class, then we can assume either that only the graph is given in the input, or that a modulator is also given. In the case of coloring chordal$-ke$ graphs, this makes no difference as finding the modulator of such a graph (i.e., the at most k edges that can make the graph chordal) is in FPT [4,9]. On the other hand, the parameterized complexity of finding the modulator of a chordal$+ke$ graph is open. Thus in our algorithm for coloring chordal$+ke$ graphs, we assume that the modulator is given in the input.

3   Tree Decomposition

A graph is *chordal* if it does not contain a cycle of length greater than 3 as an induced subgraph. This section summarizes some well-known properties of chordal graphs. First, chordal graphs can be also characterized as the intersection graphs of subtrees of a tree (see e.g. [6]):

Theorem 1. *The following two statements are equivalent:*

1. *$G(V, E)$ is chordal.*
2. *There exists a tree $T(U, F)$ and a subtree $T_v \subseteq T$ for each $v \in V$ such that $u, v \in V$ are neighbors in $G(V, E)$ if and only if $T_u \cap T_v \neq \emptyset$.*

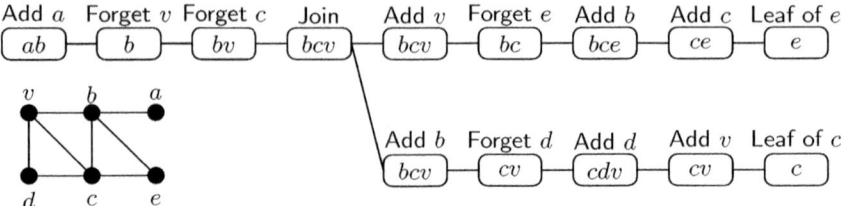

Fig. 1. Nice tree decomposition of a chordal graph.

The tree T together with the subtrees T_v is called the *tree decomposition* of G. A tree decomposition of G can be found in polynomial time (see [6,14]).

We assume that T is a rooted tree with some root $r \in U$. For clarity, we will use the word 'vertex' when we refer to the graph $G(V, E)$, and 'node' when referring to $T(U, F)$. For a node $x \in U$, denote by V_x those vertices whose subtree contains x or a descendant of x. The subgraph of G induced by V_x will be denoted by $G_x = G[V_x]$. For a node $x \in U$ of T, denote by K_x the union of v's where $x \in V(T_v)$. Clearly, the vertices of K_x are in V_x, and they form a clique in G_x, since the corresponding trees intersect in T at node x. An important property of the tree decomposition is the following: for every node $x \in U$, the clique K_x separates $V_x \setminus K_x$ and $V \setminus V_x$. That is, among the vertices of V_x, only the vertices in K_x can be adjacent to $V \setminus V_x$.

A tree decomposition will be called *nice* [10], if it satisfies the following additional requirements (see Figure 1):

- Every node $x \in U$ has at most two children.
- If $x \in U$ has two children $y, z \in U$, then $K_x = K_y = K_z$ (x is a *join* node).
- If $x \in U$ has only one child $y \in U$, then either $K_x = K_y \cup \{v\}$ (x is an *add* node) or $K_x = K_y \setminus \{v\}$ (x is a *forget* node) for some $v \in V$.
- If $x \in U$ has no children, then K_x contains exactly one vertex (x is a *leaf* node).

By splitting the nodes of the tree in an appropriate way, a tree decomposition of G can be transformed into a nice tree decomposition in polynomial time.

A vertex v can have multiple add nodes, but at most one forget node (the vertices in clique K_r of the root r have no forget nodes, but every other vertex has exactly one). For a vertex v, its subtree T_v is the subtree rooted at the forget node of v (if it exists, otherwise at the root) and whose leaves are exactly the add nodes and leaf nodes of v.

4   PrExt on Chordal Graphs

In this section we show that PRExT can be solved in polynomial time for chordal graphs if the number of colors used in the precoloring is bounded by a constant k. The algorithm presented below is a straightforward application of the tree decomposition described in Section 3. The running time of the algorithm is

$O(kn^{k+2})$, hence it is not uniformly polynomial. However, in Theorem 3 it is shown that the problem is W[1]-hard, hence we cannot hope to find a uniformly polynomial algorithm.

Theorem 2. *The* PREXT *problem can be solved in* $O(kn^{k+2})$ *time for chordal graphs, if the number of colors in the precoloring is at most k.*

Proof. It can be assumed that the colors used in the precoloring are the colors 1, 2, ..., k. For each node x of the nice tree decomposition of the graph, we solve several subproblems using dynamic programming. Each subproblem is described by a vector $[\alpha_1, \ldots, \alpha_k]$, where each α_i is either a vertex of K_x, or the symbol $\star$. We say that such a vector is *feasible* for node x, if there is a precoloring extension for G_x with the following property: if α_i $(1 \le i \le k)$ is $\star$, then color i does not appear on the clique K_x, otherwise it appears on vertex $\alpha_i \in K_x$. Notice that in a feasible vector a vertex can appear at most once (but the star can appear several times), thus in the following we consider only such vectors.

Clearly, the precoloring can be extended to the whole graph if and only if the the root node r has at least one feasible vector. The algorithm finds the feasible vectors for each node of T. We construct the feasible vectors for the nodes in a bottom-up fashion. First, they are easy to determine for the leaves. Moreover, they can be constructed for an arbitrary node if the feasible vectors for the children are already available. The techniques are standard, details omitted. □

To prove that PREXT with fixed number of precolored vertices is W[1]-hard for interval graphs, we use reduction from the edge disjoint paths problem, which is the following:

Edge disjoint paths
Input: A directed graph $G(V, E)$, with k pairs of vertices (s_i, t_i).
Parameter: The number of pairs k.
Question: Is there a set of k pairwise edge disjoint directed paths P_1, ..., P_k such that path P_i goes from s_i to t_i?

Recently, Slivkins [15] proved that the edge disjoint paths problem is W[1]-hard for directed acyclic graphs.

Theorem 3. PREXT *with fixed number of precolored vertices is* W[1]*-hard for interval graphs.*

Proof. The proof is by a parameterized reduction from the directed acyclic edge disjoint path problem. Given a directed acyclic graph $G(V, E)$ and terminal pairs s_i, t_i $(1 \le i \le k)$, we construct an interval graph with $k' = 2k$ precolored vertices in such a way that the interval graph has a precoloring extension if and only if the disjoint paths problem can be solved. Let 1, 2, ..., n be the vertices of G in a topological ordering. For each edge $\overrightarrow{xy}$ of G we add an interval $[x, y]$. For each terminal pair s_i, t_i we add two intervals $[0, s_i)$ and $[t_i, n+1)$, and precolor these intervals with color i.

Denote by $\ell(x)$ the number of intervals whose *right* end point is x (i.e., the intervals that arrive to x from the left), and by $r(x)$ the number of intervals whose left end point is x. In other words, $\ell(x)$ is the number of edges entering x plus the number of demands starting in x. If $\ell(x) < r(x)$, then add $r(x) - \ell(x)$ new intervals $[0, x)$ to the graph, if $\ell(x) > r(x)$, then add $\ell(x) - r(x)$ new intervals $[x, n+1)$. A consequence of this is that each point of $[0, n+1)$ is contained in the same number (denote it by c) of intervals: at each point the number of intervals ending equals the number of intervals starting. We claim that the interval graph has a precoloring extension with c colors if and only if the disjoint paths problem has a solution.

Assume first that there are k disjoint paths joining the terminal pairs. For each edge $\overrightarrow{xy}$, if it is used by the ith terminal pair, then color the interval $[x, y)$ with color i. Notice that the intervals we colored with color i do not intersect each other, and their union is exactly $[s_i, t_i)$. Therefore, considering also the two intervals $[0, s_i)$ and $[s_i, n+1)$ precolored with color i, each point of $[0, n+1)$ is covered by exactly one interval with color i. Therefore each point is contained in exactly $c - k$ intervals that do not have a color yet. This means that the uncolored intervals induce an interval graph where every point is in exactly $c - k$ intervals, and it is well-known that such an interval graph has clique number $c - k$ and can be colored with $c - k$ colors. Therefore the precoloring can be extended using $c - k$ colors in addition to the k colors used in the precoloring.

Now assume that the precoloring can be extended using c colors. Since each point in the interval $[0, n+1)$ is covered by exactly c intervals, therefore each point is covered by an interval of color i. Thus if an interval with color i ends at point x, then an interval with color i has to start at x. Since the interval $[0, s_i)$ has color i, there has to be an interval $[s_i, s_{i,1})$ with color i. Similarly, there has to be an interval $[s_{i,1}, s_{i,2})$ with color i, etc. Continuing this way, we will eventually arrive to an interval $[s_{i,p}, t_i)$. By the way the intervals were constructed, the edges $\overrightarrow{s_i s_{i,1}}$, $\overrightarrow{s_{i,1} s_{i,2}}$, ..., $\overrightarrow{s_{i,p} t_i}$ form a path from s_i to t_i. It is clear that the paths for different values of i are disjoint since each interval has only one color. Thus we constructed a solution to the disjoint paths problem, as required. $\square$

5   Reductions

In this section we give reductions between PREXT on $\mathscr{F}$ and coloring $\mathscr{F} + kv$, $\mathscr{F} + ke$ graphs. It turns out that if $\mathscr{F}$ is closed under disjoint union and attaching pendant vertices, then

$$\text{coloring } \mathscr{F} + ke \text{ graphs } \preceq \text{ PREXT on } \mathscr{F} \text{ with fixed } |W| \preceq$$
$$\text{coloring } \mathscr{F} + kv \text{ graphs } \preceq \text{ PREXT on } \mathscr{F} \text{ with fixed } |C_W|$$

When coloring $\mathscr{F} + ke$ or $\mathscr{F} + kv$ graphs, we assume that the modulator of the graph is given in the input. The proof of the following four results will appear in the full version:

Theorem 4. *For every class $\mathscr{F}$ of graphs, coloring $\mathscr{F}+ke$ graphs can be reduced to* PREXT *with fixed number of precolored vertices, if the modulator of the graph is given in the input.*

Theorem 5. *Let $\mathscr{F}$ be a class of graphs closed under attaching pendant vertices. Coloring $\mathscr{F} + kv$ graphs can be reduced to* PREXT *with fixed number of colors in the precoloring, if the modulator of the graph is given in the input.*

Theorem 6. *If $\mathscr{F}$ is a hereditary graph class closed under disjoint union, then* PREXT *in $\mathscr{F}$ with fixed number of precolored vertices can be reduced to the coloring of $\mathscr{F} + kv$ graphs.*

Theorem 7. *If $\mathscr{F}$ is a hereditary graph class closed under joining graphs at a vertex, then* PREXT *on $\mathscr{F}$ with a fixed number of colors in the precoloring can be reduced to the coloring of $\mathscr{F} + kv$ graphs.*

When reducing the coloring of $\mathscr{F} + ke$ or $\mathscr{F} + kv$ graphs to PREXT, the idea is to consider each possible coloring of the special vertices and solve each possibility as a PREXT problem. In the other direction, we use the special edges and vertices to build gadgets that force the precolored vertices to the required colors.

Concerning chordal graphs, putting together Theorems 2–6 gives

Corollary 1. *Coloring chordal+ke and chordal+kv graphs can be done in polynomial time for fixed k. However, coloring interval+kv (hence chordal+kv) graphs is* W[1]-*hard.*   □

In Section 6, we improve on this result by showing that coloring chordal+ke graphs is fixed-parameter tractable.

6   Coloring Chordal+ke Graphs

In Section 5 we have seen that coloring a chordal+ke graph can be reduced to the solution of PREXT problems on a chordal graph, and by Theorem 2, each such problem can be solved in polynomial time. Therefore chordal+ke graphs can be colored in polynomial time for fixed k, but with this algorithm the exponent of n in the running time depends on k. In this section we prove that coloring chordal+ke graphs is fixed-parameter tractable by presenting a uniformly polynomial time algorithm for the problem.

Let H be a chordal+ke graph, and denote by G the chordal graph obtained by deleting the special edges of G. We proceed similarly as in Theorem 2. First we construct a nice tree decomposition of H. A subgraph G_x of G corresponds to each node x of the nice tree decomposition. Let H_x be the graph G_x plus the special edges induced by the vertex set of G_x. For each subgraph H_x, we try to find a proper coloring. In fact, for every node x we solve several subproblems:

each subproblem corresponds to finding a coloring of H_x with a given property (to be defined later). The main idea of the algorithm is that the number of subproblems considered at a node can be reduced to a function of k.

Before presenting the algorithm, we introduce a technical tool that will be useful. For each node x of the nice tree decomposition, the graph H_x^* is defined by adding a clique of $|C| - |K_x|$ vertices u_1, u_2, ..., $u_{|C|-|K_x|}$ to the graph H_x, and connecting each new vertex to each vertex of K_x. The clique K_x together with the new vertices form a clique of size $|C|$, this clique will be called K_x^*. Instead of the colorings of H_x, we will consider the colorings of H_x^*. Although H_x^* is a supergraph of H_x, it is not more difficult to color than H_x: the new vertices are only connected to K_x, hence in every coloring of H_x there remains $|C| - |K_x|$ colors from C to color these vertices. However, considering the colorings of H_x^* instead of the colorings of H_x will make the arguments cleaner. The reason for this is that in every C-coloring of H_x^* every color of C appears on the clique K_x^* exactly once, which makes the description of the colorings more uniform.

Another technical trick is that we will assume that every special vertex is contained in exactly one special edge (recall that a vertex is called special if it is the end point of a special edge.) A graph can be transformed to such a form without changing the chromatic number, details omitted. The idea is to replace a special vertex with multiple vertices, and add some simple gadgets that force these vertices to have the same color. Since each special vertex is contained in only one special edge, thus each special vertex w has a unique *pair*, which is the other vertex of the special edge incident to w.

Now we define the subproblems associated with node x. A set system is defined where each set corresponds to a type of coloring that is possible on H_x^*. Let W be the set of special vertices, we have $|W| \leq 2k$. Let W_x be the special vertices contained in the subgraph H_x^*. In the following, we consider sets over $K_x^* \times W$. That is, each element of the set is a pair (v, w) with $v \in K_x^*$, $w \in W$.

Definition 1. *To each C-coloring ψ of H_x^*, we associate a set $S_x(\psi) \subseteq K_x^* \times W$ such that $(v, w) \in S_x(\psi)$ $(v \in K_x^*, w \in W_x)$ if and only if $\psi(v) = \psi(w)$. The set system $\mathscr{S}_x$ over $K_x^* \times W$ contains a set S if and only if there is a coloring ψ of H_x^* such that $S = S_x(\psi)$.*

The set $S_x(\psi)$ describes ψ on H_x^* as it is seen from the "outside", i.e., from $H \setminus H_x^*$. In H_x^* only K_x^* and W_x are connected to the outside. Since K_x^* is a clique of size $|C|$, every color appears on exactly one vertex, this is the same for every coloring. Seen from the outside, the only difference between the colorings is how the colors are assigned to W_x. The set $S_x(\psi)$ captures this information.

Subgraph H_x^* (hence H_x) is C-colorable if and only if the set system $\mathscr{S}_x$ is not empty. Therefore to decide the C-colorability of H, we have to check whether $\mathscr{S}_r$ is empty, where r is the root of the nice tree decomposition.

Before proceeding further, we need some new definitions.

Definition 2. *A set $S \subseteq K_x^* \times W$ is regular, if for every $w \in W$, there is at most one element of the form (v, w) in S. Moreover, we also require that if*

$v \in K_x^* \cap W$ then $(v, v) \in S$. The set S contains *vertex* w, *if there is an element* (v, w) *in* S *for some* $v \in K_x^*$.

For a coloring ψ of H_x^*, set $S_x(\psi)$ is regular and contains only vertices from W_x.

Definition 3. *For a set $S \in K_x^* \times W$, its* blocker $B(S)$ *is a subset of $K_x^* \times W$ such that $(v, w) \in B(S)$ if and only if $(v, w') \in S$ for the pair w' of w. We say that sets S_1 and S_2 form a* non-blocking pair *if $B(S_1) \cap S_2 = \emptyset$ and $S_1 \cap B(S_2) = \emptyset$.*

If ψ is a coloring of H_x^*, then the set $B(S_x(\psi))$ describes the requirements that have to be satisfied if we want to extend ψ to the whole graph. For example, if $(v, w) \in S_x(\psi)$, then this means that $v \in K_x^*$ has the same color as special vertex w. Now $(v, w') \in B(S_x(\psi))$ for the pair w' of w. This tells us that we *should not* color w' with the same color as v, because in this case the pairs w and w' would have the same color.

To be a non-blocking pair, it is sufficient that one of $B(S_1) \cap S_2$ and $S_1 \cap B(S_2)$ is empty:

Lemma 1. *For two sets $S_1, S_2 \in K_x \times W$, we have that $B(S_1) \cap S_2 = \emptyset$ if and only if $S_1 \cap B(S_2) = \emptyset$.*

Proof. Suppose that $B(S_1) \cap S_2 = \emptyset$, but $(v, w) \in S_1 \cap B(S_2)$ (the other direction follows by symmetry). Since $(v, w) \in B(S_2)$, this means that $(v, w') \in S_2$ where w' is the pair of w. But in this case $(v, w) \in S_1$ implies that $(v, w') \in B(S_1)$, contradicting $B(S_1) \cap S_2 = \emptyset$. □

The following lemma motivates the definition of the non-blocking pair, it turns out to be very relevant to our problem. If x is a join node, then we can give a new characterization of $\mathscr{S}_x$, based on the set systems of its children.

Lemma 2. *If x is a join node with children y and z, then*

$$\mathscr{S}_x = \{S_y \cup S_z : S_y \in \mathscr{S}_y \text{ and } S_z \in \mathscr{S}_z \text{ form a non-blocking pair}\}.$$

Proof. If $S \in \mathscr{S}_x$, then there is a corresponding coloring ψ of H_x^*. Coloring ψ induces a coloring ψ_y (resp. ψ_z) of H_y^* (resp. H_z^*). Let S_y (resp. S_z) be the set that corresponds to coloring ψ_y (resp. ψ_z). We show that S_y and S_z form a non-blocking pair, and $S = S_y \cup S_z$. By Lemma 1, it is enough to show that $S_y \cap B(S_z) = \emptyset$. Suppose that $S_y \cap B(S_z)$ contains the element (v, w) for some $v \in K_y^* = K_x^*$ and $w \in W_y$. By the definition of S_y, this means that $\psi_y(v) = \psi_y(w)$. Since $(v, w) \in B(S_z)$, thus $(v, w') \in S_z$ for the pair $w' \in W$ of w. Therefore $\psi_z(v) = \psi_z(w')$ follows. However, $\psi_y(v) = \psi_z(v)$, hence $\psi_y(w) = \psi_z(w')$, which is a contradiction, since w and w' are neighbors, and ψ is a proper coloring of H_x^*. Now we show that $S = S_y \cup S_z$. It is clear that $(v, w) \in S_y$ implies $(v, w) \in S$, hence $S_y \cup S_z \subseteq S$. Moreover, suppose that $(v, w) \in S$. Without loss of generality, it can be assumed that w is contained in H_y^*. This implies that $(v, w) \in S_y$, as required.

Now let $S_y \in \mathscr{S}_y$ and $S_z \in \mathscr{S}_z$ be a non-blocking pair, it has to be shown that $S = S_y \cup S_z$ is in $\mathscr{S}_x$. Let ψ_y (resp. ψ_z) be the coloring corresponding to S_y (resp. S_z). In general, ψ_y and ψ_z might assign different colors to the vertices of $K_x^* = K_y^* = K_z^*$. However, since K_x^* is a clique and every color appears exactly once on it, by permuting the colors in ψ_y, we can ensure that ψ_y and ψ_z agree on K_x^*. We claim that if we merge ψ_y and ψ_z, then the resulting coloring ψ is a proper coloring of H_x^*. The only thing that has to be verified is whether ψ assigns different colors to the end vertices of these special edges that are contained completely neither in H_y^* nor H_z^*. Suppose that special vertices $w \in W_y \setminus W_z$ and $w' \in W_z \setminus W_y$ are pairs, but $\psi(w) = \psi(w')$. We know that $(v, w) \in S_y$ for some $v \in K_y^*$, and similarly $(v', w') \in S_z$. By definition, this means that $\psi_y(v) = \psi_y(w)$ and $\psi_z(v') = \psi(w')$. Since ψ_y and ψ_z assign the same colors to the vertices of the clique K_x^*, thus this is only possible if $v = v'$, implying $(v, w') \in S_z$. However, $B(S_y)$ also contains (v, w') contradicting the assumption that $B(S_y) \cap S_z = \emptyset$. Now it is straightforward to verify that the set corresponding to ψ is $S = S_y \cup S_z$, proving that $S \in \mathscr{S}_x$. $\qquad\square$

Lemma 2 gives us a way to obtain the system $\mathscr{S}_x$ if x is a join node and the systems for the children are known. It can be shown for add nodes and forget nodes as well that their set systems can be constructed if the set systems are given for their children. However, we do not prove this here, since this observation does not lead to a uniformly polynomial algorithm. The problem is that the size of $\mathscr{S}_x$ can be $O(n^k)$, therefore it cannot be represented explicitly. On the other hand, in the following we show that it is not necessary to represent the whole set system, most of the sets can be thrown away, it is enough to retain only a constant number of sets.

We will replace $\mathscr{S}_x$ by a system $\mathscr{S}_x^*$ representative for $\mathscr{S}_x$ that has constant size. Representative systems and their use in finding disjoint sets were introduced by Monien [13] (and subsequently used also in [1]). Here we give a definition adapted to our problem:

Definition 4. *A subsystem $\mathscr{S}_x^* \subseteq \mathscr{S}_x$ is representative for $\mathscr{S}_x$ if the following holds: for each regular set $U \subseteq K_x \times W$ that does not contain vertices in $W_x \setminus K_x^*$, if $\mathscr{S}_x$ contains a set S disjoint from $B(U)$, then $\mathscr{S}_x^*$ also contains a set S' also disjoint from $B(U)$. We say that the subsystem $\mathscr{S}_x^*$ is minimally representative for $\mathscr{S}_x$, if it is representative for $\mathscr{S}_x$, but it is not representative after deleting any of the sets from $\mathscr{S}_x^*$.*

That is, if $\mathscr{S}_x$ can present a member avoiding all the forbidden colorings described by $B(U)$, then $\mathscr{S}_x^*$ can present such a member as well. For technical reasons, we are interested only in requirements $B(U)$ with U as described above.

The crucial idea is that the size of a minimally representative system can be bounded by a function of k independent of n (if the size of each set in $\mathscr{S}_x$ is at most $2k$). This is a consequence of the following version of Bollobás' inequality:

Theorem 8 (Bollobás [3]). *Let (A_1, B_1), (A_2, B_2), $\ldots$, (A_m, B_m) be a sequence of pairs of sets over a common ground set X such that $A_i \cap B_j = \emptyset$ if and only if $i = j$. Then*

$$\sum_{i=1}^{m} \binom{|A_i| + |B_i|}{|A_i|}^{-1} \leq 1.$$

Lemma 3. *If $\mathscr{S}_x^*$ is minimally representative for $\mathscr{S}_x$, then $|\mathscr{S}_x^*| \leq \binom{4k}{2k}$.*

Proof. Let $\mathscr{S}_x^* = \{A_1, A_2, \ldots, A_m\}$. Since $\mathscr{S}_x^*$ is minimally representative for $\mathscr{S}_x$, therefore for every $1 \leq i \leq m$, there is a regular set $B_i = B(U_i) \subseteq K_x \times W$ satisfying Definition 4 such that $\mathscr{S}_x$ has a set disjoint from B_i, but A_i is the only set in $\mathscr{S}_x^*$ disjoint from B_i (otherwise A_i could be safely removed from $\mathscr{S}_x^*$). This means that $A_i \cap B_i = \emptyset$, and $A_j \cap B_i \neq \emptyset$ for every $i \neq j$. Therefore $(A_1, B_1), (A_2, B_2), \ldots, (A_m, B_m)$ satisfy the requirements of Theorem 8, hence

$$1 \geq \sum_{i=1}^{m} \binom{|A_i| + |B_i|}{|A_i|}^{-1} \geq \sum_{i=1}^{m} \binom{|W_x| + |W|}{|W_x|}^{-1} \geq m \binom{4k}{2k}^{-1}.$$

Therefore $m \leq \binom{4k}{2k}$, and the lemma follows.   □

Lemma 3 shows that one can obtain a constant size representative system by throwing away sets until the system becomes a minimally representative. Another way of obtaining a constant size system is to use the data structure of Monien [13] for finding and storing representative systems. Using that method, we can obtain a representative system of size at most $2k^{2k}$. This can be somewhat larger than $\binom{4k}{2k}$ given by Lemma 3, but it is also good for our purposes.

We show that instead of determining the set system $\mathscr{S}_x$ for each node, it is sufficient to find a set system $\mathscr{S}_x^*$ representative for $\mathscr{S}_x$. That is, if for each child y of x we are given a system $\mathscr{S}_y^*$ representative for $\mathscr{S}_y$, then we can construct a system $\mathscr{S}_x^*$ representative for $\mathscr{S}_x$. For a join node x, one can find a set system $\mathscr{S}_x^*$ representative for $\mathscr{S}_x$ by a characterization analogous to Lemma 2:

Lemma 4. *Let x be a join node with children y and z, and let $\mathscr{S}_y^*$ be representative for $\mathscr{S}_y$, and $\mathscr{S}_z^*$ representative for $\mathscr{S}_z$. Then the system*

$$\mathscr{S}_x^* = \{S_y \cup S_z : S_y \in \mathscr{S}_y^* \text{ and } S_z \in \mathscr{S}_z^* \text{ form a non-blocking pair}\}$$

is representative for $\mathscr{S}_x$.

Proof. Since $\mathscr{S}_y^* \subseteq \mathscr{S}_y$ and $\mathscr{S}_z^* \subseteq \mathscr{S}_z$, by Lemma 2 it follows that $\mathscr{S}_x^* \subseteq \mathscr{S}_x$. Therefore we have to show that for every regular set U not containing vertices from $W_x \setminus K_x^*$, if there is a set $S \in \mathscr{S}_x$ disjoint from $B(U)$, then there is a set $S' \in \mathscr{S}_x^*$ also disjoint from $B(U)$. Let ψ be the coloring corresponding to set S, and let ψ_y (resp. ψ_z) be the coloring of H_y^* (resp. H_z^*) induced by ψ. Let $S_y \in \mathscr{S}_y$ and $S_z \in \mathscr{S}_z$ be the sets corresponding to ψ_y and ψ_z. We have seen in the proof of Lemma 2 that S_y and S_z form a non-blocking pair and $S = S_y \cup S_z$, hence S_y is disjoint from $B(U) \cup B(S_z) = B(U \cup S_z)$. Note that U does not contain vertices from $W_x \setminus K_x^*$, and S_z contains only vertices from H_z^*, hence $U \cup S_z$ is regular, and does not contain vertices from $W_y \setminus K_y^*$. Since

$\mathscr{S}_y^*$ is representative for $\mathscr{S}_y$, there is a set $S_y' \in \mathscr{S}_y^*$ that is also disjoint from $B(U \cup S_z)$. By Lemma 1, $S_y' \cap B(S_z) = \emptyset$ implies that $B(S_y') \cap S_z = \emptyset$, hence S_z is disjoint from $U \cup B(S_y') = B(U \cup S_y')$. Since $\mathscr{S}_z^*$ is representative for $\mathscr{S}_z$, there is a set $S_z' \in \mathscr{S}_z^*$ that is also disjoint from $B(U \cup S_y')$. Applying again Lemma 1, we get that S_y' and S_z' form a non-blocking pair, hence $S' = S_y' \cup S_z'$ is in S_x^*. Since S' is disjoint from $B(U)$, thus $\mathscr{S}_x^*$ contains a set disjoint from $B(U)$.   □

If x is an add node or forget node with children y and a system $\mathscr{S}_y^*$ representative for S_y is given, then we can construct a system $\mathscr{S}_x^*$ that is representative for $\mathscr{S}_x$. The construction is conceptually not difficult, but requires a tedious discussion. We omit the details.

Therefore starting from the leaves, the systems $\mathscr{S}_x^*$ can be constructed using bottom up dynamic programming. After constructing $\mathscr{S}_x^*$, we use the data structure of Monien to reduce the size of $\mathscr{S}_x^*$ to a constant. This will ensure that each step of the algorithm can be done in uniformly polynomial time. By checking whether $\mathscr{S}_r^*$ is empty for the root r, we can determine whether the graph has a C-coloring. This proves the main result of the section:

Theorem 9. *Coloring chordal+ke graphs is in* FPT *if the modulator of the graph is given in the input.*

References

1. N. Alon, R. Yuster, and U. Zwick. Finding and counting given length cycles. *Algorithmica*, 17(3):209–223, 1997.
2. M. Biró, M. Hujter, and Zs. Tuza. Precoloring extension. I. Interval graphs. *Discrete Math.*, 100(1-3):267–279, 1992.
3. B. Bollobás. On generalized graphs. *Acta Math. Acad. Sci. Hungar*, 16:447–452, 1965.
4. L. Cai. Fixed-parameter tractability of graph modification problems for hereditary properties. *Inform. Process. Lett.*, 58(4):171–176, 1996.
5. L. Cai. Parameterized complexity of vertex colouring. *Discrete Appl. Math.*, 127:415–429, 2003.
6. M. C. Golumbic. *Algorithmic graph theory and perfect graphs*. Academic Press, New York, 1980.
7. M. Hujter and Zs. Tuza. Precoloring extension. II. Graph classes related to bipartite graphs. *Acta Mathematica Universitatis Comenianae*, 62(1):1–11, 1993.
8. M. Hujter and Zs. Tuza. Precoloring extension. III. Classes of perfect graphs. *Combin. Probab. Comput.*, 5(1):35–56, 1996.
9. H. Kaplan, R. Shamir, and R. E. Tarjan. Tractability of parameterized completion problems on chordal, strongly chordal, and proper interval graphs. *SIAM J. Comput.*, 28(5):1906–1922, 1999.
10. T. Kloks. *Treewidth*, volume 842 of *Lecture Notes in Computer Science*. Springer-Verlag, Berlin, 1994. Computations and approximations.
11. D. Marx. Precoloring extension on chordal graphs, 2004. Submitted.
12. D. Marx. Precoloring extension on unit interval graphs, 2004. Submitted.
13. B. Monien. How to find long paths efficiently. In *Analysis and design of algorithms for combinatorial problems (Udine, 1982)*, volume 109 of *North-Holland Math. Stud.*, pages 239–254. North-Holland, Amsterdam, 1985.

14. D. J. Rose, R. E. Tarjan, and G. S. Lueker. Algorithmic aspects of vertex elimination on graphs. *SIAM J. Comput.*, 5(2):266–283, 1976.
15. A. Slivkins. Parameterized tractability of edge-disjoint paths on directed acyclic graphs. In *Algorithms – ESA 2003, 11th Annual European Symposium*, volume 2832 of *Lecture Notes in Comput. Sci.*, pages 482–493. Springer, Berlin, 2003.

On Decidability of MSO Theories of Representable Matroids

Petr Hliněný[1][*] and Detlef Seese[2]

[1] Dept. of Computer Science, VŠB – Technical University of Ostrava,
17. listopadu 15, 70833 Ostrava, Czech Republic
petr.hlineny@vsb.cz
[2] Institute AIFB, University Karlsruhe (TH),
D-76128 Karlsruhe, Germany
seese@aifb.uni-karlsruhe.de

Abstract. We show that, for every finite field $\mathbb{F}$, the class of all $\mathbb{F}$-representable matroids of branch-width at most a constant t has a decidable MSO theory. In the other direction, we prove that every class of $\mathbb{F}$-representable matroids with a decidable MSO theory must have uniformly bounded branch-width.

Keywords: matroid, branch-width, MSO theory, decidability.
2000 Math subject classification: 03B25, 05B35, 68R05, 68R10, 03B15.

1 Introduction

Monadic second order logic, which extends first order logic by allowing quantification over monadic predicates, is famous for their high expressive power in combination with a manageable model theory (see e.g. [12]). For this reason it has found many applications in different areas, as e.g. decidability, model checking, data bases, and computational complexity.

Of special importance in this area are classes of graphs (or other structures) of bounded tree-width, branch-width, or clique-width, since for these classes MSO logic posseses besides the good model theory also very good algorithmic properties. On the structural side, strong interest in tree-width has been motivated by the (now famous) Graph Minor project [19] of Robertson and Seymour which, besides many deep theoretical results, revolutionized the area of algorithm design in computer science. In particular, many problems which are NP-hard for arbitrary structures, could be solved in polynomial and often even linear time if they are restricted to structures of bounded tree-width or bounded clique-width (see e.g. [1], or [8],[7]).

Interestingly, general algorithmic results on efficient computability over structures of bounded tree-width (branch-width, clique-width, etc.) come hand in hand with related logic results on decidability of theories. For example, for each

---

[*] Other affiliation: Institute of Mathematics and Computer Science, Matej Bel University, Banská Bystrica, Slovakia. Supported by grant VEGA 1/1002/04.

R. Downey, M. Fellows, and F. Dehne (Eds.): IWPEC 2004, LNCS 3162, pp. 96–107, 2004.

$k > 0$, the monadic second order theory of the class of all graphs of tree-width at most k, or of clique-width at most k, respectively, is decidable (see [2,6], or [22,23]).

Here we shall concentrate on matroids, as a strong generalization of graphs. Nowadays, one can witness in the matroid community a great effort to extend the above mentioned Robertson-Seymour's theoretical work on graph minors as far as possible to matroids, followed by important new structural results about representable matroids, eg. [10,11]. Inspired by those advances, we focus on extending the research of related complexity and logic questions from graphs to matroids, building on recent works [13,14] of the first author.

Since this paper is intended for general computer-science and logic audiences, we provide some basic definitions concerning matroid structure, and decidability and interpretability of theories from mathematical logic, in the next three sections. We bring up the MSO theory of matroids in Section 5, and present some related recent results there; like we show that the MSO theory of the class of all matroids of branch-width at most k is decidable, for every $k > 0$. We present our main result in Section 6, which extends results by the second author from [22]. We prove that, for every finite field $\mathbb{F}$, a class of $\mathbb{F}$-representable matroids with a decidable MSO theory must have uniformly bounded branch-width.

2   Basics of Matroids

We refer to Oxley [16] for matroid terminology. A *matroid* is a pair $M = (E, \mathcal{B})$ where $E = E(M)$ is the ground set of M (elements of M), and $\mathcal{B} \subseteq 2^E$ is a nonempty collection of *bases* of M. Moreover, matroid bases satisfy the "exchange axiom"; if $B_1, B_2 \in \mathcal{B}$ and $x \in B_1 - B_2$, then there is $y \in B_2 - B_1$ such that $(B_1 - \{x\}) \cup \{y\} \in \mathcal{B}$. We consider only finite matroids. Subsets of bases are called *independent sets*, and the remaining sets are *dependent*. Minimal dependent sets are called *circuits*. All bases have the same cardinality called the *rank* $\mathrm{r}(M)$ of the matroid. The *rank function* $\mathrm{r}_M(X)$ in M is the maximal cardinality of an independent subset of a set $X \subseteq E(M)$.

If G is a (multi)graph, then its *cycle matroid* on the ground set $E(G)$ is denoted by $M(G)$. The independent sets of $M(G)$ are acyclic subsets (forests) in G, and the circuits of $M(G)$ are the cycles in G. Another example of a matroid is a finite set of vectors with usual linear dependency. If $\mathbf{A}$ is a matrix, then the matroid formed by the column vectors of $\mathbf{A}$ is called the *vector matroid* of $\mathbf{A}$, and denoted by $M(\mathbf{A})$. The matrix $\mathbf{A}$ is a *representation* of a matroid $M \simeq M(\mathbf{A})$. We say that the matroid $M(\mathbf{A})$ is $\mathbb{F}$-*represented* if $\mathbf{A}$ is a matrix over a field $\mathbb{F}$. (Fig. 1.) A *graphic matroid*, i.e. such that it is a cycle matroid of some multigraph, is representable over any field.

The *dual* matroid M^* of M is defined on the same ground set E, and the bases of M^* are the set-complements of the bases of M. A set X is *coindependent* in M if it is independent in M^*. An element e of M is called a *loop* (a *coloop*), if $\{e\}$ is dependent in M (in M^*). The matroid $M \setminus e$ obtained by *deleting* a non-coloop element e is defined as $(E - \{e\}, \mathcal{B}^-)$ where $\mathcal{B}^- = \{B : B \in \mathcal{B}, e \notin B\}$.

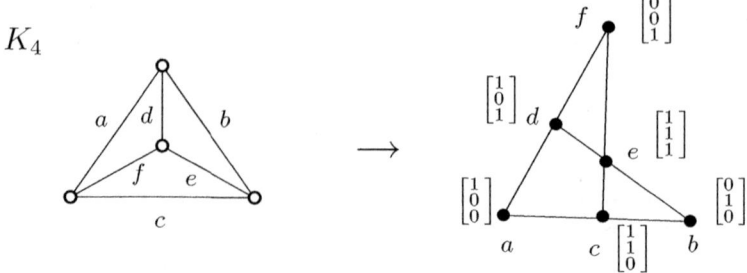

Fig. 1. An example of a vector representation of the cycle matroid $M(K_4)$. The matroid elements are depicted by dots, and their (linear) dependency is shown using lines.

The matroid M/e obtained by *contracting* a non-loop element e is defined using duality $M/e = (M^* \setminus e)^*$. (This corresponds to contracting an edge in a graph.) A *minor* of a matroid is obtained by a sequence of deletions and contractions of elements. Since these operations naturally commute, a minor M' of a matroid M can be uniquely expressed as $M' = M \setminus D/C$ where D are the coindependent deleted elements and C are the independent contracted elements. The following claim is folklore in matroid theory:

Lemma 2.1. *Let $N = M \setminus D/C$. Then a set $X \subseteq E(N)$ is dependent in N if and only if there is a dependent set $Y \subseteq E(M)$ in M such that $Y - X \subseteq C$.*

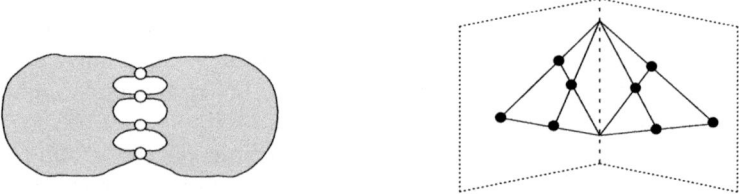

Fig. 2. An illustration to a 4-separation in a graph, and to a 3-separation in a matroid.

Another important concept is matroid connectivity, which is close, but somehow different, to traditional graph connectivity. The *connectivity function* λ_M of a matroid M is defined for all subsets $A \subseteq E$ by

$$\lambda_M(A) = r_M(A) + r_M(E - A) - r(M) + 1\,.$$

Here $r(M) = r_M(E)$. A subset $A \subseteq E$ is *k-separating* if $\lambda_M(A) \leq k$. A partition $(A, E - A)$ is called a *k-separation* if A is k-separating and both $|A|, |E - A| \geq k$.

Geometrically, the spans of the two sides of a k-separation intersect in a subspace of rank less than k. See in Fig. 2. In a corresponding graph view, the connectivity function $\lambda_G(F)$ of an edge subset $F \subseteq E(G)$ equals the number of vertices of G incident both with F and with $E(G) - F$. (Then $\lambda_G(F) = \lambda_{M(G)}(F)$ provided both sides of the separation are connected in G.)

3   Tree-Width and Branch-Width

The notion of graph *tree-width* is well known. Let Q_n denote the $n \times n$-grid graph, i.e. the graph on $V(Q_n) = \{1, 2, \ldots, n\}^2$ and $E(Q_n) = \{\{(i,j)(i',j')\} : 1 \leq i, j, i', j' \leq n, \{|i - i'|, |j - j'|\} = \{0, 1\}\}$. We say that a class $\mathcal{G}$ of graphs has *bounded tree-width* if there is a constant k such that any graph $G \in \mathcal{G}$ has tree-width at most k. A basic structural result on tree-width is given in [20]:

Theorem 3.1. (Robertson, Seymour) *A graph class $\mathcal{G}$ has bounded tree-width if and only if there exists a constant m such that no graph $G \in \mathcal{G}$ has a minor isomorphic to Q_m.*

The same paper [20] also presents a similar, but less known, parameter called branch-width, and proves that branch-width is within a constant factor of tree-width on graphs.

Assume that λ is a symmetric function on the subsets of a ground set E. (Here $\lambda \equiv \lambda_G$ is the connectivity function of a graph, or $\lambda \equiv \lambda_M$ of a matroid.) A *branch decomposition* of λ is a pair (T, τ) where T is a sub-cubic tree ($\Delta(T) \leq 3$), and τ is a bijection of E into the leaves of T. For e being an edge of T, the *width* of e in (T, τ) equals $\lambda(A) = \lambda(E - A)$, where $A \subseteq E$ are the elements mapped by τ to leaves of one of the two connected components of $T - e$. The width of the branch decomposition (T, τ) is maximum of the widths of all edges of T, and *branch-width* of λ is the minimal width over all branch decompositions of λ.

Recall the definitions of graph and matroid connectivity functions from Section 2. Then branch-width of $\lambda \equiv \lambda_G$ is called *branch-width of a graph G*, and that of $\lambda \equiv \lambda_M$ is called *branch-width of a matroid M*. (See examples in Fig. 3.) We remark that it is possible to define matroid tree-width [15] which is within a constant factor of branch-width, but this is not a straightforward extension of traditional graph tree-width. Considering branch-width on matroids, the following recent result [11] is crucial for our paper:

Theorem 3.2. (Geelen, Gerards, Whittle) *For every finite field $\mathbb{F}$; a class $\mathcal{N}$ of $\mathbb{F}$-representable matroids has bounded branch-width if and only if there exists a constant m such that no matroid $N \in \mathcal{N}$ has a minor isomorphic to $M(Q_m)$.*

4   Decidability of Theories

We will use the following notion of a theory. Let $\mathcal{K}$ be a class of structures and let L be a suitable logic for $\mathcal{K}$. A sentence is a set of well-formed L-formulas

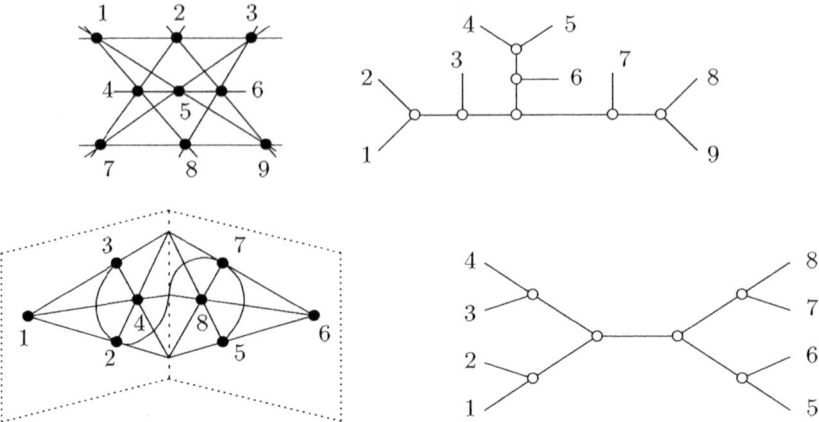

Fig. 3. Two examples of width-3 branch decompositions of the Pappus matroid (top left, in rank 3) and of the binary affine cube (bottom left, in rank 4). The lines in matroid pictures show dependencies among elements.

without free variables. The set of all L-sentences true in $\mathcal{K}$ is denoted as L-theory of $\mathcal{K}$. We use $\mathrm{Th}_L(\mathcal{K})$ as a short notation for this theory. Hence, a theory can be viewed as the set of all properties, expressible in L, which all structures of $\mathcal{K}$ possess. In case that $\mathcal{K} = \{G\}$ we write $\mathrm{Th}_L(G)$ instead of $\mathrm{Th}_L(\mathcal{K})$. Using this definition we obtain $\mathrm{Th}_L(\mathcal{K}) = \bigcap\{\mathrm{Th}_L(G) : G \in \mathcal{K}\}$. We write $\mathrm{Th}(\mathcal{K})$, $\mathrm{Th}_{MSO}(\mathcal{K})$ if L is first order logic, or monadic second order logic (abbreviated as MSO logic), respectively.

For graphs there are actually two variants of MSO logic, commonly denoted by MS_1 and MS_2. In MS_1, set variables only denote sets of vertices. In MS_2, set variables can also denote sets of edges of the considered graph. In other words the difference between both logics is that in MS_1 the domain of the graph consists of the vertices only and the relation is just the usual adjacency between vertices, while in MS_2 the domain is two-sorted and contains vertices as well as edges and the relation is the incidence relation. The expressive power of both logics was studied by Courcelle in [4].

A theory is said to be *decidable* if there is an algorithm deciding, for an arbitrary sentence $\varphi \in L$, whether $\varphi \in \mathrm{Th}_L(\mathcal{K})$ or not, i.e. whether φ is true in all structures of $\mathcal{K}$. Otherwise this theory is said to be *undecidable*. More information concerning the terminology from logic needed in this section can be found in classical textbooks as [9]. A good introduction into the decidability of theories can be found in [18] (see also [12] for monadic theories).

To prove decidability of theories the method of model interpretability, introduced in [17] is often the best tool of choice. To describe the idea of the method assume that two classes of structures $\mathcal{K}$ and $\mathcal{K}'$ are given, and that L and L', respectively, are corresponding languages for the structures of these classes. The basic idea of the interpretability of theory $\mathrm{Th}_L(\mathcal{K})$ into $\mathrm{Th}_{L'}(\mathcal{K}')$ is to transform formulas of L into formulas of L', by translating the nonlogical symbols of L by formulas of L', in such a way that truth is preserved in a certain way. Here we assume that the logics underlying both languages are the same. Otherwise, one has to translate also the logical symbols.

We explain this translation in a simple case of relational structures. First one chooses an L'-formula $\alpha(x)$ intended to define in each L'-structure $G \in \mathcal{K}'$ a set of individuals $G[\alpha] := \{a : a \in dom(G)$ and $G \models \alpha(a)\}$, where $dom(G)$ denotes the domain (set of individuals) of G. Then one chooses for each s-ary relational sign R from L an L'-formula $\beta_R(x_1, \ldots, x_s)$, with the intended meaning to define a corresponding relation $G[\beta_R] := \{(a_1, \ldots, a_s) : a_1, \ldots, a_s \in dom(G)$ and $G \models \beta_R(a_1, \ldots, a_s)\}$. All these formulas build the formulas of the interpretation $I = (\alpha(x), \beta_R(x_1, \ldots, x_s), \ldots)$.

With the help of these formulas one can define for each L'-structure G a structure $G^I := (G[\alpha], G[\beta_R], \ldots)$, which is just the structure defined by the chosen formulas in G. Sometimes G^I is also denoted as $I(G)$ and I is called an (L, L')-interpretation of G^I in G. In case that both L and L' are MSO languages, this interpretation is also denoted as MSO-interpretation. Using these formulas there is also a natural way to translate each L-formula φ into an L'-formula φ^I. This is done by induction on the structure of formulas. The atomic formulas are simply substituted by the corresponding chosen formulas with the corresponding substituted variables. Then one may proceed via induction as follows:

$$(\neg\chi)^I := \neg(\chi^I), \quad (\chi_1 \wedge \chi_2)^I := (\chi_1)^I \wedge (\chi_2)^I,$$

$$\left(\exists x\, \chi(x)\right)^I := \exists x \left(\alpha(x) \wedge \chi^I(x)\right),$$

$$(x \in X)^I := x \in X, \quad \left(\exists X\, \chi(X)\right)^I := \exists X\, \chi^I(X).$$

The resulting translation is called an interpretation with respect to L and L'. Its concept could be briefly illustrated with a picture:

$$
\begin{array}{ccc}
\varphi \in L & \quad I \quad & \varphi^I \in L' \\
H \in \mathcal{K} & \longrightarrow & G \in \mathcal{K}' \\
& & \\
G^I \simeq H & \xleftarrow{\quad I \quad} & G
\end{array}
$$

For theories, interpretability is now defined as follows. Let $\mathcal{K}$ and $\mathcal{K}'$ be classes of structures and L and L' be corresponding languages. Theory $\mathrm{Th}_L(\mathcal{K})$ is said to be interpretable in $\mathrm{Th}_{L'}(\mathcal{K}')$ if there is an (L, L')-interpretation I translating each L-formula φ into an L'-formula φ^I, and each L'-structure $G \in \mathcal{K}'$ into an L-structure G^I as above, such that the following two conditions are satisfied:

(i) For every structure $H \in \mathcal{K}$, there is a structure $G \in \mathcal{K}'$ such that $G^I \cong H$,

(ii) for every $G \in \mathcal{K}'$, the structure G^I is isomorphic to some structure of $\mathcal{K}$.

It is easy to see that interpretability is transitive. The key result for interpretability of theories is the following theorem [17]:

Theorem 4.1. (Rabin) *Let $\mathcal{K}$ and $\mathcal{K}'$ be classes of structures, and L and L' be suitable languages. If $\mathrm{Th}_L(\mathcal{K})$ is interpretable in $\mathrm{Th}_{L'}(\mathcal{K}')$, then undecidability of $\mathrm{Th}_L(\mathcal{K})$ implies undecidability of $\mathrm{Th}_{L'}(\mathcal{K}')$.*

5   MSO Theory of Matroids

Considering logic point of view, a matroid M on a ground set E is the collection of all subsets 2^E together with a unary predicate *indep* such that $indep(F)$ if and only if $F \subseteq E$ is independent in M. (One may equivalently consider a matroid with a unary predicate for bases or for circuits, see in [13].) We shortly write MS_M to say that the language of *MSO logic is applied to (independence) matroids*. If $\mathcal{N}$ is a class of independence matroids, then the MS_M theory of $\mathcal{N}$ is denoted by $\mathrm{Th}_{MSO}(\mathcal{N})$.

To give readers a better feeling for the expressive power of MS_M on a matroid, we write down a few basic matroid predicates now.

- We write $basis(B) \equiv indep(B) \wedge \forall D\big(B \not\subseteq D \vee B = D \vee \neg\, indep(D)\big)$ to express the fact that a basis is a maximal independent set.
- Similarly, we write $circuit(C) \equiv \neg\, indep(C) \wedge \forall D\big(D \not\subseteq C \vee D = C \vee indep(D)\big)$, saying that C is dependent, but all proper subsets of C are independent.
- A cocircuit is a dual circuit in a matroid (i.e. a bond in a graph). We write $cocircuit(C) \equiv \forall B\big[\, basis(B) \rightarrow \exists x(x \in B \wedge x \in C)\big] \wedge \forall X\big[X \not\subseteq C \vee X = C \vee \exists B\big(\, basis(B) \wedge \forall x(x \notin B \vee x \notin X)\big)\big]$ saying that a cocircuit C intersects every basis, but each proper subset of C is disjoint from some basis.

It is shown that the language of MS_M is at least as powerful as that of MS_2 on graphs. Let $G \uplus H$ denotes the graph obtained from disjoint copies of G and H by adding all edges between them. The following statement is proved in [13]:

Theorem 5.1. (Hliněný) *Let G be a loopless multigraph, and let M be the cycle matroid of $G \uplus K_3$. Then any MSO sentence (in MS_2) about an incidence graph G can be expressed as a sentence about M in MS_M.*

In other words, the MSO theory of (loopless) incidence multigraphs is interpretable in a certain subclass of 3-connected graphic matroids.

The next result we are going to mention speaks about (restricted) recognizability of MS_M-definable matroid properties via tree automata. To formulate this, we have to introduce briefly the concept of *parse trees* for representable matroids of bounded branch-width, which has been first defined in [13]. For a finite field $\mathbb{F}$, an integer $t \geq 1$, and an arbitrary $\mathbb{F}$-represented matroid M of branch-width at most $t + 1$; a t-boundaried parse tree $\bar{T}$ over $\mathbb{F}$ is a rooted ordered binary tree, whose leaves are labeled with elements of M, and the inner nodes are labeled with symbols of a certain finite alphabet (depending on $\mathbb{F}$ and t). Saying roughly, symbols of the alphabet are "small configurations" in the projective geometry over $\mathbb{F}$. The parse tree $\bar{T}$ uniquely determines an $\mathbb{F}$-representation (up to projective transformations) of the matroid $P(\bar{T}) \simeq M$. See [13] for more details and the result:

Theorem 5.2. (Hliněný) *Let $\mathbb{F}$ be a finite field, $t \geq 1$, and let ϕ be a sentence in the language of MS_M. Then there exists a finite tree automaton $\mathcal{A}_t^\phi$ such that the following is true: A t-boundaried parse tree $\bar{T}$ over $\mathbb{F}$ is accepted by $\mathcal{A}_t^\phi$ if and only if $P(\bar{T}) \models \phi$. Moreover, the automaton $\mathcal{A}_t^\phi$ can be constructed (algorithmically) from given $\mathbb{F}$, t, and ϕ.*

Corollary 5.3. *Let* $\mathbb{F}$ *be a finite field,* $t \geq 1$, *and let* $\mathcal{B}_t$ *be the class of all matroids representable over* $\mathbb{F}$ *of branch-width at most* $t + 1$. *Then the theory* $\mathrm{Th}_{MSO}(\mathcal{B}_t)$ *is decidable.*

Proof. Assume we are given an MS_M-sentence ϕ. We construct the automaton $\mathcal{A}_t^\phi$ from Theorem 5.2. Moreover, there is an (easily constructible [13]) automaton $\mathcal{V}_t$ accepting valid t-boundaried parse trees over $\mathbb{F}$. Then $\mathcal{B}_t \not\models \phi$ if and only if there is a parse tree accepted by $\mathcal{V}_t$, but not accepted by $\mathcal{A}_t^\phi$. We thus, denoting by $-\mathcal{A}_t^\phi$ the complement of $\mathcal{A}_t^\phi$, construct the cartesian product automaton $\mathcal{A} = (-\mathcal{A}_t^\phi) \times \mathcal{V}_t$ accepting the intersection of the languages of $-\mathcal{A}_t^\phi$ and of $\mathcal{V}_t$. Then we check for emptiness of $\mathcal{A}$ using standard tools of automata theory.   □

6   Large Grids and Undecidability

We need the following result, which was proved first in a more general form in [21] (see also [22]).

Theorem 6.1. (Seese) *Let* $\mathcal{K}$ *be a class of adjacency graphs such that for every integer* $k > 1$ *there is a graph* $G \in \mathcal{K}$ *such that* G *has the* $k \times k$ *grid* Q_k *as an induced subgraph. Then the* MS_1 *theory of* $\mathcal{K}$ *is undecidable.*

Using Theorems 3.1, 6.1 and interpretation, one concludes [22]:

Theorem 6.2. (Seese) *a) If a family* $\mathcal{G}$ *of planar graphs has a decidable* MS_1 *theory, then* $\mathcal{G}$ *has bounded tree-width.*
b) If a graph family $\mathcal{G}$ *has a decidable* MS_2 *theory, then* $\mathcal{G}$ *has bounded tree-width.*

Related results can be found also in [5] and [6]. The troubles, why part (a) of this theorem has to be formulated for planar graphs, lie in the fact that MS_1 logic (unlike MS_2) lacks expressive power to handle minors in arbitrary graphs. However, that is not a problem with our MS_M logic, cf. Theorem 5.1 or [14], and hence we can extend the (now stronger) part (b) to representable matroids as follows:

Theorem 6.3. *Let* $\mathbb{F}$ *be a finite field, and let* $\mathcal{N}$ *be a class of matroids that are representable by matrices over* $\mathbb{F}$. *If the (monadic second-order)* MS_M *theory* $\mathrm{Th}_{MSO}(\mathcal{N})$ *is decidable, then the class* $\mathcal{N}$ *has bounded branch-width.*

The key to the proof of this theorem is given in Theorem 3.2, which basically states that the obstructions to small branch-width on matroids are the same as on graphs. Unfortunately, the seemingly straightforward way to prove Theorem 6.3 — via direct interpretation of graphs (Theorem 6.2) in the class of graphic minors of matroids in $\mathcal{N}$, is not so simple due to technical problems with (low) connectivity. That is why we give here a variant of this idea bypassing Theorem 6.2, and using an indirect interpretation of (graph) grids in matroid grid minors.

Remark. A restriction to $\mathbb{F}$-representable matroids in Theorem 6.3 is not really necessary; it comes more from the context of the related matroid structure research. According to [11], it is enough to assume that no member of $\mathcal{N}$ has a $U_{2,m}$- or $U_{2,m}^*$-minor (i.e. an m-point line or an m-point dual line) for some constant m.

We begin the proof of Theorem 6.3 with an interpretation of the MS_M theory of all minors of the class $\mathcal{N}$. To achieve this goal, we use a little technical trick first. Let a *DC-equipped matroid* be a matroid M with two distinguished unary predicates D and C on $E(M)$ (with intended meaning as a pair of sets $D, C \subseteq E(M)$ defining a minor $N = M \setminus D/C$).

Lemma 6.4. *Let $\mathcal{N}$ be a class of matroids, and let $\mathcal{N}_{DC}$ denote the class of all DC-equipped matroids induced by members of $\mathcal{N}$. If $\mathrm{Th}_{MSO}(\mathcal{N})$ is decidable, then so is $\mathrm{Th}_{MSO}(\mathcal{N}_{DC})$.*

Proof. We may equivalently view the distinguished predicates D, C as free set variables in MS_M. Let $\phi(D, C)$ be an MS_M formula, and $N \in \mathcal{N}$. Then, by standard logic arguments, $N_{DC} \models \phi(D, C)$ for all DC-equipped matroids N_{DC} induced by N if and only if $N \models \forall D, C\, \phi(D, C)$. Hence $\mathcal{N}_{DC} \models \phi(D, C)$ if and only if $\mathcal{N} \models \forall D, C\, \phi(D, C)$. Since $\forall D, C\, \phi(D, C)$ is an MSO formula if ϕ is such, the statement follows. □

Lemma 6.5. *Let $\mathcal{N}$ be a class of matroids, and $\mathcal{N}_m$ be the class of all minors of members of $\mathcal{N}$. Then $\mathrm{Th}_{MSO}(\mathcal{N}_m)$ is interpretable in $\mathrm{Th}_{MSO}(\mathcal{N}_{DC})$.*

Proof. We again regard the distinguished predicates D, C of $\mathcal{N}_{DC}$ as free set variables in MS_M. Let us consider a matroid $N_1 \in \mathcal{N}_m$ such that $N_1 = N \setminus D_1/C_1$ for $N \in \mathcal{N}$. We are going to use a "natural" interpretation of N_1 in the DC-equipped matroid N_{DC} which results from N with a particular equipment $D = D_1$, $C = C_1$. (Notice that both theories use the same language of MSO logic, and the individuals of N_1 form a subset of the individuals of N.) Let ψ be an MS_M formula. The translation ψ^I of ψ is obtained inductively:

– For each (bound) element variable x in ψ; it is replaced with

$$\exists x\, \theta(x) \quad \longrightarrow \quad \exists x\, \big(x \notin C \wedge x \notin D \wedge \theta(x)\big).$$

– For each (bound) set variable X in ψ; it is replaced with

$$\exists X \theta(X) \quad \longrightarrow \quad \exists X \forall z\, \big((z \notin X \vee z \notin C) \wedge (z \notin X \vee z \notin D) \wedge \theta(X)\big).$$

– Every occurence of the *indep* predicate in ψ is rewritten as (cf. Lemma 2.1)

$$indep^I(X) \equiv \forall Y\, \big(\,indep(Y) \vee \exists z(z \in Y \wedge z \notin X \wedge z \notin C)\big).$$

Consider now the structure N^I defined by $indep^I$ in $N_{DC} \in \mathcal{N}_{DC}$. By Lemma 2.1, a set $X \subseteq E(N^I) = E(N_1)$ is independent in N^I if and only if X is independent in N_1, and hence N^I is a matroid isomorphic to $N_1 = N \setminus D/C \in \mathcal{N}_m$. Moreover, it is immediate from the construction of ψ^I that $N_1 \models \psi$ iff $N_{DC} \models \psi^I$. Thus, I is an interpretation of $\mathrm{Th}_{MSO}(\mathcal{N}_m)$ in $\mathrm{Th}_{MSO}(\mathcal{N}_{DC})$. □

Next, we define, for a matroid M, a *4CC-graph of M* as the graph G on the vertex set $E(M)$, and edges of G connecting those pairs of elements $e, f \in E(M)$, such that there are a 4-element circuit C and a 4-element cocircuit C' in M containing both $e, f \in C \cap C'$. (This is *not* the usual way of interpretation in which the ground set of a matroid is formed by graph edges.) The importance of our definition is in that 4CC-graphs "preserve" large grids:

Lemma 6.6. *Let $m \geq 6$ be even, and $M = M(Q_m)$. Denote by G the 4CC-graph of M. Then G has an induced subgraph isomorphic to Q_{m-2}.*

Proof. Recall that circuits in a cycle matroid of a graph correspond to graph cycles, and cocircuits to graph bonds (minimal edge cuts). The only 4-element cycles in a grid clearly are face-cycles in the natural planar drawing of Q_m. The only edge cuts with at most 4 edges in Q_m are formed by sets of edges incident with a single vertex in Q_m, or by edges that are "close to the corners".

Let $E' \subseteq E(Q_m)$ denote the edge set of the subgraph induced on the vertices (i, j) where $1 < i, j < m$. Let G' denotes the corresponding subgraph of G induced on E'. Choose $x \in E'$, and assume up to symmetry $x = \{(i, j), (i', j')\}$ where $i' = i + 1$ and $j' = j$. According to the above arguments, the only neighbours of x in G' are in the set

$$E' \cap \big\{ \{(i, j-1), (i, j)\}, \{(i, j), (i, j+1)\}, \{(i', j'-1), (i', j')\}, \{(i', j'), (i', j'+1)\} \big\} .$$

We now define "coordinates" for the elements $x \in E'$ as follows

$$x = \{(i, j), (i', j')\}, \ i \leq i', j \leq j' : \qquad k_x = i + j, \qquad \ell_x = i + j' - 2j .$$

As one may easily check from the above description of neighbours, two elements $x, y \in E'$ are adjacent in G' if and only if $\{|k_x - k_y|, |\ell_x - \ell_y|\} = \{0, 1\}$. Hence the elements $x \in E'$ such that $\frac{m}{2} + 1 < k_x, \ell_x < \frac{m}{2} + m - 1$ induce in G' a grid isomorphic to Q_{m-2}. □

Now we are to finish a chain of interpretations from Theorem 6.1 to a proof of our Theorem 6.3.

Lemma 6.7. *Let $\mathcal{M}$ be a matroid family, and let $\mathcal{F}_4$ denote the class of all adjacency graphs which are 4CC-graphs of the members of $\mathcal{M}$. Then the MS_1 theory of $\mathcal{F}_4$ is interpretable in the MS_M theory $\mathrm{Th}_{MSO}(\mathcal{M})$.*

Proof. Let us take a graph $G \in \mathcal{F}_4$ which is a 4CC-graph of a matroid $M \in \mathcal{M}$. Now G is regarded as an adjacency graph structure, and so the individuals (the domain) of G are the vertices $V(G)$. These are to be interpreted in the ground set $E(M)$, the domain of M. Let ψ be an MS_1 formula. The translation ψ^I in MS_M of ψ is obtained simply by replacing every occurence of the adj predicate in ψ with

$$adj^I(x, y) \equiv \exists C, C'$$

$$\left(|C| = |C'| = 4 \wedge circuit(C) \wedge cocircuit(C) \wedge x, y \in C \wedge x, y \in C'\right),$$

where the matroid MS_M predicates $circuit$ and $cocircuit$ are defined in Section 5, and $|X| = 4$ has an obvious interpretation in the FO logic.

Consider the structure G^I defined by the predicate adj^I on the domain $E(M)$. It is $G^I \simeq G$ by definition, for all pairs G, M as above. Moreover, adj^I is defined in the MSO logic. Hence we have got an interpretation I of $\mathrm{Th}_{MSO_1}(\mathcal{F}_4)$ in $\mathrm{Th}_{MSO}(\mathcal{M})$. □

Proof of Theorem 6.3. We prove the converse direction of the implication. Assume that a matroid class $\mathcal{N}$ does not have bounded branch-width, and denote by $\mathcal{N}_m$ the class of all matroids which are minors of some member of $\mathcal{N}$. By Theorem 3.2, for every integer $m > 1$, there is a matroid $N \in \mathcal{N}_m$ isomorphic to the cycle matroid of the grid $N \simeq M(Q_m)$. Now denote by $\mathcal{F}_4$ the class of all graphs which are 4CC-graphs of members of $\mathcal{N}_m$. Then, using Lemma 6.6, there exist members of $\mathcal{F}_4$ having induced subgraphs isomorphic to the grid Q_k, for every integer $k > 1$.

Hence the class $\mathcal{K} = \mathcal{F}_4$ satisfies the assumptions of Theorem 6.1, and so the MS_1 theory of $\mathcal{F}_4$ is undecidable. So is the MS_M theory $\mathrm{Th}_{MSO}(\mathcal{N}_m)$ using the interpretation in Lemma 6.7, and Theorem 4.1. We analogously apply the interpretation in Lemma 6.5 to $\mathrm{Th}_{MSO}(\mathcal{N}_m)$, and conclude that also $\mathrm{Th}_{MSO}(\mathcal{N}_{DC})$ is undecidable, where $\mathcal{N}_{DC}$ is the class of all DC-equipped matroids induced by $\mathcal{N}$ as above. Finally, Lemma 6.4 implies that the MS_M theory $\mathrm{Th}_{MSO}(\mathcal{N})$ is undecidable, as needed. □

Acknowledgement

It is a pleasure for us to express our thanks to M. Fellows, M. Hallett, R. Niedermeier and N. Nishimura, the organizers of the Dagstuhl Seminar in 2003 on "'Fixed Parameter Algorithms"' for inviting us to this interesting workshop, where the idea for this paper was born. Special thanks go to Mike Fellows and Rod Downey for stimulating discussions on the subject.

References

1. S. Arnborg, J. Lagergren, D. Seese, *Easy Problems for Tree-Decomposable Graphs*, Journal of Algorithms 12 (1991), 308–340.
2. B. Courcelle, *The decidability of the monadic second order theory of certain sets of finite and infinite graphs*, LICS'88, Logic in Computer Science, Edinburg, 1988.
3. B. Courcelle, *The Monadic Second-Order Logic of Graphs I. Recognizable sets of Finite Graphs*, Information and Computation 85 (1990), 12–75.

4. B. Courcelle, *The monadic second order logic of graphs VI: on several representa-tions of graphs by relational structures*, Discrete Appl. Math. 54 (1994) 117–149. Erratum 63 (1995) 199–200.

5. B. Courcelle, *The monadic second-order logic of graphs XIV: Uniformly sparse graphs and edge set quantification*, Theoretical Computer Science 299 (2003) 1–36.

6. B. Courcelle, *The Monadic Second-Order Logic of Graphs XV: On a Conjecture by D. Seese*, manuscript, LaBri, Bordeaux 1 University, March 8, 2004, 1–40.

7. B. Courcelle, J.A. Makowsky, U. Rotics, *Linear Time Solvable Optimization Prob-lems on Graphs of Bounded Clique-Width*, Theory Comput. Systems 33 (2000) 125–150.

8. B. Courcelle, M. Mosbah, *Monadic second-order evaluations on tree-decomposable graphs*, Theoret. Comput. Sci. 109 (1993), 49–82.

9. H.-D. Ebbinghaus, J. Flum, W. Thomas, Mathematical Logic, Springer Verlag, 2nd edition, 1994.

10. J.F. Geelen, A.H.M. Gerards, G.P. Whittle, *Branch-Width and Well-Quasi-Ordering in Matroids and Graphs*, J. Combin. Theory Ser. B 84 (2002), 270–290.

11. J.F. Geelen, A.H.M. Gerards, G.P. Whittle, *Excluding a Planar Graph from a GF(q)-Representable Matroid*, manuscript, 2003.

12. Y. Gurevich, *Monadic second-order theories*, In: Chapter XIII from J. Barwise, S. Feferman: Model-Theoretic Logics; Springer-Verlag New York 1985, 479–506.

13. P. Hliněný, *Branch-Width, Parse Trees, and Monadic Second-Order Logic for Ma-troids*, submitted. Extended abstract in: STACS 2003, Lecture Notes in Computer Science 2607, Springer Verlag (2003), 319–330.

14. P. Hliněný, *On Matroid Properties Definable in the MSO Logic*, In: Math Founda-tions of Computer Science MFCS 2003, Lecture Notes in Computer Science 2747, Springer Verlag Berlin (2003), 470–479.

15. P. Hliněný, G.P. Whittle, *Matroid Tree-Width*, submitted, 2003. Extended abstract in: Eurocomb'03, ITI Series 2003–145, Charles University, Prague, Czech Republic, 202–205.

16. J.G. Oxley, Matroid Theory, Oxford University Press, 1992.

17. M.O. Rabin, *A simple method for undecidability proofs and some applications*, Logic, Methodology and Philosophy of Sciences 1 (ed. Bar-Hillel), North-Holland, Amsterdam 1964, 58–68.

18. M.O. Rabin, *Decidable theories*, In: Chapter C.3 of Handbook of Mathematical Logic, J. Barwise, North-Holland Publ.Co. 1977, 595–629.

19. N. Robertson, P.D. Seymour, *Graph Minors – A Survey*, Surveys in Combinatorics, Cambridge Univ. Press 1985, 153–171.

20. N. Robertson, P.D. Seymour, *Graph Minors X. Obstructions to Tree-Decompo-sition*, J. Combin. Theory Ser. B 52 (1991), 153–190.

21. D. Seese, *Ein Unentscheidbarkeitskriterium*, Wissenschaftliche Zeitschrift der Humboldt-Universitaet zu Berlin, Math.-Nat. R. XXIV (1975) 6, 772–778.

22. D. Seese, *The structure of the models of decidable monadic theories of graphs*, Annals of Pure and Aplied Logic 53 (1991), 169–195.

23. D. Seese, *Interpretability and Tree Automata: a simple way to solve Algorithmic Problems on Graphs closely related to Trees*, In: Tree Automata and Languages, M. Nivat and A. Podelski (editors), Elsevier Sceince Publishers 1992, 83–114.

On Miniaturized Problems in Parameterized Complexity Theory

Yijia Chen and Jörg Flum

Abteilung für Mathematische Logik, Universität Freiburg, Eckerstr. 1,
79104 Freiburg, Germany
chen@zermelo.mathematik.uni-freiburg.de
Joerg.Flum@math.uni-freiburg.de

Abstract. We introduce a general notion of miniaturization of a problem that comprises the different miniaturizations of concrete problems considered so far. We develop parts of the basic theory of miniaturizations in this general framework. Using the appropriate logical formalism, we show that the miniaturization of a definable problem in W[t] lies in W[t], too. In particular, the miniaturization of the dominating set problem is in W[2].

1 Introduction

Parameterized complexity theory provides a framework for a refined complexity analysis of algorithmic problems that are intractable in general. Central to the theory is the notion of *fixed-parameter tractability*, which relaxes the classical notion of tractability, polynomial time computability, by admitting algorithms whose runtime is exponential, but only in terms of some *parameter* that is usually expected to be small. Let FPT denote the class of all fixed-parameter tractable problems. A well-known example of a problem in FPT is the vertex cover problem, the parameter being the size of the vertex cover we ask for.

As a complexity theoretic counterpart, a theory of *parameterized intractability* has been developed. In classical complexity, the notion of NP-completeness is central to a nice and simple theory for intractable problems. Unfortunately, the world of parameterized intractability is more complex: there is a big variety of seemingly different classes of parameterized intractability. For a long while, the smallest complexity class of parameterized intractable problems considered in the literature was W[1], the first class of the so-called W-hierarchy. (In particular, FPT $\subseteq$ W[1]; and FPT $\neq$ W[1] would imply PTIME $\neq$ NP.)

Recently, this situation has changed: In [4], Downey et al. consider various problems in W[1] that, apparently, are not W[1]-hard. Most of them are "miniaturizations" of well-studied problems in parameterized complexity theory; for example, mini-CIRCUIT SAT is the problem that takes a circuit C of size $\leq k \cdot \log m$, where k is the parameter and m in unary is part of the input, and asks whether C is satisfiable. This problem is called a miniaturization of CIRCUIT SAT, as the size ($\leq k \cdot \log m$) of C is small compared with m (under

R. Downey, M. Fellows, and F. Dehne (Eds.): IWPEC 2004, LNCS 3162, pp. 108–120, 2004.
© Springer-Verlag Berlin Heidelberg 2004

the basic assumption of parameterized complexity that the parameter k is small too). In [4], Downey et al. introduce the class MINI[1] as the class of parameterized problems fpt-reducible to mini-CIRCUIT SAT. MINI[1] now provides very nice connections between classical complexity and parameterized complexity as it is known that FPT = MINI[1] if and only if n variable 3-SAT can be solved in time $2^{o(n)}$. This equivalence stated in [4] is based on a result of Cai and Juedes [1].

Besides this "miniaturization route", a second route to MINI[1] has been considered by Fellows in [7]; he calls it the "renormalization route" to MINI[1]. He "renormalizes" the parameterized VERTEX COVER problem and considers the so-called $k \cdot \log n$ VERTEX COVER problem: It takes as input a graph G and as parameter a natural number k; it asks if G has vertex cover of size $k \cdot \log n$. This problem turns out to be MINI[1]-complete (cf. [7]).

Before outlining the purpose and the contents of this paper let us give two quotations, the first one from Fellows' paper [7] and the second one from Downey's paper [3]:

> Dozens of renormalized FPT problems and miniaturized arbitrary problems are now known to be MINI[1]-complete. However, what is known is quite problem specific.

> Can the hierarchy [starting with MINI[1]] be extended?

Among others, in this paper we try to develop the theory of miniaturized problems on a more abstract level and we address the problems mentioned in these quotations. Concerning the second problem, even though we introduce a hierarchy of complexity classes, we conjecture, among others encouraged by the results of this paper, that the world of parameterized intractability in W[1] is so rich that, probably, there are various more or less natural hierarchies in W[1].

We sketch the content of the different sections. In Section 2 we give the necessary preliminaries. In particular, we introduce the notion of a size function, a polynomial time function $\| \ \|$ defined on the inputs x of a given problem with the property that the length $|x|$ of x is polynomially bounded in $\|x\|$. For example, for a graph $G = (V, E)$, natural choices could be $|V|$, the number of vertices, or $|V|+|E|$, the number of vertices and edges, or $\Theta(|V|+|E| \cdot \log |V|)$, the total length of its binary description; but, in general, $|E|$ is not a size function.

In Section 3, for a given size function $\| \ \|$, we define the concept of the miniaturization mini$^{\| \ \|}$-Q of an arbitrary problem Q. Now, a proof essentially due to Cai and Juedes [1] goes through for this concept showing that mini$^{\| \ \|}$-Q is fixed-parameter tractable just in case Qx is solvable in time $2^{o(\|x\|)}$. In Proposition 2 we extend the well-known fact that a linear reduction from Q to Q' yields an fpt-reduction from the miniaturization of Q to that of Q' and essentially show that the existence of a linear reduction from Q to Q' is *equivalent* to the existence of an fpt-reduction of the miniaturization of Q to that of Q' that is linear with respect to the parameters. Perhaps therefore, there are so many not fpt-equivalent miniaturizations.

There is a way of defining parameterized problems by means of first-order formulas with a free *set* variable X that has been dubbed *Fagin-definability* in [8], since it is related to Fagin's theorem characterizing NP as the class of Σ_1^1-definable problems. For example, the parameterized clique problem is Fagin-definable by the formula

$$\forall y \forall z((Xy \wedge Xz \wedge y \neq z) \rightarrow Eyz).$$

In [5], Downey et al. showed that $W[t]$, the tth class of the W-hierarchy, contains all parameterized problems Fagin-defined by Π_t-formulas and conversely, there are $W[t]$-complete problems Fagin-defined by Π_t-formulas. In general, the miniaturization may increase the computational complexity of a problem; e.g., the parameterized vertex cover problem is fixed-parameter tractable while its miniaturization is not (unless $\mathrm{MINI}[1] = \mathrm{FPT}$). But the miniaturization of every Fagin-definable problem in $W[t]$ lies in $W[t]$, too. We prove this result in Section 4 for $t = 1$ generalizing an idea used in [4] for the independent set problem. Some miniaturized problems considered in the literature can be regarded as miniaturization of *unweighted* Fagin-definable problems, a concept we introduce in this paper. In Section 5 we prove that in a certain sense weighted and unweighted definable problems have the same computational complexity. Using this result, we get the result on Fagin-definable problems for arbitrary t.

As mentioned above, Π_1-formulas of the form $\varphi(X) = \forall x_1 \ldots \forall x_t \psi(X)$ with a set variable X and with quantifier-free $\psi(X)$ are used to obtain the Fagin-definable problems in $W[1]$. We obtain a hierarchy of classes within $W[1]$ taking the length t of the block of quantifiers into consideration. We study the basic properties of this hierarchy; in particular, we show that the (appropriate) miniaturization of t-SAT is complete in the tth class of this hierarchy. Recall that Impagliazzo and Paturi [10] have shown that, assuming the exponential-time hypothesis, the complexity of t-SAT increases with t.

So far, when comparing the complexity of miniaturized and other parameterized problems, we used many-one reductions (more precisely, FPT many-one reductions). In some papers, Turing reductions have been considered. As we show in Section 7, most problems studied in this paper are Turing equivalent.

Finally, in Section 8 we deal with renormalizations. Besides the renormalization of the vertex cover problem introduced in Fellows [7], we consider a slightly different renormalization and also show its fpt-equivalence to the miniaturization. We shall see that this result cannot be extended to arbitrary Fagin-definable problems, in particular not to the clique problem.

Due to space limitations, we have to defer the proofs to the full version of the paper.

2   Preliminaries

We use standard terminology (cf. [6,8]) and therefore, only briefly recall some notations, definitions, and results and introduce the concept of size function.

2.1 First-Order Logic and Propositional Logic

A *vocabulary* τ is a finite set of relation symbols. A *(relational) structure* $\mathcal{A}$ of vocabulary τ (or, simply structure), consists of a *finite* set A called the *universe*, and an interpretation $R^{\mathcal{A}} \subseteq A^r$ of each r-ary relation symbol $R \in \tau$.

For example, let $\tau_{\mathrm{circ}} := \{E, I, O, G_\wedge, G_\vee, G_\neg\}$, where E is a binary relation symbol and I, O, $G_\wedge$, $G_\vee$, $G_\neg$ are unary relation symbols. We view Boolean circuits as τ_{circ}-structures

$$\mathcal{C} = (C, E^{\mathcal{C}}, I^{\mathcal{C}}, O^{\mathcal{C}}, G_\wedge^{\mathcal{C}}, G_\vee^{\mathcal{C}}, G_\neg^{\mathcal{C}}),$$

where $(C, E^{\mathcal{C}})$ is the directed acyclic graph underlying the circuit, $I^{\mathcal{C}}$ is the set of all input nodes, $O^{\mathcal{C}}$ just contains the output node, $G_\wedge^{\mathcal{C}}$, $G_\vee^{\mathcal{C}}$, and $G_\neg^{\mathcal{C}}$ are the sets of and-gates, or-gates (and-gates and or-gates of arbitrary arity), and negation-gates, respectively. The *weight* of a truth value assignment to the input nodes of $\mathcal{C}$ is the number of input nodes set to TRUE by the assignment.

Often for graphs we shall use the more common notation $G = (V, E)$ (or, $G = (V(G), E(G))$), where V is the set of vertices and E the set of edges.

We define the *size* $\|\mathcal{A}\|_0$ of a τ-structure $\mathcal{A}$ to be the number

$$\|\mathcal{A}\|_0 := |A| + \sum_{R \in \tau} \mathrm{arity}(R) \cdot |R^{\mathcal{A}}| \cdot \log |A|.$$

In fact, the length of a reasonable encoding of $\mathcal{A}$ as a string is $\Theta(\|\mathcal{A}\|_0)$.

For $t \geq 1$, by Π_t we denote the class of all first-order formulas of the form

$$\forall x_{11} \ldots \forall x_{1k_1} \exists x_{21} \ldots \exists x_{2k_2} \ldots Q x_{t1} \ldots Q x_{tk_t}\ \psi,$$

where $Q = \forall$ if t is odd and $Q = \exists$ otherwise, and where ψ is quantifier-free. Σ_t-formulas are defined analogously starting with a block of existential quantifiers.

Formulas of propositional logic are built up from *propositional variables* $X_1, X_2, \ldots$ by taking conjunctions, disjunctions, and negations. We distinguish between *small conjunctions*, denoted by $\wedge$, which are just conjunctions of two formulas, and *big conjunctions*, denoted by $\bigwedge$, which are conjunctions of arbitrary finite sets of formulas. Analogously, we distinguish between *small disjunctions*, denoted by $\vee$, and *big disjunctions*, denoted by $\bigvee$.

For $t \geq 0$ and $d \geq 1$ we define the sets $\Gamma_{t,d}$ and $\Delta_{t,d}$ of propositional formulas by induction on t (here, by $(\lambda_1 \wedge \ldots \wedge \lambda_r)$ we mean the iterated small conjunction $((\ldots((\lambda_1 \wedge \lambda_2) \wedge \lambda_3)\ldots) \wedge \lambda_r)$:

$$\Gamma_{0,d} := \{(\lambda_1 \wedge \ldots \wedge \lambda_r) \mid \lambda_1, \ldots, \lambda_r \text{ literals and } r \leq d\},$$
$$\Delta_{0,d} := \{(\lambda_1 \vee \ldots \vee \lambda_r) \mid \lambda_1, \ldots, \lambda_r \text{ literals and } r \leq d\},$$
$$\Gamma_{t+1,d} := \{\textstyle\bigwedge \Pi \mid \Pi \subseteq \Delta_{t,d}\},$$
$$\Delta_{t+1,d} := \{\textstyle\bigvee \Pi \mid \Pi \subseteq \Gamma_{t,d}\}.$$

Let CNF denote the class of all propositional formulas in conjunctive normal form. A formula is in d-CNF, if it is a conjunction of disjunctions of at most

d literals. Often, we identify a formula $\alpha \in d$-CNF with a τ_d-structure $\mathcal{A}(\alpha)$. Here $\tau_d := \{N, C\}$, where N is binary and C is d-ary, and $\mathcal{A}(\alpha)$ has the set $\{X, \neg X \mid X \text{ variable of } \alpha\}$ as universe and

$$N^{\mathcal{A}(\alpha)} := \{(X, \neg X) \mid X \text{ variable of } \alpha\};$$
$$C^{\mathcal{A}(\alpha)} := \{(\lambda_1, \ldots, \lambda_d) \mid (\lambda_1 \vee \ldots \vee \lambda_d) \text{ is a clause of } \alpha\},$$

w.l.o.g. we assume each clause has exactly d literals. A propositional formula α is *k-satisfiable*, if there is an assignment satisfying α of weight k (i.e., setting exactly k variables of α to TRUE).

2.2   Size Functions

Let Σ be an alphabet. For a string $x \in \Sigma^*$, we denote its length by $|x|$.

Definition 1. *A function* $\| \ \| : \Sigma^* \to \mathbb{N}$ *is a* size function, *if it is computable in polynomial time and if, for some* $c \in \mathbb{N}$,

$$|x| \le \|x\|^c$$

holds for all $x \in \Sigma^*$.

In particular, $|\ |$ is a size function. The function $\| \ \|_0$ (cf. Section 2.1) defined for τ-structures (more precisely, for the encodings of τ-structures by strings) is a size function. We introduce further size functions for τ-structures:

$$\|\mathcal{A}\|_+ := |A| + \sum_{R \in \tau} \text{arity}(R) \cdot |R^{\mathcal{A}}|,$$
$$\|\mathcal{A}\|_- := |A|.$$

Note that

- for a graph G with n vertices and m edges: $\|G\|_- = n$ and $\|G\|_+ = \Theta(n + m)$;
- for a circuit $\mathcal{C}$ with n gates and m lines: $\|\mathcal{C}\|_- = n$ and $\|\mathcal{C}\|_+ = \Theta(n + m)$;
- for a propositional formula $\alpha \in d$-CNF with n variables and m clauses: $\|\alpha\|_- (:= \|\mathcal{A}(\alpha)\|_-) = \Theta(n)$ and $\|\alpha\|_+ (:= \|\mathcal{A}(\alpha)\|_+) = \Theta(n + m)$. [1]

2.3   Parameterized Complexity

A *parameterized problem* is a set $Q \subseteq \Sigma^* \times \Pi^*$, where Σ and Π are finite alphabets. If $(x, y) \in \Sigma^* \times \Pi^*$ is an instance of a parameterized problem, we refer to x as the *input* and to y as the *parameter*.

---

[1] Note that for arbitrary propositional formulas the number of variables does not define a size function; for formulas α in d-CNF we obtain a size function, since we identify α with $\mathcal{A}(\alpha)$. Equivalently, we could restrict ourselves to formulas α that contain every clause at most once.

A parameterized problem $Q \subseteq \Sigma^* \times \Pi^*$ is *fixed-parameter tractable*, if there are a computable function $f : \mathbb{N} \to \mathbb{N}$, a polynomial p, and an algorithm that, given a pair $(x, y) \in \Sigma^* \times \Pi^*$, decides if $(x, y) \in Q$ in at most $f(|y|) \cdot p(|x|)$ steps.

FPT denotes the complexity class consisting of all fixed-parameter tractable parameterized problems.

Of course, we may view any parameterized problem $Q \subseteq \Sigma^* \times \Pi^*$ as a classical problem, say, in the alphabet obtained from $\Sigma \cup \Pi$ by adding new symbols '(', ',', ')'.

We mainly consider parameterized problems $Q \subseteq \Sigma^* \times \mathbb{N}$; so we give most definitions and state most results only for this case.

To illustrate our notation, let us give two examples of parameterized problems, the *weighted satisfiability problem* WSAT(Θ) for a class Θ of propositional formulas and the parameterized model-checking problem p-MC($\Sigma_1[\tau]$) for Σ_1-sentences of vocabulary τ:

WSAT(Θ)	p-MC($\Sigma_1[\tau]$)
Input: α in Θ. *Parameter:* $k \in \mathbb{N}$. *Problem:* Decide if α is k-satisfiable.	*Input:* A τ-structure $\mathcal{A}$. *Parameter:* $\varphi \in \Sigma_1[\tau]$. *Problem:* Decide if $\mathcal{A}$ satisfies φ.

For computable functions $f, g : \mathbb{N} \to \mathbb{N}$, we write $f \in o^{\mathrm{eff}}(g)$, if $f \in o(g)$ holds in an effective way, i.e., if there is a computable function h such that, given any $\ell \in \mathbb{N}$ with $\ell > 0$, we have $f(m)/g(m) \leq 1/\ell$ for all $m \geq h(\ell)$.

Essentially the idea underlying the proof of Proposition 2 can be used to show that the effective versions of two notions of *subexponential time*, namely $\mathbf{DTIME}(2^{o(n)})$ and $\bigcap_{\epsilon > 0} \mathbf{DTIME}(2^{\epsilon \cdot n})$ coincide:

Proposition 1. *For a classical problem $Q \subseteq \Sigma^*$ the following are equivalent.*

1. *$Q \in \mathbf{DTIME}(2^{o^{\mathrm{eff}}(n)})$, i.e., $Q \in \mathbf{DTIME}(2^{o(g)})$ for some function $g \in o^{\mathrm{eff}}(\mathrm{id})$, where id denotes the identity function on $\mathbb{N}$.*
2. *For every rational $\epsilon > 0$, there is an algorithm $\mathbb{A}_\epsilon$ deciding Q in time $O(2^{\epsilon \cdot n})$. Moreover, $\mathbb{A}_\epsilon$ can be computed from ϵ.*

Lemma 1 (Cai and Juedes [1]). *Let $Q \subseteq \Sigma^* \times \mathbb{N}$ be a parameterized problem. Assume that there is an algorithm solving Qxk in time $2^{h(f(k) \cdot \log |x|)}$ for some computable function $h \in o^{\mathrm{eff}}(\mathrm{id})$. Then Qxk is solvable in time $g(k) + O(|x|)$ for some computable function g. In particular, $Q \in$ FPT.*

Complementing the notion of fixed-parameter tractability, there is a theory of parameterized intractability. It is based on the notion of fpt-*reduction* (the natural notion of many-one reduction of parameterized complexity theory, see [6,8]).

We write $Q \leq^{\mathrm{fpt}} Q'$, if there is an fpt-reduction from Q to Q', and $Q =^{\mathrm{fpt}} Q'$ if $(Q \leq^{\mathrm{fpt}} Q'$ and $Q' \leq^{\mathrm{fpt}} Q)$. We set

$$[Q]^{\mathrm{fpt}} := \{Q' \mid Q' \leq^{\mathrm{fpt}} Q\} \quad \text{and} \quad [\mathrm{C}]^{\mathrm{fpt}} := \bigcup_{Q \in \mathrm{C}} [Q]^{\mathrm{fpt}}$$

for a class C of parameterized problems. For $t \geq 1$, the class $W[t]$ is defined by

$$W[t] := \left[\{\mathrm{WSAT}(\Gamma_{t,d}) \mid d \in \mathbb{N}\}\right]^{\mathrm{fpt}}.$$

Clearly, $\mathrm{FPT} \subseteq W[1] \subseteq W[2] \ldots$ and it is conjectured that $\mathrm{FPT} \neq W[1]$ (which would imply $\mathrm{PTIME} \neq \mathrm{NP}$). It is well-known that

$$W[1] = \left[\{p\text{-}\mathrm{MC}(\Sigma_1[\tau]) \mid \tau \text{ a vocabulary}\}\right]^{\mathrm{fpt}}.$$

3 The Miniaturization of an Arbitrary Problem

In this section, for a classical problem Q and a size function $\| \ \|$, we introduce its miniaturization $\mathrm{mini}^{\| \ \|}\text{-}Q$, a parameterized problem, and study the relationship between the complexity of Q and $\mathrm{mini}^{\| \ \|}\text{-}Q$.

Definition 2. *Let $Q \subseteq \Sigma^*$ and let $\| \ \| : \Sigma^* \to \mathbb{N}$ be a size function. The parameterized miniaturization $\mathrm{mini}^{\| \ \|}\text{-}Q$ with respect to $\| \ \|$ is the parameterized problem:*

$\mathrm{mini}^{\| \ \|}\text{-}Q$
  Input: $n, k \in \mathbb{N}$ in unary, and $x \in \Sigma^*$.
  Parameter: k.
  Problem: Decide if $\|x\| \leq k \cdot \log n$ and $x \in Q$.

Remark 1. a) Let $Q \subseteq \Sigma^*$ and $\| \ \|$ be a size function with $|x| \leq \|x\|^c$. The condition $\|x\| \leq k \cdot \log n$ can be checked in time polynomial in k and n only. Therefore, often the problem $\mathrm{mini}^{\| \ \|}\text{-}Q$ is presented in the more appealing form:

$\mathrm{mini}^{\| \ \|}\text{-}Q$
  Input: $n, k \in \mathbb{N}$ in unary, $x \in \Sigma^*$ with $\|x\| \leq k \cdot \log n$.
  Parameter: k.
  Problem: Decide if $x \in Q$.

---

b) Arguing similarly as in part a), one shows that if $\mathrm{mini}^{\| \ \|}\text{-}Q$ is in FPT, then there is an algorithm solving $\mathrm{mini}^{\| \ \|}\text{-}Q$ (on instance n, k, x) in $\leq f(k) \cdot p(n)$ steps for some computable function f and some polynomial p.
c) If $Q \in \mathrm{PTIME}$, then $\mathrm{mini}^{\| \ \|}\text{-}Q \in \mathrm{FPT}$.

The following result relates the fixed-parameter tractability of $\mathrm{mini}^{\| \ \|}\text{-}Q$ with the solvability of Q in subexponential time. Its proof uses an idea of [1] in the form presented in [4].

Proposition 2. *For $Q \subseteq \Sigma^*$ and any size function $\| \ \| : \Sigma^* \to \mathbb{N}$ the following are equivalent:*

1. *Q is solvable in time $2^{o^{\mathrm{eff}}(\|x\|)}$, i.e., in time $2^{h(\|x\|)}$ for some $h \in o^{\mathrm{eff}}(\mathrm{id})$.*
2. *$\mathrm{mini}^{\| \ \|}\text{-}Q \in \mathrm{FPT}$.*

3. *There is an algorithm that, for every instance n, k, x of* $\text{mini}^{\| \ \|}\text{-}Q$ *with* $\|x\| \leq k \cdot \log n$ *decides if* $(n, k, x) \in \text{mini}^{\| \ \|}\text{-}Q$ *in time* $f(k) + O(n)$ *for some computable* f.

The implication from (1) to (2) in the following proposition corresponds to the well-known fact that a polynomial time "linear size" reduction between two problems yields an fpt-reduction of their miniaturizations:

Proposition 3. *Let* $Q_1 \subseteq \Sigma_1^*$ *and* $Q_2 \subseteq \Sigma_2^*$ *and let* $\| \ \|_i : \Sigma_i^* \rightarrow \mathbb{N}$ *be a size function for* $i = 1, 2$. *Then, the following are equivalent:*

1. *There is a function* $f : \Sigma_1^* \rightarrow \Sigma_2^*$ *computable in time* $2^{o^{\text{eff}}(\|x\|_1)}$ *such that for all* $x \in \Sigma_1^*$, $\|f(x)\|_2 \in O(\|x\|_1)$ *and such that* f *is a reduction of* Q_1 *to* Q_2, *i.e.,*

$$x \in Q_1 \iff f(x) \in Q_2.$$

2. *There is an* fpt-reduction R *from* $\text{mini}^{\| \ \|_1}\text{-}Q_1$ *to* $\text{mini}^{\| \ \|_2}\text{-}Q_2$ *such that for any instance* (n_1, k_1, x_1) *of* $\text{mini}^{\| \ \|_1}\text{-}Q_1$ *with* $\|x_1\|_1 \leq k_1 \cdot \log n_1$ *we have* $R(n_1, k_1, x_1) = (n_2, k_2, x_2)$ *with* $k_2 = O(k_1)$ *and* $\|x_2\|_2 \leq k_2 \cdot \log n_2$.

Remark 2. a) Take as Q_1 a language in **DTIME**$(2^{O(|x|)}) \setminus$ **DTIME**$(2^{o(|x|)})$ and as Q_2 a language in **DTIME**$(2^{o^{\text{eff}}(|x|)})$ complete for EXPTIME under polynomial time reductions. In particular, there is a polynomial time reduction from Q_1 to Q_2. By Proposition 3, $\text{mini}^{\| \ \|}\text{-}Q_1 \notin$ FPT and $\text{mini}^{\| \ \|}\text{-}Q_2 \in$ FPT. Hence, there is no fpt-reduction from $\text{mini}^{\| \ \|}\text{-}Q_1$ to $\text{mini}^{\| \ \|}\text{-}Q_2$. This example shows that the condition "$\|f(x)\|_2 \in O(\|x\|_1)$" in (1) of the preceding proposition cannot be weakened to "$\|f(x)\|_2 \leq q(\|x\|_1)$ for some polynomial q".

b) For a natural number $d \geq 1$ replace the condition $k_2 \in O(k_1)$ in (2) of Proposition 3 by $k_2 \in O(k_1^d)$. Then, along the lines of the preceding proof, one can show that there is a reduction f from Q_1 to Q_2 according to (1) satisfying $\|f(x)\|_2 \in O(\|x\|_1^d)$

We close this section with some examples. Let CIRCSAT, SAT, and t-SAT denote the satisfiability problem for circuits, for propositional formulas in CNF, and for propositional formulas in t-CNF, respectively. In Section 2.1, for a circuit $\mathcal{C}$, we defined $\|\mathcal{C}\|_0$, $\|\mathcal{C}\|_+$, and $\|\mathcal{C}\|_-$. Essentially they are the (total) size of an encoding of $\mathcal{C}$, the number of gates + the number of lines of $\mathcal{C}$, and the number of gates of $\mathcal{C}$, respectively. In the following, we abbreviate $\text{mini}^{\| \ \|_+}\text{-CIRCSAT}$ and $\text{mini}^{\| \ \|_-}\text{-CIRCSAT}$ by $\text{mini}^+\text{-CIRCSAT}$ and $\text{mini}^-\text{-CIRCSAT}$, respectively. The same notations are used for other problems. Taking as Q in Proposition 2 the problem CIRCSAT, we get the following result (cf. [1,4]); it shows, for example, that $\text{mini}^+\text{-CIRCSAT} \in$ FPT is quite unlike.

Proposition 4. 1. *For* $\| \ \| \in \{\| \ \|_+, \| \ \|_-\}$: $\text{mini}^{\| \ \|}\text{-CIRCSAT} \in$ FPT *if and only if there is a subexponential algorithm for* CIRCSAT, *i.e., if there is an algorithm with running time* $2^{o^{\text{eff}}(\|\mathcal{C}\|)}$ *checking if the circuit* $\mathcal{C}$ *is satisfiable.*
2. $\text{mini}^{\| \ \|_0}\text{-CIRCSAT} \in$ FPT.

By Proposition 3, the well-known linear reductions between CircSat, Sat, and 3-Sat yield:

- mini$^+$-CircSat $=^{\text{fpt}}$ mini$^+$-Sat $=^{\text{fpt}}$ mini$^+$-3-Sat.

Denote by VC, IS, and Clique the vertex cover problem, the independent set problem, and the clique problem, respectively; e.g., the instances to Clique consist of pairs (G, r), where $G = (V, E)$ is a graph and r is a natural number with $r \leq |V|$. $(G, r) \in$ Clique if and only if there is a clique of size r in G. We let $\|(G, r)\|_- = \|G\|_-$ and $\|(G, r)\|_+ = \|G\|_+$ and use the analogous notations for VC and IS. Clearly, the mini$^+$ versions of these problems are fpt-reducible to their mini$^-$ versions. Using this fact and well-known linear reductions between the corresponding problems, we get the following fpt-reductions between their miniaturized versions:

- mini$^+$-3-Sat $\leq^{\text{fpt}}$ mini$^+$-VC $=^{\text{fpt}}$ mini$^+$-IS $\leq^{\text{fpt}}$ mini$^-$-IS $=^{\text{fpt}}$ mini$^-$-VC $=^{\text{fpt}}$ mini$^-$-Clique.

4   The Miniaturization of Fagin-Definable Problems

It is well-known that the parameterized halting problem HP for Turing machines, parameterized by the number of steps, is in W[1], even it is W[1]-complete. As pointed out in Downey [3], it is unknown whether mini$^{|\ |}$-HP $\in$ W[1]; we conjecture that this is not the case. In this section we show that the miniaturizations of Fagin-definable problems in W[1] are themselves in W[1].

We start by recalling the definition of Fagin-definable problem. Let τ be a vocabulary and C a class of τ-structures decidable in polynomial time. Let $\varphi(X)$ be a first-order formula of vocabulary τ that contains a free set variable X (that is, X is a unary second-order variable); it defines a problem $W_{\varphi(X)}$ on C:

$W_{\varphi(X)}(\text{C})$
> *Input:* A τ-structure $\mathcal{A}$ in C.
> *Parameter:* $r \in \mathbb{N}$ with $r \leq |A|$.
> *Problem:* Is there a subset S of A of cardinality r satisfying $\varphi(X)$
> in $\mathcal{A}$, i.e., with $\mathcal{A} \models \varphi(S)$?

We say that $\varphi(X)$ *Fagin-defines* $W_{\varphi(X)}(\text{C})$ *on* C and that a parameterized problem $Q \subseteq$ C is *Fagin-definable*, if $Q = W_{\varphi(X)}(\text{C})$ for some $\varphi(X)$.

For example, the parameterized vertex cover problem, the independent set problem, and the dominating set problem are Fagin-defined on the class Graph of all graphs by $\forall y \forall z (Eyz \rightarrow (Xy \lor Xz))$, by $\forall y \forall z ((Xy \land Xz) \rightarrow \neg Eyz)$, and by $\forall y \exists z (Xz \land (y = z \lor Eyz))$, respectively.

If C is the class of all τ-structures, we denote $W_{\varphi(X)}(\text{C})$ by $W_{\varphi(X)}$. For notational simplicity, we formulate most results for Fagin-definable problems $W_{\varphi(X)}$. Their extensions to Fagin-definable problems $W_{\varphi(X)}(\text{C})$ are easy.

So far, we defined the miniaturization for classical problems only. Here and later, when speaking of the miniaturization of a parameterized problem Q, we consider Q as a classical problem as explained in Section 2.3. Again for a structure $\mathcal{A}$ and $r \in \mathbb{N}$ with $r \leq |A|$, we set $\|(\mathcal{A}, r)\|_- = \|\mathcal{A}\|_-$ and $\|(\mathcal{A}, r)\|_+ = \|\mathcal{A}\|_+$. For example for a formula $\varphi(X)$ of vocabulary τ, mini$^-$-$W_{\varphi(X)}$ is the problem

---

mini$^-$-$W_{\varphi(X)}$

        Input: $n, k \in \mathbb{N}$ in unary, a τ-structure $\mathcal{A}$ with $|A| \leq k \cdot \log n$,
           and $r \in \mathbb{N}$ with $r \leq |A|$.
    Parameter: k.
     Problem: Is there a subset S of A of cardinality r with $\mathcal{A} \models \varphi(S)$?

It was shown in [5] (see [8]) that $W[t] = \left[\{W_{\varphi(X)} \mid \varphi(X) \text{ a } \Pi_t\text{-formula}\}\right]^{\text{fpt}}$.

The following theorem shows that the miniaturization with respect to $\|\ \|_-$ (and hence, its miniaturization with respect to $\|\ \|_+$) of a Fagin-definable problem in W[1], lies in W[1], too. In [4], it was shown that mini$^-$-IS is reducible to the parameterized independent set problem, a problem in W[1]. There, a Turing reduction was exhibited. We generalize the idea underlying this reduction appropriately to obtain our result.

Theorem 1. *Let $\varphi(X)$ be a Π_1-formula. Then,* mini$^-$-$W_{\varphi(X)} \in$ W[1].

Above we gave definitions of VC and IS by means of Π_1-formulas, hence:

Corollary 1. mini$^-$-VC, mini$^-$-IS $\in$ W[1].

5   The Miniaturization of Unweighted Definable Problems

In general, in parameterized complexity we consider *weighted* satisfiability problems; often, in the context of miniaturizations they are considered in the *unweighted* form. In this section, for Fagin-definable problems we show that the miniaturized weighted and unweighted problems have the same computational complexity. Using this result, we prove, for every $t \geq 1$, that the miniaturization of a Fagin-definable problem in W[t] lies in W[t], too.

Definition 3. *Let τ be a vocabulary. Let $\varphi(X)$ be a first-order formula of vocabulary τ with the free set variable X; it defines a classical problem $U_{\varphi(X)}$ given by:*

$U_{\varphi(X)}$

        Input: A τ-structure $\mathcal{A}$.
     Problem: Is there a subset S of A with $\mathcal{A} \models \varphi(S)$?

Theorem 2. *For $q \geq 3$ and $t \geq 1$:*

1. $\left[\{\text{mini}^+\text{-}U_{\varphi(X)} \mid \varphi(X) \in \Pi_{1,q}\}\right]^{\text{fpt}} = \left[\{\text{mini}^+\text{-}W_{\varphi(X)} \mid \varphi(X) \in \Pi_{1,q}\}\right]^{\text{fpt}}$.

2. $\left[\{\text{mini}^-\text{-}U_{\varphi(X)} \mid \varphi(X) \in \Pi_{1,q}\}\right]^{\text{fpt}} = \left[\{\text{mini}^-\text{-}W_{\varphi(X)} \mid \varphi(X) \in \Pi_{1,q}\}\right]^{\text{fpt}}$.

3. $\left[\{\text{mini}^+\text{-}U_{\varphi(X)} \mid \varphi(X) \in \Pi_t\}\right]^{\text{fpt}} = \left[\{\text{mini}^+\text{-}W_{\varphi(X)} \mid \varphi(X) \in \Pi_t\}\right]^{\text{fpt}}$.

4. $\left[\{\text{mini}^-\text{-}U_{\varphi(X)} \mid \varphi(X) \in \Pi_t\}\right]^{\text{fpt}} = \left[\{\text{mini}^-\text{-}W_{\varphi(X)} \mid \varphi(X) \in \Pi_t\}\right]^{\text{fpt}}$.

By Theorem 1 this implies:

Corollary 2. *If* $\varphi(X) \in \Pi_1$ *then* $\mathrm{mini}^{-}\text{-}U_{\varphi(X)}, \mathrm{mini}^{+}\text{-}U_{\varphi(X)} \in \mathrm{W}[1]$.

Theorem 3. *Let* $t \geq 1$ *and* $\varphi(X) \in \Pi_t$. *Then* $\mathrm{mini}^{+}\text{-}W_{\varphi(X)}, \mathrm{mini}^{-}\text{-}W_{\varphi(X)} \in \mathrm{W}[t]$.

Corollary 3. $\mathrm{mini}^{-}\text{-}\textsc{Dominating Set} \in \mathrm{W}[2]$.

Recall that the parameterized dominating set problem (p-DS) is complete for W[2], so Corollary 3 implies that if p-DS $\in$ FPT, then the dominating set problem can be decided in time $2^{o^{\mathrm{eff}}(|V|)}$. Moreover by a refined argument, similar to that needed to prove Theorem 3, one can show the following result analogous to a result concerning the clique problem proved in [2].

Theorem 4. *If* p-DS *can be decided in time* $f(k) \cdot n^{o^{\mathrm{eff}}(k)}$ *for a computable function* f, *then both the dominating set problem and the clique problem are solvable in time* $2^{o^{\mathrm{eff}}(|V|)}$.

This improves a result stated in [7].

6   The M^{-}-hierarchy

The previous analysis suggests the definition of a hierarchy within W[1], taking the number of universal quantifiers into account: For $t \geq 1$, we introduce the class M$^{-}[t]$ by

$$\mathrm{M}^{-}[t] := \left[\{\mathrm{mini}^{-}\text{-}U_{\varphi(X)} \mid \varphi(X) \in \Pi_{1,t}\}\right]^{\mathrm{fpt}}.$$

By Corollary 2, M$^{-}[t] \subseteq$ W[1]; therefore,

$$\mathrm{M}^{-}[1] \subseteq \mathrm{M}^{-}[2] \subseteq \ldots \subseteq \mathrm{M}^{-}[t] \subseteq \ldots \mathrm{W}[1]. \tag{1}$$

By Theorem 2, for $t \geq 3$,

$$\mathrm{M}^{-}[t] = \left[\{\mathrm{mini}^{-}\text{-}W_{\varphi(X)} \mid \varphi(X) \in \Pi_{1,t}\}\right]^{\mathrm{fpt}}.$$

Theorem 5. *For all* $t \geq 2$, $\mathrm{mini}^{-}\text{-}t\text{-}\textsc{Sat}$ *is complete in* M$^{-}[t]$.

It is well-known that 2-SAT is in PTIME, thus $\mathrm{mini}^{-}\text{-}2\text{-}\textsc{Sat} \in$ FPT by part c) of Remark 1. Hence, (M$^{-}[1] =$)M$^{-}[2] =$ FPT by the preceding theorem.

Is the hierarchy in (1) (starting with $t = 2$) strict? By the preceding theorem, we know that $\mathrm{mini}^{-}\text{-}t\text{-}\textsc{Sat}$ is M$^{-}[t]$-complete; hence, in connection with this problem one should mention the result of Impagliazzo and Paturi [10] that, assuming the exponential-time hypothesis, the complexity of t-SAT increases with t.

7   MINI[1] and Turing Reductions

In [4], the class MINI[1] (sometimes denoted by M[1]) was introduced:

$$\text{MINI}[1] := [\text{mini}^+\text{-CIRCSAT}]^{\text{fpt}}.$$

Since mini$^+$-CIRCSAT $=^{\text{fpt}}$ mini$^+$-3-SAT $\leq^{\text{fpt}}$ mini$^-$-3-SAT, we know by Theorem 5 that MINI[1] $\subseteq$ M$^-$[3].

Sometimes (e.g., in [4,7]), in connection with the class MINI[1], Turing reductions (more precisely, parameterized Turing reductions) have been considered; the fpt-reductions considered so far in this paper were many-one reductions.

From the point of view of Turing reductions nearly all problems considered here have the same complexity. In fact, it has been implicitly shown in Impagliazzo, Paturi and Zane [11] that for $t \geq 3$, there is a parameterized Turing reduction from mini$^-$-t-SAT to mini$^+$-t-SAT, hence, these two problems are Turing equivalent. In particular, if we denote by MINI$_T$[1] and M$_T^-$[t] the closure under parameterized Turing reductions of MINI[1] and M$^-$[t], respectively, we have MINI$_T$[1] = M$_T^-$[3], but also M$_T^-$[t] = M$_T^-$[3] for all $t \geq 3$.

8   Renormalizations

In the context of miniaturized problems two "renormalizations" of VC have been considered, $(k \cdot \log n)^+$-VC and $(k \cdot \log n)^-$-VC. Let $\| \ \|$ be an arbitrary size function on the class of graphs. Define $(k \cdot \log n)^{\| \ \|}$-VC by

---

$(k \cdot \log n)^{\| \ \|}$-VC
> *Input:* $G = (V, E)$.
> *Parameter:* $k \in \mathbb{N}$.
> *Problem:* Does G have a vertex cover of size $k \cdot \log \|G\|$?

Clearly, $(k \cdot \log n)^-$-VC and $(k \cdot \log n)^+$-VC denote $(k \cdot \log n)^{\| \ \|-}$-VC and $(k \cdot \log n)^{\| \ \|+}$-VC, respectively. We show that both problems, $(k \cdot \log n)^-$-VC and $(k \cdot \log n)^+$-VC, are fpt-equivalent to mini$^-$-VC. The equivalence of the first and the third problem is stated in [7]. There, no proof is given and hence, we do not know if [7] refers to Turing reductions or to many-one reductions.

Theorem 6. $(k \cdot \log n)^-$-VC, $(k \cdot \log n)^+$-VC, *and* mini$^-$-VC *are fpt-equivalent.*

Remark 3. In the same way, one could define the $k \cdot \log n$ renormalizations of various parameterized problems, for example, of all Fagin-definable ones. Since we have no substantial results for the general case, we introduced the notion of renormalization for the vertex cover problem directly. In fact, the preceding theorem does not generalize to CLIQUE:

Claim: If FPT $\neq$ W[1], then $(k \cdot \log n)^-$-CLIQUE $\not\leq^{\text{fpt}}$ mini$^-$-CLIQUE.

For the reader familiar with the terminology of [9], we state the following generalization of this claim:

For $t \geq 2$, if $W[t] \neq FPT$, then $(k \cdot \log n)^- W_{\varphi(X)} \not\leq^{fpt} mini^- W_{\varphi(X)}$ for every generic $\Pi_{t/1}$-formula $\varphi(X)$.

In particular, for the dominating set problem, we have

$$\text{If } W[2] \neq FPT, \text{ then } (k \cdot \log n)^- \text{-DS} \not\leq^{fpt} mini^- \text{-DS}.$$

9   Conclusions

We have introduced a general notion of miniaturization of a problem that comprises the different miniaturizations of concrete problems considered so far. Using the appropriate logical formalism, we were able to show that the miniaturizations of definable problems in $W[t]$ are in $W[t]$, too. Based on this logical formalism we introduced a hierarchy of complexity classes in $W[1]$.

References

1. L. Cai and D. Juedes. Subexponential parameterized algorithms collapse the W-hierarchy. In *Proceedings of ICALP'01, LNCS 2076*, 2001.
2. J. Chen, X. Huang, I. Kanj and G. Xia. Linear FPT Reductions and Computational Lower Bounds. In *Proceedings of STOC'04*, 2004.
3. R. Downey. Parameterized complexity for the sceptic. In *Proceedings of the 18th IEEE Conference on Computational Complexity*, pages 147–168, 2003.
4. R. Downey, V. Estivill, M. Fellows, E. Prieto, and F. Rosamond. Cutting up is hard to do: the parameterized complexity of k-cut and related problems. In *Proceedings of CATS'03, ENTCS*, 78(0), 2003.
5. R.G. Downey, M.R. Fellows, and K. Regan. Descriptive complexity and the W-hierarchy. In P. Beame and S. Buss, editors, *Proof Complexity and Feasible Arithmetic*, volume 39 of *AMS-DIMACS Volume Series*, pages 119–134. AMS, 1998.
6. R.G. Downey and M.R. Fellows. *Parameterized Complexity*. Springer-Verlag, 1999.
7. M. Fellows. *New directions and new challenges in algorithm design and complexity, parameterized.*. In *Proceedings of WADS 2003, LNCS 2748*, 2003.
8. J. Flum and M. Grohe. Fixed-parameter tractability, definability, and model checking. *SIAM Journal on Computing*, 31(1):113–145, 2001.
9. J. Flum, M. Grohe, and M. Weyer. Bounded fixed-parameter tractability and $\log^2 n$ nondeterministic bits . In *Proceedings of ICALP'04*, 2004.
10. R. Impagliazzo and R. Paturi. Complexity of k-SAT. *JCSS*, 62(2): 367-375, 2001.
11. R. Impagliazzo, R. Paturi, and F. Zane. Which problems have strongly exponential complexity? *JCSS*, 63(4):512-530, 2001.

Smaller Kernels for Hitting Set Problems of Constant Arity⋆

Naomi Nishimura[1], Prabhakar Ragde[1], and Dimitrios M. Thilikos[2]

[1] School of Computer Science, University of Waterloo, Waterloo, Ontario, Canada
{nishi,plragde}@uwaterloo.ca
[2] Departament de Llenguatges i Sistemes Informàtics, Universitat Politècnica de
Catalunya, Campus Nord – Mòdul C5, c/Jordi Girona Salgado 1-3,
08034 Barcelona, Spain
sedthilk@lsi.upc.es

Abstract. We demonstrate a kernel of size $O(k^2)$ for 3-HITTING SET (HITTING SET when all subsets in the collection to be hit are of size at most three), giving a partial answer to an open question of Niedermeier by improving on the $O(k^3)$ kernel of Niedermeier and Rossmanith. Our technique uses the Nemhauser-Trotter linear-size kernel for VERTEX COVER, and generalizes to demonstrating a kernel of size $O(k^{r-1})$ for r-HITTING SET (for fixed r).

Keywords: hitting set, fixed parameter algorithms, kernels

1 Introduction

Kernelization is a central technique in the development of fixed-parameter tractable algorithms. Intuitively, a kernelization for a parameterized problem $\mathcal{P}$ is a polynomial-time algorithm that, given any instance $\mathcal{I}$, either determines that it is a "no" instance, or finds another instance $\mathcal{I}'$ (with perhaps a modified parameter value) such that $\mathcal{I}'$ is a "yes" instance if and only if $\mathcal{I}$ is, and the size of $\mathcal{I}'$ (which is the kernel) is bounded by some function of k. Since many problems have exponential-time brute-force algorithms, applying one to $\mathcal{I}'$ yields a fixed-parameter tractable algorithm for $\mathcal{P}$. Although technically any problem admitting a fixed-parameter tractable algorithm has a kernel, it is more useful to discover kernelization techniques as an aid to finding fixed-parameter tractable algorithms than to extract kernels from such algorithms.

This paper describes an algorithm for finding kernels for variations of the HITTING SET problem, one of the original NP-complete problems from Karp's 1972 paper [6]. In order to describe these variations, we need some terminology. Given two sets S and T, we say S *hits* T (or T *is hit by* S) if $S \cap T \neq \emptyset$. We

---

⋆ The two first authors were supported by the Natural Sciences and Engineering Research Council of Canada (NSERC). The third author was partially supported by the EU within the 6th Framework Programme under contract 001907 (DELIS) and by the Spanish CICYT project TIC-2002-04498-C05-03 (TRACER).

R. Downey, M. Fellows, and F. Dehne (Eds.): IWPEC 2004, LNCS 3162, pp. 121–126, 2004.
© Springer-Verlag Berlin Heidelberg 2004

apologize for the violent metaphor, which we inherited; our preference would be to use *touches* or *meets*.

r-HITTING SET
Input: A collection $\mathcal{C}$ of subsets of a finite set S such that for every $T \in \mathcal{C}$, $|T| \leq r$.
Parameter: A non-negative integer k.
Question: Is there a subset H of S such that $|H| \leq k$ and for every $T \in \mathcal{C}$, H hits T?

We describe an instance of r-HITTING SET by a tuple $\mathcal{I} = (S, \mathcal{C}, k)$. For any $r \geq 2$, r-HITTING SET is NP-complete, since for $r = 2$ it is simply the VERTEX COVER problem (S is the set of vertices and $\mathcal{C}$ the set of edges).

Because of the set-theoretic nature of r-HITTING SET, kernels for it have a slightly stronger property than described in the general introduction above. Given a collection $\mathcal{C} \subseteq 2^S$ and a subset S' of S, we denote by $\mathcal{C}|_{S'}$ the collection of sets in $\mathcal{C}$ restricted to S', that is, the set $\{T \cap S' \mid T \in \mathcal{C}, T \cap S' \neq \emptyset\}$. A kernel for an instance of r-HITTING SET is a subset K of S such that any solution to the instance $\mathcal{I}' = (K, \mathcal{C}|_K, k)$ of r-HITTING SET is a solution to $\mathcal{I}$.

If we can find a kernel of size $f(k)$ for r-HITTING SET, we can solve any instance in time $O(f(k)^k n)$ by simply trying all subsets of the kernel to see if they are solutions to the restricted problem. As we mentioned above, this approach works for other kernelizable problems, though in the case of r-HITTING SET an approach based on bounded search trees is faster [9], and kernelization serves mainly to bring the running time of the search algorithm down from $O(c^k n)$ to $O(c^k f(k) + n)$ (for some constant c). The ultimate goal is to find a linear-sized kernel for problems, because this brings down the cost of the brute-force search, and implies the existence of a constant-factor approximation algorithm.

2   Reducing Kernel Size

The first instance of kernelization most students of parameterized complexity [5] see is the $O(k^2)$ kernel for VERTEX COVER due to S. Buss [3, cited in [2]]. But VERTEX COVER is one of the few problems known to have a linear-sized kernel, provided by the following theorem due to Nemhauser and Trotter [7] (with improved running time due to Chen et al. [4]).

Theorem 1. *There is an algorithm running in time $O(kn + k^3)$ that, for any instance $\mathcal{I} = (S, \mathcal{C}, k)$ of VERTEX COVER with input size n, either computes a kernel $K \subseteq S$, $|K| \leq 2k$, or proves that $\mathcal{I}$ is a "no" instance.*

The proof of Theorem 1 makes elegant use of the linear programming relaxation of the integer program for VERTEX COVER, but we do not need to know anything about the proof; we will use the algorithm as a black-box subroutine in computing our kernel for r-HITTING SET. To illustrate the method, we first demonstrate a kernel of size $6k^2$ for 3-HITTING SET. Theorem 2 below improves the $O(k^3)$ kernel for 3-HITTING SET given by Niedermayer and Rossmanith [9], which is basically a generalization of Buss's $O(k^2)$ kernel for VERTEX COVER. We save a factor of k by using the fact that Nemhauser and Trotter give a kernel of size $O(k)$, not $O(k^2)$, for r-HITTING SET in the case $r = 2$, though an entirely different technique is employed to make use of this fact. Niedermeier [8, p. 31] lists as an open problem the use of Nemhauser-Trotter techniques to improve the size of the kernel for r-HITTING SET.

Theorem 2. *There is an algorithm running in time $O(kn + k^4)$ that, for any instance $\mathcal{I} = (S, \mathcal{C}, k)$ of 3-HITTING SET with input size n, either computes a kernel $K \subseteq S$, $|K| \leq 6k^2$, or proves that $\mathcal{I}$ is a "no" instance.*

Proof. Our algorithm actually does something slightly stronger; it computes a set $F \subseteq S$ of elements that must be in any solution to $\mathcal{I}$, and a set $M \subseteq S$ of elements that may be in the solution, with $|F \cup M| \leq 6k^2$. (Many kernelization algorithms do this, including that of Theorem 1, though we do not use this fact.)

We start by forming a collection $\mathcal{G} \subseteq \mathcal{C}$ using the following greedy algorithm: Start with $\mathcal{G}$ empty, and repeatedly choose an arbitrary set $C \in \mathcal{C}$. Add C to $\mathcal{G}$, and delete from $\mathcal{C}$ any set with a nonempty intersection with C. Repeat until $\mathcal{C}$ is empty. This takes $O(n)$ time.

If $|\mathcal{G}|$ contains more than k sets, $\mathcal{I}$ is a "no" instance, because any two sets in $\mathcal{G}$ are disjoint, and so more than k elements are required to hit them all. If $|\mathcal{G}| \leq k$, we proceed to construct the kernel. Let E be the set of elements appearing in the sets in $\mathcal{G}$, that is, $E = \bigcup_{G \in \mathcal{G}} G$. E must hit every set in $\mathcal{C}$ (if it did not hit some set, the set would not have been deleted in the algorithm that created $\mathcal{G}$, which is a contradiction), and $|E| \leq 3k$. Thus E is a hitting set, but if a hitting set of size k exists, it may not be contained within E. We will use the elements of E to construct our kernel.

For each $e \in E$, we define $\mathcal{C}_e$ to be the collection of sets in $\mathcal{C}$ containing the element e, that is, $\mathcal{C}_e = \{T \mid e \in T, T \in \mathcal{C}\}$. Think of $\mathcal{C}_e$ as a subproblem induced by the element e. We also define $\mathcal{C}'_e$ to be the sets in $\mathcal{C}_e$ but with the element e removed, that is, $\mathcal{C}'_e = \{T \setminus \{e\} \mid T \in \mathcal{C}_e\}$. Note that $\mathcal{I}_e = (S \setminus \{e\}, \mathcal{C}'_e, k)$ is an instance of VERTEX COVER, since every set in $\mathcal{C}'_e$ has size at most two. Since every $T \in \mathcal{C}_e$ must be hit, we will either choose e to be in the hitting set, or we find a hitting set for $\mathcal{I}_e$. The former can be achieved by adding e to F, and the latter by adding to M a kernel for the instance $\mathcal{I}_e$ of VERTEX COVER. Doing this for every element $e \in E$ will give us the sets F and M that we seek.

Applying Theorem 1, either $\mathcal{I}_e$ is a "no" instance, or we can find a kernel K_e of size at most $2k$ for $\mathcal{I}_e$ in time $O(kn_e + k^3)$ (where n_e is the size of the instance $\mathcal{I}_e$). If $\mathcal{I}_e$ is a "no" instance, e must be in any solution H for $\mathcal{I}$. Suppose it is not. Then in order for H to hit all the sets in $\mathcal{C}_e$, H would have to be a solution to $\mathcal{I}_e$. But since $\mathcal{I}_e$ is a "no" instance, this would make the size of H greater than k, which is a contradiction to H being a solution for $\mathcal{I}$. Thus if $\mathcal{I}_e$ is a "no" instance, we add e to F. If instead it is a "yes" instance, we add K_e to M.

Since, for each of the at most $3k$ elements of E, we either added one element to F or at most $2k$ elements to M, $|F \cup M| \leq 6k^2$, as required.

We claim that the set $K = F \cup M$ is a kernel for $\mathcal{I}$. To see this, suppose that $\mathcal{I}'$ is the problem defined by the candidate kernel K, that is, $\mathcal{I}' = (K, \mathcal{C}|_K, k)$.

If $\mathcal{I}'$ is a "no" instance, then $\mathcal{I}$ must be as well, since if H is a solution for $\mathcal{I}$, $H \cap K$ will be a solution for $\mathcal{I}'$. If $\mathcal{I}'$ is a "yes" instance, then let $H' \subseteq K$ be a solution. Since every element added to F must be in any solution, we know that $F \subseteq H'$. We need to show that H' is a solution for $\mathcal{I}$. To do this, we take an arbitrary set T in $\mathcal{C}$, and show that H' hits it.

If F hits T, we are done, so suppose it does not. Since E hits T, there must be some element e in $E \cap T$; if e is in the hitting set H', then T is hit by H', since it is hit by e. If e is not in H', then since F does not hit T, e is not in F. It follows that $\mathcal{I}_e$ is a "yes" instance, and $K_e \subseteq M$.

Since K_e is a kernel for $\mathcal{I}_e$, K_e must hit $T \setminus \{e\}$ (because K_e contains a solution for $\mathcal{I}_e$, which is a "yes" instance). Since $T \cap K_e$ is nonempty, it is a set in $\mathcal{C}|_K$. Since H' is a hitting set for $\mathcal{C}|_K$, it must hit $T \cap K_e$, and therefore it hits T, as required.

The running time of the procedure is $O(n)$ to find E plus $\sum_{e \in E} O(kn_e + k^3)$ to find F and M, where n_e is the number of sets in $\mathcal{C}$ that contain e. Since every set in $\mathcal{C}$ contains at most three elements, $\sum_e n_e \leq 3n$, and since $|E| \leq 3k$, the total running time is $O(kn + k^4)$.   □

In proving the above theorem, we used a subroutine for kernelizing VERTEX COVER to create an algorithm for kernelizing 3-HITTING SET. We can continue this process, using a subroutine for kernelizing $(r - 1)$-HITTING SET to create an algorithm for kernelizing r-HITTING SET.

Theorem 3. *For fixed $r \geq 2$, if there is an algorithm running in time $O(kn + k^r)$ that finds an $O(k^{r-2})$-size kernel for "yes" instances of $(r - 1)$-HITTING SET, then there is an algorithm running in time $O(kn + k^{r+1})$ that finds an $O(k^{r-1})$-size kernel for "yes" instances of r-HITTING SET.*

Proof. The proof of this theorem is a generalization of the proof of Theorem 2, which we can describe more succinctly now. Let $\mathcal{I} = (S, \mathcal{C}, k)$ be an instance of r-HITTING SET. As before, we form a set F of elements that must be in any solution to $\mathcal{I}$, and a set M of elements that might be, with $F \cup M$ being our kernel of the desired size. Let $\mathcal{G}$ be a maximal pairwise disjoint collection of sets from $\mathcal{C}$ chosen using a greedy algorithm running in time $O(n)$. If $|\mathcal{G}| > k$, there is no solution to $\mathcal{I}$. If $|\mathcal{G}| \leq k$, we let $E = \bigcup_{C \in \mathcal{G}} C$. E must hit every set in $\mathcal{C}$, and $|E| \leq rk$.

For each $e \in E$, we define $\mathcal{C}_e = \{T \mid e \in T, T \in \mathcal{C}\}$, and $\mathcal{C}'_e = \{T \setminus \{e\} \mid T \in \mathcal{C}_e\}$. Then $\mathcal{I}_e = (S \setminus \{e\}, \mathcal{C}'_e, k)$ is an instance of $(r - 1)$-HITTING SET. By the statement of the theorem, either $\mathcal{I}_e$ is a "no" instance, or we can find a kernel K_e of size $O(k^{r-2})$ for $\mathcal{I}_e$, where n_e is the size of $\mathcal{I}_e$. In the former case, e must be in any solution H for $\mathcal{I}$, because if it is not, H would have to be a solution to $\mathcal{I}_e$ in order for it to hit all the sets in $\mathcal{C}_e$; thus we can add e to F. In the latter case, we add K_e to M. Since we are adding at most rk sets of size $O(k^{r-2})$ to either F or M, $|F \cup M| = O(k^{r-1})$.

Then $K = F \cup M$ is a kernel for $\mathcal{I}$. To see this, we define $\mathcal{I}' = (K, \mathcal{C}|_K, k)$. If $\mathcal{I}'$ is a "no" instance, so is $\mathcal{I}$. If $\mathcal{I}'$ is a "yes" instance, let H' be a solution for it; $F \subseteq H'$. If T is any set in $\mathcal{C}$, we must show that H' hits T. Either F hits T, or, since E hits T, there exists $e \in E \cap T$; if $e \in H'$, then T is hit by H'. If $e \notin H'$, $\mathcal{I}_e$ is a "yes" instance. Since K_e is a kernel for $\mathcal{I}_e$, it hits $T \setminus \{e\}$, so $T \cap K_e$ is a set in $\mathcal{C}|_K$. Since H' is a hitting set for $\mathcal{C}|_K$, it must hit $T \cap K_e$ and thus T.

The running time of the procedure is $O(n) + \sum_e O(kn_e + k^r)$; since every set contains at most r elements, $\sum_e n_e \leq rn$, and the total running time is $O(kn + k^{r+1})$. $\qquad\square$

The method of Niedermeier and Rossmanith also generalizes easily to provide a kernel of size $O(k^r)$ for r-HITTING SET (though this is not explicitly mentioned in their paper). Unfortunately, the constants hidden in the O-notation increase with r in our case, but not for the Niedermeier-Rossmanith kernel. Thus Theorem 3 is not of much practical interest for larger values of r; even in the case $r = 3$, the Niedermeier-Rossmanith kernel is smaller for $k < 6$. Niedermeier [8, p. 34] mentions that Nemhauser-Trotter kernelization seems to perform well in practice, and this suggests that our kernelization for 3-HITTING SET may also have practical merit. However, recent empirical work [1] describes instances where LP-based methods such as Nemhauser-Trotter do poorly. Clearly, much more investigation remains to be done.

3   Conclusion

Since HITTING SET is W[2]-complete when the size of sets in the collection is not bounded in size [5], it is unlikely that we will find a linear-sized kernel for this problem. However, the statement "r-HITTING SET has a kernel of size $f(r)k$" for some $f(r) = 2^{\Omega(r)}$ is not inconsistent with what we believe about separations among parameterized complexity classes. It would be interesting to either prove this statement, or to demonstrate it false by proving lower bounds on the size of kernels for this and other FPT problems.

4   Acknowledgements

Part of this work was done while the authors were attending the 2004 Workshop on Fixed-Parameter Tractability Methods in Geometry and Games at the Bellairs Research Institute of McGill University in Holetown, Barbados. We wish to thank Sue Whitesides for inviting us. The second author also wishes to thank Arju and Zuki for swimming safely so that he could pace on the beach and think.

References

1. F. N. Abu-Khzam, R. L. Collins, M. R. Fellows, M. A. Langston, W. H. Suters, and C. T. Symons, Kernelization algorithms for the vertex cover problem: theory and experiments. Proceedings, *Workshop on Algorithm Engineering and Experiments (ALENEX)*, 2004.
2. Jonathan F. Buss and Judy Goldsmith. Nondeterminism within P. *SIAM Journal on Computing*, 22(3):560–572, 1993.

3. Sam Buss. private communication, 1989.
4. J. Chen, I.A. Kanj, and W. Jia. Vertex cover: further observations and further improvements. *Journal of Algorithms*, 41:280–301, 2001.
5. R.G. Downey and M.R. Fellows. *Parameterized Complexity*. Springer, 1999.
6. Richard M. Karp. Reducibility among combinatorial problems. In R. E. Miller and J. W. Thatcher, editors, *Complexity of Computer Computations*, pages 85–103. Plenum Press, 1972.
7. G. L. Nemhauser and L. E. Trotter Jr. Vertex packings: Structural properties and algorithms. *Mathematical Programming*, 8:232–248, 1975.
8. Rolf Niedermeier. *Invitation to fixed-parameter algorithms*. PhD thesis, Universität Tübingen, 2002. Habilitation thesis.
9. Rolf Niedermeier and Peter Rossmanith. An efficient fixed parameter algorithm for 3-hitting set. *Journal of Discrete Algorithms*, 2(1):93–107, 2003.

Packing Edge Disjoint Triangles: A Parameterized View

Luke Mathieson, Elena Prieto, and Peter Shaw

School of Electrical Engineering and Computer Science
The University of Newcastle
Australia

Abstract. The problem of packing k edge-disjoint triangles in a graph has been thoroughly studied both in the classical complexity and the approximation fields and it has a wide range of applications in many areas, especially computational biology [BP96]. In this paper we present an analysis of the problem from a parameterized complexity viewpoint. We describe a fixed-parameter tractable algorithm for the problem by means of kernelization and crown rule reductions, two of the newest techniques for fixed-parameter algorithm design. We achieve a kernel size bounded by $4k$, where k is the number of triangles in the packing.

1   Introduction

The problem of finding the maximum number of vertex or edge disjoint cycles in an undirected graph G has applications in many different fields, for instance in computational biology [BP96]. The problem is defined as follows: Let $G = (V, E)$ be a simple graph. A triangle T in G is any induced subgraph of G having precisely 3 edges and 3 vertices. A graph $G = (V, E)$ is said to have a k packing of triangles if there exist k disjoint copies $T_1, ..., T_k$ of T in the vertex set of G. The packing is called vertex-disjoint if $T_1, ..., T_k$ share no vertices and is called edge-disjoint if we allow $T_1, ..., T_k$ to have some vertices in common but no edges exist in $T_i \cap T_j$ when $i \neq j$. In this paper we look at the parameterized version of the edge-disjoint case. More formally, we will study in detail the parameterized complexity of the following problem:

k-EDGE DISJOINT TRIANGLE PACKING (ETP)
Instance: A graph $G = (V, E)$, a positive integer k
Parameter: k
Question: Are there at least k edge disjoint instances of T in G?

This problem is NP-hard for general graphs [HOL81] and has also been shown to be NP-hard for planar graphs, even if the maximum degree is 5 or more. Regarding approximability, ETP is known to be APX-hard [K94]. A general result of [HS89] leads to a polynomial time $(3/2 + \epsilon)$ approximation algorithm for any $\epsilon > 0$ for this problem. If G is planar Baker [B94] gives a polynomial time approximation scheme for the vertex-disjoint case which can be extended to

R. Downey, M. Fellows, and F. Dehne (Eds.): IWPEC 2004, LNCS 3162, pp. 127–137, 2004.
© Springer-Verlag Berlin Heidelberg 2004

solve ETP. Caprara and Rizzi give a polynomial time algorithm for ETP when the maximum degree of G is ≤ 4 and prove that the problem remains APX-hard even for planar graphs with maximum degree 5 or more [CR01]. The similar problem of finding k vertex-disjoint triangles (NTP) has also been studied from a parameterized complexity point of view achieving a cubic kernel [FHRST04].

2   Parameterized Complexity and Kernelization: The Method of Coordinatized Kernels

2.1   The Method

Definition 1. *A parameterized problem L is kernelizable if there is a parametric transformation of L to itself that satisfies:*

1. *The running time of the transformation of (x, k) into (x', k'), where $|x| = n$, is bounded by a polynomial $q(n, k)$ (so that in fact this is a polynomial time transformation of L to itself, considered classically, although with the additional structure of a parametric reduction),*
2. *$k' \leq k$, and*
3. *$|x'| \leq h(k)$, where h is an arbitrary function.*

We define the size of the kernel for L to be $|h(k)|$.

It is important to note the following result by [DFS99]:

Lemma 1. *A parameterized problem L is in FPT if and only if it is kernelizable.*

Proof of Lemma 1. Based on ideas from [CCDF97]   □
From this definition the goal would obviously be to try to produce reduction rules to reduce the size of the instance. Then, simply trying an 'intelligent' brute force to the remaining instance would suffice to solve the problem. Obviously, the smaller the kernel is, the better algorithm we get.

In *the method of coordinatized kernels*, the general aim is to prove a kernelization lemma. In the case of solving a graph maximization problem such as k-EDGE DISJOINT TRIANGLE PACKING, this lemma has the following form:

Lemma 2. *(General Kernelization Lemma) If $G = (V, E)$ is an irreducible instance of a problem Π and G has more than $f(k)$ vertices, then (G, k) is a Yes-instance for the problem.*

To prove such a lemma, we still have to set up a list of reduction rules R (leading to a specific notion of irreducibility) and a suitable bounding function $f(k)$. These reduction rules will reduce the size of the instance without destroying any optimal solution. The rules can always be performed in polynomial time.
Our techniques for deriving a smaller kernel are based on:

– the polynomial time reduction rules mentioned above, and
– combinatorial extremal arguments.

The combinatorial extremal arguments help us obtain structural claims in the context of a so-called *boundary lemma*. This boundary lemma takes the following form:

Lemma 3. *(General Boundary Lemma) Let G be R-irreducible and let (G, k) be a* No*-instance and $(G, k + 1)$ be a* No*-instance for Π. Then the number of vertices of G is less than $g(k)$ for some function g.*

Once the boundary lemma is proved, the kernelization lemma follows by simple contradiction.

This *method of coordinatized kernels* has already been used to tackle many other graph maximization problems such as k-NONBLOCKER [DFFPR03], k-MAX LEAF [FMRS00] and k-STAR PACKING [FPS03] proving to be equally as successful as with k-EDGE DISJOINT TRIANGLE PACKING. But its usefulness has proven also important with minimization problems such as k-GRAPH MODIFICA-TION PROBLEM [FL03] and even with non-graph problems like SET SPLITTING [DFR03].

2.2   The Crown Rule

Traditionally, most reduction rules used to reduce the size of the problem kernel were limited to reduction rules that were based on the structure of a fixed number of vertices. For example, given an instance (G, k) of the VERTEX COVER problem which has a vertex v of degree 1 with $N(v) = u$, we can simplify the problem by reducing the instance from (G, k) to (G', k') where $G' = G \backslash \{u, v\}$ and $k' = k - 1$ without losing any information on the solution set of the problem.

However, other methods to simplify the problem exist that are not based on the structure of a fixed number of vertices. In [CFJ03] another reduction rule is introduced that uses a structural observation based on the following definition,

Definition 2. *A Crown Decomposition of a graph $G = (V, E)$ is a tripartition of the vertices of the graph into three sets H, C and X with the following properties:*

1. *C is an independent set,*
2. *H is a cutset in V such that there are no edges between C and X.*
3. *H is matched into C.*

The *Crown Decomposition* is applied to both SAVING-k-COLORS and VERTEX COVER. An instance of the VERTEX COVER problem (G, k) that contains a crown decomposition, can be reduced to (G', k') where $G' = G \setminus H \setminus C$ and $k' = k - |H|$ by removing the two sets H (the Head) and C (the Crown) of vertices, as there is an optimal cover for (G, k) which contains all the vertices in H and none of the vertices in C [CFJ03].

Here the Crown Decomposition is used to generalize the simple reduction rule for a degree 1 vertex (which is the same for both problems) to one that admits a global structural generalization [F03], as each edge in the matching must be covered by the vertex that is connected to the rest of the graph. Further, it can be shown that by repeatedly applying the crown reduction, a linear kernel can be obtained in polynomial time. For both problems, VERTEX COVER and SAVING-k-COLORS, a $3k$ kernel can be achieved [CFJ03].

It has been shown that the structural nature of crown decomposition based rules allows them to be applied faster than other computationally expensive techniques such as Linear Programming or Network Flow algorithm [ALSS03]. In addition, the simple structure of the crown decomposition implies that it can be adapted to other problems. This has already become a useful tool to obtain significantly smaller (often linear) sized problem kernels.

The following theorem given in [CFJ03] can be used to bound the size of the graph based on the size of the independent set for which we can find a Crown. This is expressed here in a slightly more general form.

Lemma 4. *If a graph $G = (V, E)$ has an independent set $I \subseteq V(G)$ such that $|N(I)| \leq |I|$ then G has a crown.*

In this paper we describe how a slightly modified version of the crown decomposition can be used to obtain a linear kernel for the ETP problem using an auxiliary graph module. Other modifications of the crown rule can be found in [FPS03] [FHRST04] [PS03].

3   The Method on k-EDGE DISJOINT TRIANGLE PACKING

As mentioned in section 2.1, a list of reduction rules for the problem must be produced. These reduction rules are all computable in polynomial time on the size of the instance and transform G into another graph G' with the property that G has a solution of size k if and only if G' has a solution of size k' where $k' \leq k$ and in general $|G'| \leq |G|$.

The kernelization and boundary lemmas will be proved in section 3.2.

3.1   Reduction Rules

For each Reduction Rule described below we assume that none of the previous Reduction Rules can be applied to (G, k) anymore.

Reduction Rule 1 *If there exist $u, v, w \in V$ such that u and v have degree 2, and $(u, v), (u, w), (v, w) \in E$, then G is a* Yes-*instance of k-EDGE DISJOINT TRIANGLE PACKING if and only if $(G' = G \setminus \{u, v\}, k - 1)$ is a* Yes-*instance.*

> **Proof of Reduction Rule 1.** The vertices u, v and w form a dangling triangle which will always be part of any optimal solution of k-EDGE DISJOINT TRIANGLE PACKING.   □

Reduction Rule 2 *If a vertex v is not part of any triangle then (G, k) is a* Yes-*instance of k-EDGE DISJOINT TRIANGLE PACKING if and only if $(G' = G \setminus \{v\}, k)$ is a* Yes-*instance of k-EDGE DISJOINT TRIANGLE PACKING.*

Proof of Reduction Rule 2. Trivial, as v will never participate in any triangle packing. □

Reduction Rule 3 *If an edge e is not part of any triangle then (G, k) is a Yes-instance of k-*EDGE DISJOINT TRIANGLE PACKING *if and only if $(G' = G \setminus \{e\}, k)$ is a Yes-instance of k-*EDGE DISJOINT TRIANGLE PACKING.

Proof of Reduction Rule 3. Trivial, as e will never participate in any triangle packing. □

We say that an edge (u, v) is spanned by a vertex w if (u, w) and (v, w) are in E. A set V' of vertices spans a set E' of edges if for every $e \in E'$ there exists a vertex in V' spanning e. We will denote E' as $S(V')$.

Reduction Rule 4 *(Fat-Head Crown Rule) Assume the following structure:*

1. *C, an independent set of vertices.*
2. *H, a set of edges spanned by C, i.e. $S(C)$.*
3. *X, the rest.*
4. *The endpoints of the edges in H, $V(H)$ form a cutset such that there are no edges from C to $X \setminus V(H)$.*
5. *f, an injective function mapping each edge $(h_1, h_2) \in H$ to a unique pair of edges $(h_1, u), (h_2, u)$ where $u \in C$.*

*Then (G, k) is a Yes-instance of k-*EDGE DISJOINT TRIANGLE PACKING*if and only if $(G \setminus (C \cup H), k - |H|)$ is a Yes-instance of k-*EDGE DISJOINT TRIANGLE PACKING.

Proof of Reduction Rule 4.
First note that $V(H)$ forms a cutset with no edges from C to $X \setminus V(H)$ as every edge in the graph must be part of a triangle and if there existed an edge from C to $X \setminus V(H)$ that edge would not be part of a triangle and thus the graph would not be reduced under Reduction Rule 3.
We will prove the two implications of the if and only if separately:
$\Rightarrow$: f assigns a triangle to each edge in H. This edge can only be used in a single triangle in the packing. Further, the maximum number of triangles we can possibly add to the packing with the elements of H is $|H|$. With our construction we achieve this goal using only elements in C. Note that the elements in C are independent, and therefore can only form triangles with the edges in H. It is clear that by choosing the triangles formed with C and H we can do no worse than choosing triangles formed with H and X, as using the vertices in C allows the possibility of extra triangles to be formed in X.
$\Leftarrow$: Assume we have a set of $|H|$ non-edges[1] between vertices in the graph. If we add an independent set C of vertices spanning these non-edges plus all the edges missing then we will get a new solution with precisely $|H|$ more triangles. □

[1] A non-edge is a pair of vertices u, v in G such that $(u, v) \notin E$

Lemma 5. *Let I be an independent set of vertices in $G = (V, E)$ such that $|I| \geq |S(I)|$. Then there exists a fat-head crown, i.e. a partition (C, H, X) of G and a function f constituting a fat-head crown as described in Reduction Rule 4.*

Proof of Lemma 5. Let G be a graph with an independent set I such that $|I| \geq |S(I)|$. From $G = (V, E)$ we construct a graph model $G' = (V', E')$ by altering G as follows:

- For each edge $e \in S(I)$ add a vertex v to V'.
- For every u in I *spanning* e, add (u, v) to E'.
- Add an edge from v to the endpoints of e.
- Remove the edges from the endpoints of e to u from E'.

By our construction it is clear that if $|I| \geq |S(I)|$ in G then $|I'| \geq |N(I')|$ in G' and by Lemma 4 there exists a crown in G', where I' is the vertices of G' corresponding to I in G.

If G' has a crown decomposition (C', H', X'), then G has a fat-head crown (C, H, X), where:

- $C = C'$ is an independent set by definition.
- H is the set of edges represented by the vertices in H' excluding its endpoints.
- f corresponds to the matching between C' and H' (by the nature of the construction of G').
- X is the rest of the graph.

Thus, if $|I| \geq |S(I)|$ there exists a fat-head crown decomposition. $\square$

3.2   Kernelization and Boundary Lemmas

Lemma 6. *(Kernelization Lemma) If $G = (V, E)$ is reduced under reduction rules 1-4, and $|V| > 4k$, then G is a Yes-instance for k-EDGE DISJOINT TRIANGLE PACKING.*

Before we prove this lemma we need the following result.

Lemma 7. *(Boundary Lemma) If $G = (V, E)$ is a Yes-instance of the k-EDGE DISJOINT TRIANGLE PACKINGproblem for parameter k, a No-instance for parameter $k + 1$, and G is reduced under reduction rules 1-4, then $|V| \leq 4k$.*

Proof of Lemma 7. (By minimum counterexample). Assume for contradiction that there exists a counterexample G', such that G' is reduced and contains a k-EDGE DISJOINT TRIANGLE PACKING, but there is no packing of the edges of G allowing $k + 1$ triangles. Assume further that $|V(G)| > 4k$.

Let us consider the following witness structure as a partition of V:

- P, an edge-disjoint triangle packing of size k plus all edges between triangles in the packing (*ghost edges*). Note that P contains less than $3k$ vertices.
- O, the rest (*outsiders*).

STRUCTURAL CLAIMS:

Claim 1 *There are no triangles in O.*

Proof of Claim 1. If there existed a triangle in O, then we could increase the size of the packing by 1, contradicting our assumption that G is a No-instance for $(k+1)$-EDGE DISJOINT TRIANGLE PACKING.   □

Claim 2 $< O >_{|edges} = \emptyset$ *(i.e. O is an independent set).*

Proof of Claim 2. Assume there exists an edge (u_1, u_2) for some $u_1, u_2 \in O$. Then by reduction rule 3, it must be part of some triangle, and by claim 1, the triangle is not in O. So there exists a vertex $v \in P$ that forms this triangle. Then we have a triangle that is not in the packing, so we could increase out packing by 1, contradicting our assumptions. □

Claim 3 *For all $u \in O$, u does not span any ghost edges in P.*

Proof of Claim 3. Assume there was such a vertex. Then we would have a triangle that is not included in the packing, and we could increase the packing by 1, contradicting our assumptions.   □

Claim 4 *For every triangle $(u, v, w) \in P$, if one edge is spanned by $x \in O$, then there is no $y \in O, y \neq x$ such that y spans another edge in (u, v, w).*

Proof of Claim 4. Assume that we had such an arrangement. Then, without loss of generality, assume x spans the edge (u, v) and y spans the edge (v, w). We can replace $(u, v, w) \in P$ with two triangles (u, v, x) and (v, w, y), and thus increase the size of the packing by 1, contradicting our assumptions.   □

Claim 5 *If a vertex $x \in O$ spans more than one edge in a triangle[2] $(u, v, w) \in P$, then it is the only vertex that spans any edge in (u, v, w).*

Proof of Claim 5. This is simply a special case of claim 4.   □

From now on we consider a partition of P into three subsets P_0, P_1 and P_A defined as follows:

- P_0: triangles in P with no vertices in O spanning any of their edges.
- P_1: triangles in P which have exactly one edge spanned by vertices in O.
- P_A: triangles in P which have all edges spanned by some vertex in O.

---

[2] Note that if a vertex spans more than one edge of a triangle, it must span all three

We define the number of triangles in P_0 to be s, the number of triangles in P_1 to be l and the number of triangles in P_A to be $k - l - s$.
By this definition we can claim:

Claim 6 *Let $O_0 = N(P_0)_{|_O}$, then $|O_0| = 0$.*

Claim 7 *Let $O_A = N(P_A)_{|_O}$, then $|O_A| \leq k - l - s (\leq k - s)$.*

Proof of Claim 7. By claim 5, each triangle in P_A must have precisely one neighbor in O, also note that a vertex in O may completely span two triangles P_A. Then as $|P_A| = k - l - s$ (by definition), $|O_A| \leq k - l - s$. $\square$

Claim 8 *Let $O_1 = N(P_1)_{|_{O - O_A}}$, then $|O_1| \leq l$.*

Proof of Claim 8. By definition of P_1, only one edge in each triangle in P_1 is spanned by vertices in O. There are precisely l triangles in P_1, and by lemma 5, if $|O_1| > l$, we would have reduced by reduction rule 2.2 (the crown rule). $\square$

Claim 9 $|O| = |O_A \cup O_1| = |O_A| + |O_1| - |O_A \cap O_1| \leq l + k - l - s = k - s \leq k$.

Proof of Claim 9. As the number of vertices in O_0 is 0 (claim 6), $|O| = |O_A \cup O_1|$. The rest follows trivially from the definitions of set operations and claims 7 and 8. $\square$

Claim 10 $|V| \leq 3k + k = 4k$.

Proof of Claim 10. By definition $|V| = |P| + |O|$. $\square$

Thus from claim 10, we have that the total size of $|V|$ is at most $4k$ contradicting our initial assumptions and thus we have the proof of the boundary lemma. $\square$

Proof of Lemma 6. Assume in contradiction to the stated theorem that there exists a reduced graph G of size $|V(G)| > 4k$ but has no k-Edge Disjoint Triangle Packing.
Let $k' < k$ be the largest k' for which G is a Yes-instance. By the Boundary Lemma 7 we know that $|V(G)| \leq 4k' < 4k$. This contradicts the assumption. $\square$

4   Description of the Algorithm and Running Time Analysis

In the previous section we obtained a kernel for the problem of size $4k$ on the number of vertices. Now, we are in a position to apply the following algorithm to this reduced instance to achieve the claimed running time. The algorithm is nothing but a brute-force approach on the $4k$ kernel: it first tries all possible collections of vertices that could produce a k-EDGE DISJOINT TRIANGLE PACKING and then, for each one of these collections, tries all subsets of $3k$ edges. The number of vertices which could possibly give a k-EDGE DISJOINT TRIANGLE PACKING less than $3k$ as shown in section 3.2. The number of possible combinations of $3k$ edges, considering that we have at most $3k$ vertices in a collection, will be always less than $\binom{(3k)^2}{3k}$.

Step 1. Generate all subsets of vertices that are
        candidates to form an edge-disjoint triangle
        packing.
Step 2. For each of these subsets find all possible
        k-triangles by testing every subset of edges
        of size $3k$.
Step 3. If the answer is Yes for one of the subsets of
        edges in step 2, answer Yes and halt.
Step 4. Else answer No and halt.

The running time of this algorithm is the following:

$$\sum_{i=0}^{3k}\binom{4k}{i}\binom{i^2}{3k} \leq \binom{(3k)^2}{3k}\sum_{i=0}^{3k}\binom{4k}{i} \leq (3k)^{\frac{9k}{2}}\sum_{i=0}^{4k}\binom{4k}{i}$$

$$= (3k)^{\frac{9k}{2}}\cdot 2^{4k} = 2^{\log{(3k)}^{\frac{9k}{2}}}\cdot 2^{4k} \leq 2^{\frac{9k}{2}\cdot\log 3k+\frac{9k}{2}} = \mathcal{O}(2^{\frac{9k}{2}\log k+\frac{9k}{2}})$$

It is to be noted that we could achieve single exponential running time for k-EDGE DISJOINT TRIANGLE PACKING. The algorithm to obtain this running time uses a color coding technique, also known as hashing [AYZ95]. The idea is to use dynamic programming on a $3k$ coloring of the edges of the graph. For each of the colorings we generate a table which has an entry for every set S of colors whose size is a multiple of 3. The entry for S,

$$answer(S) = \begin{cases} 1 & \text{if } \exists\, S' \subseteq S, \text{ such that } |S'| = |S| - 3, S' = S \setminus \{u, v, w\} \\ & \text{and } answer(S') = 1 \\ 0 & \text{otherwise} \end{cases}$$

The running time of this algorithm is $\mathcal{O}(2^{3k}\cdot \#\mathcal{H})$, where $\#\mathcal{H}$ is the number of Hash Functions, which is of the order of c^k. Unfortunately c is a huge constant, making this algorithm unfeasible in practical terms, for more details on this type of technique for packing problems we refer the reader to [FK+04].

5   Conclusions and Further Research

To pack k edge-disjoint triangles, we utilize two of the newest techniques in parameterized complexity, namely *the method of coordinatized kernels* and *crown type reduction rules*.

In section 3.1 the use of polynomial time reduction rules, especially the crown rule, prunes the instance to the manageable size of $4k$ vertices. Here k is the number of triangles we are trying to pack, which is smaller than $|E|$. The method provides us with a systematic tool to analyze the kernel size obtained by these reduction rules.

We achieve a running time for the triangle packing case that could be extended to cycles of length greater than three using the same technique. Probably a modification of the crown reduction rule we use in this paper would be needed in the more general case of k-EDGE DISJOINT CYCLE PACKING:

> k-EDGE DISJOINT CYCLE PACKING
> Instance: A graph $G = (V, E)$, a positive integer k
> Question: Are there at least k edge disjoint instances of C_r in G?

The choice of parameter in this case is not as clear. We could either choose only k as in the case of k-EDGE DISJOINT TRIANGLE PACKING or both k and r, the length of the cycle. The problem is still NP-complete [HS89] and APX-hard [K94] and has a great number of applications in computational biology [BP96].

References

[ALSS03] F. Abu-Khazam, M. Langston, P. Shanbhag, C. Symons. High-Performance Tools for Fixed-Parameter Tractable Implementations, *29th Workshop on Graph Theoretic Concepts in Computer Science, Workshop on Fixed Parameter Tractability* (2003).

[AYZ95] N. Alon, R. Yuster, U. Zwick. *Color-Coding*, Journal of the ACM, Volume 42(4), pages 844–856 (1995).

[B94] B. Baker. *Approximation algorithms for NP-complete problems on planar graphs.* Journal of the ACM, Volume 41(1) (1994).

[BP96] V. Bafna and P. Pevzner, *Genome Rearrangements and Sorting by Reversals,* SIAM J. Comput. 25(2) (1996).

[CCDF97] L. Cai, J. Chen, R. Downey and M. Fellows. *Advice Classes of Parameterized Tractability.* Annals of Pure and Applied Logic 84, pages 119–138 (1997).

[CFJ03] B. Chor, M. Fellows. D. Juedes. *'Saving k Colors in Time $O(n^{5/2})$.* Manuscript in preparation.

[CR01] A. Caprara, R. Rizzi. *Packing Triangles in Bounded Degree Graphs.* Inf. Process. Lett. 84(4), pages 175–180 (2002).

[DF99] R. Downey and M. Fellows. *Parameterized Complexity* Springer-Verlag (1999).

[DFFPR03] F. Dehne, M. Fellows, H. Fernau, E. Prieto, F.Rosamond *A Description of the Method of Coordinatized Kernels Illustrated by* NONBLOCKER. Manuscript in preparation.

[DFR03] F. Dehne, M. Fellows, F.Rosamond. *An FPT Algorithm for Set Splitting.* Proceedings 29th Workshop on Graph Theoretic Concepts in Computer Science, Springer Verlag LNCS 2880, (2003).

[DFS99] R. Downey, M. Fellows, U. Stege. *Parameterized Complexity: A Framework for Systematically Confronting Computational Intractability.* AMS-DIMACS Series in Discrete Mathematics and Theoretical Computer Science, Volume 49, pages 49-99 (1999).

[F03] M. Fellows. *Blow-ups, Win/Wins and Crown Rules: Some New Directions in FPT.* Proceedings 29th Workshop on Graph Theoretic Concepts in Computer Science, Springer Verlag LNCS 2880, pages 1-12, (2003).

[FHRST04] M. Fellows, P. Heggernes, F. Rosamond, C. Sloper, J.A. Telle *Exact Algorithms for Finding k Disjoint Triangles in an Arbitrary Graph.* Proceedings 30th Workshop on Graph Theoretic Concepts in Computer Science (WG '04), Springer Verlag LNCS, (2004).

[FK+04] M. Fellows, C. Knauer, N. Nishimura, P. Radge, F. Rosamond, U. Stege, D. Thilikos, S. Whitesides. *Faster fixed-parameter tractable algorithms for matching and packing problems.* Accepted in 12th Annual European Symposium on Algorithms (ESA '04), Springer Verlag LNCS, (2004).

[FL03] M. Fellows, H. Fernau, M. Langston, E. Prieto, P. Shaw, C. Symons, U. Stege. *The Graph Modification Problem: Edge Editing to k Cliques.* Manuscript in preparation.

[FMRS00] M. Fellows, C. McCartin, F. Rosamond, and U.Stege. *Coordinatized Kernels and Catalytic Reductions: An Improved FPT Algorithm for Max Leaf Spanning Tree and Other Problems,* Foundations of Software Technology and Theoretical Computer Science, (2000).

[FPS03] M. Fellows, E. Prieto, C. Sloper. *Looking at the Stars.* Manuscript in preparation.

[HOL81] I. Holyer, *The NP-completeness of some edge-partition problems,* SIAM J. Comput. 10, pages 713-717 (1981).

[HS89] C. A. J. Hurkens and A. Schrijver. *On the size of systems of sets every t of which have an SDR, with an application to the worst-case ratio of heuristics for packing problems.* SIAM J. Discret. Math. 2(1), pages 68–72 (1989).

[K94] V. Kann. *Maximum bounded H-matching is MAX-SNP-complete.* Information Processing Letters 49, pages 309–318 (1994).

[PS03] E. Prieto and C. Sloper. Either/Or: Using Vertex Cover Structure in designing FPT-algorithms - the case of k-Internal Spanning Tree, *Proceedings of WADS. Workshop on Algorithms and Data Structures, Ottawa, Canada,* LNCS 2748 pages 474–483 (2003).

Looking at the Stars

Elena Prieto[1] and Christian Sloper[2]

[1] School of Electrical Engineering and Computer Science,
The University of Newcastle
NSW, Australia
elena@cs.newcastle.edu.au
[2] Department of Informatics,
University of Bergen
Norway
sloper@ii.uib.no

Abstract. The problem of packing k vertex-disjoint copies of a graph H into another graph G is NP-complete if H has more than two vertices in some connected component. In the framework of parameterized complexity we analyze a particular family of instances of this problem, namely the packing of stars. We prove that packing k copies of $H = K_{1,s}$ is fixed-parameter tractable and give a quadratic kernel for the general case. When we consider the special case of $s = 2$, i.e. H being a star with two leaves, we give a linear kernel and an algorithm running in time $\mathcal{O}^*(2^{5.3k})$.

1 Introduction

The problem of MAXIMUM H-MATCHING, also called MAXIMUM H-PACKING, is of practical interest in the areas of scheduling [BM02], wireless sensor tracking [BK01], wiring-board design and code optimization [HK78] and many others.

The problem is defined as follows: Let $G = (V, E)$ be a graph and $H = (V_H, E_H)$ be a fixed graph with at least three vertices in some connected component. An H-packing for G is a collection of disjoint subgraphs of G, each isomorphic to H. In an optimization sense, the problem that we want to solve would be to find the maximum number of vertex disjoint copies of H in G. The problem is NP-complete [HK78] when the graph H has at least three vertices in some connected component. Note that in the case where H is the complete graph on two nodes H-packing is the very well studied (and polynomial time solvable) problem MAXIMUM MATCHING. MAXIMUM H-PACKING has been thoroughly studied in terms of approximation. The problem has been proved to be MAX-SNP-complete [K94] and approximable within $|V_H|/2 + \varepsilon$ for any $\varepsilon > 0$ [HS89]. Several restrictions have also been considered (planar graphs, unit disk graphs etc.) in terms of the complexity of their approximation algorithms. For a review of these we refer the reader to [AC99]. In parameterized complexity a recent result by [FHRST04] gives a general algorithm for packing an arbitrary graph H into G. Their result gives a $2^{\mathcal{O}(|H|k \log k + k|H| \log |H|)}$ algorithm for the general case.

We discuss the parameterized complexity of the MAXIMUM H-PACKING problem for the case when H belongs to the restricted family of graphs $\mathcal{F} = K_{1,s}$, a star with s leaves. More formally:

R. Downey, M. Fellows, and F. Dehne (Eds.): IWPEC 2004, LNCS 3162, pp. 138–148, 2004.
© Springer-Verlag Berlin Heidelberg 2004

$K_{1,s}$-PACKING
INSTANCE: Graph $G = (V, E)$, a positive integer k
QUESTION: Are there at least k vertex disjoint instances of $K_{1,s}$ in G?

This problem has already been studied within the framework of classical complexity theory [HK86]. In their paper, Hell and Kirkpatrick studied the complexity of packing complete bipartite graphs into general graphs. We include a brief introduction to this topic in Section 2. In Section 3 we show that the general problem is tractable if parameterized, and that we can obtain a quadratic kernel. In Section 4 we show that the special case of packing $K_{1,2}$'s has a linear kernel, and in Section 5 we give a quick algorithm for both the general and special case. In contrast [FHRST04] obtains only a $\mathcal{O}(k^3)$ for packing a graph with three vertices, namely K_3. Finally we conclude with some directions for further research.

2   A Brief Introduction to Parameterized Complexity

In classical complexity theory, a decision problem is specified by two elements: the input to the problem and the question to be answered. In parameterized complexity a third element of information is added, the parameter. Also, in classical complexity theory, the notion of 'good' behavior is related to polynomial time solvability, i.e. if there is an algorithm solving the problem in polynomial time on the size of the instance. In parameterized complexity a problem is said to have a "good" behavior if the combinatorial explosion can be limited to the parameter, i.e. it is fixed-parameter tractable.

2.1   The Two Races

Currently, in FPT algorithm design there are two races [F03]. Firstly, to obtain the best running time, is the obvious race as it derives straight from the definition of fixed-parameter tractability. Here the emphasis is put in the function f. We will adopt the $\mathcal{O}^*(f(k))$-notation for running time, where the polynomial part of the function is ignored. Generally the polynomials will be negligible compared to $f(k)$.

Improvements in the algorithms in this race may arise from better preprocessing rules, better branching strategies or more sophisticated methods of analysis.

Another race is the kernelization race, which is the one we contribute to (mainly) in this paper. The kernel of a problem is defined as follows:

Definition 1. *A parameterized problem L is kernelizable if there is a parametric transformation of L to itself that satisfies:*

1. *The running time of the transformation of (x, k) into (x', k'), where $|x| = n$, is bounded by a polynomial $q(n, k)$ (so that in fact this is a polynomial time transformation of L to itself, considered classically, although with the additional structure of a parametric reduction),*
2. *$k' \leq k$, and*
3. *$|x'| \leq h(k')$, where h is an arbitrary function.*

From this definition the goal would obviously be to try to produce better reduction rules to reduce the size of the instance as much as possible. It is important to note the following result by [DFS00], which gives us the link between the two race:

Lemma 1. *A parameterized problem L is in* FPT *if and only if it is kernelizable.*

The two races are worth playing as they may lead to substantial improvements on the quality of the algorithms we design and also to new strategies for practical implementations of these algorithms.

2.2 Preliminaries

We assume simple, undirected, connected graphs $G = (V, E)$ where $|V| = n$. The neighbors of a vertex v are denoted as the set $N(v)$, and the neighbors of a set $S \subseteq V$, $N(S) = \bigcup_{v \in S} N(v) \setminus S$.

The induced subgraph of $S \subseteq V$ is denoted $G[S]$.

We use the simpler $G \setminus v$ to denote $G = (V \setminus \{v\}, E)$ for v a vertex and $G \setminus e$ to denote $G = (V, E \setminus \{e\})$ for e an edge. Likewise $G \setminus V'$ denotes $G = (V \setminus V', E)$ and $G \setminus E'$ denotes $G = (V, E \setminus E')$ where V' is a set of vertices and E' is a set of edges.

We say that $K_{1,s}$ is a s-star or a star of size s. P_i denotes a path of i vertices and $(i-1)$ edges.

3 Parameterized Complexity of STAR PACKING

In this section we are going to prove a series of polynomial time preprocessing rules (reduction rules) and eventually show that we can obtain a kernel of quadratic size on the parameter k for the parameterized version of $K_{1,s}$-packing.

We use the following natural parametrization of $K_{1,s}$-PACKING:

k-$K_{1,s}$-PACKING
INSTANCE: Graph $G = (V, E)$
PARAMETER: k
QUESTION: Are there k vertex disjoint instances of $K_{1,s}$ in G?

To remove vertices of high degree and remove useless edges between vertices of low degree we introduce the following reduction rules.

Lemma 2. *Let G be a graph such that $\exists v \in V, deg(v) > k(s+1) - 1$. G has a k-$K_{1,s}$-packing if and only if $G' = G \setminus v$ has a $(k-1)$-$K_{1,s}$-packing.*

Proof. If G has a k-$K_{1,s}$-packing then it is obvious that G' has a $(k-1)$-$K_{1,s}$ as v cannot participate in two different stars.

If G' has a $(k-1)$-$K_{1,s}$-packing we can create a k-$K_{1,s}$-packing by adding v. The $k-1$ stars already packed cannot use more than $(s+1)(k-1)$ of v's neighbors, leaving s for v.

Lemma 3. *Let G be a graph where $\exists u, v \in V(G), uv \in E(G)$ and $deg(u) \le deg(v) < s$. G has a k-packing if and only if $G' = G \setminus uv$ contains a k-packing.*

Proof. If G has a k-$K_{1,s}$-packing then it is obvious that G' has a k-$K_{1,s}$-packing as uv can never participate in a $K_{1,s}$.

If G' has a k-$K_{1,s}$-packing it is obvious that G has a k-$K_{1,s}$-packing as well.

To prove that the fixed parameter version of k-STAR PACKING is tractable we will use a new technique first seen in [FM+00]. This technique borrows ideas from extremal graph theory. In essence we look for a minimum counterexample represented by a witness structure and with the aid of reduction rules produce a series of structural claims that help us find a bound on the size of the instance produced by those reduction rules. This technique has proven to be a very potent tool in FPT -algorithm design, both because of its strength and also because of its ease of use.

We try to establish a bound where it is guaranteed that any graph larger than this bound has a k-$K_{1,s}$-PACKING, thus implying a kernel for the problem. We will prove this fact, labelling it *kernelization lemma*, at the end of this section.

Let a graph be *reduced* when lemmas 2 and 3 are no longer applicable. In this sense both these lemmas will be commonly referred to as *reduction rules*.

Lemma 4. *(Boundary Lemma)*
  If a graph G is reduced and has a k-$K_{1,s}$-packing, but no $(k + 1)$-$K_{1,s}$-packing then $|V(G)| \le k(s^3 + ks^2 + ks + 1)$.

Proof. Assume there exists a counterexample G, such that G is reduced and contains a k-$K_{1,s}$-packing W, but no $(k+1)$-$K_{1,s}$-packing and size $|V(G)| > k(s^3+ks^2+ks+1)$.

Let Q be $V \setminus W$. Let Q_i be the vertices in Q that have degree i in the subgraph induced by Q. We will now prove a series of claims that bound the number of vertices in Q.

Claim 1. $\forall i \ge s, Q_i = \emptyset$
  Proof of Claim 1. This is clear as otherwise W could not be maximal.
  □

Claim 2. A $K_{1,s}$-star $S \in W$ has at most $s^2 + k(s + 1) - 1$ neighbors in Q.

  Proof of Claim 2. Assume in contradiction that there is a reduced graph with a star S that has more than $s^2 + k(s+1) - 1$ neighbors in Q. Let $v \subseteq S$ be the vertex in S that has the highest number of neighbors in Q. From Lemma 2 we have that v has at most $k(s+1) - 1$ neighbors. Let u be the vertex in $S \setminus v$ that has the highest number x of vertices in $Q \setminus N(v)$, it is clear that $x < s$ otherwise u and v can form two new stars, contradicting the assumption that G has no $(k+1)$-packing. From this we conclude that at most one vertex in S has s or more neighbors in Q. This gives us that each star in W has at most $s^2 + k(s + 1) - 1$ neighbors in Q.   □

Claim 3. W has at most $k \cdot (s^2 + k(s + 1) - 1)$ neighbors in Q.

This follows from Claim 2.

Let $R = V \setminus (W \cup N(W))$ i.e. the set of vertices of Q which do not have neighbors in W.

Claim 4. R is an independent set in G.

> *Proof of Claim 4.* Assume in contradiction that there exists two vertices $u, v \in R$ that have an edge $uv \in E$ between them. By Claim 1 we know that both u and v have degree less than s, but then by Lemma 3 we know that uv has been reduced, contradicting the assumption. □

Claim 4 ensures us that all vertices in R have an edge to one or more vertex in Q. By Claim 1 we know that each of the vertices in $Q \setminus R$ have at most $s - 1$ such neighbors and thus by Claim 3 we know that the total size of R is at most $(s - 1) \cdot |Q \setminus R|$.

In total, G has size $|V(G)| = |W| + |Q| \leq k(s + 1) + s \cdot k \cdot (s^2 + k(s + 1) - 1) = k(s^3 + ks^2 + ks + 1)$ contradicting the assumption that the graph had more than $k(s^3 + ks^2 + ks + 1)$ vertices. This concludes the proof of the boundary lemma.   □

Having proved the boundary lemma we can now state that any reduced instance that is still 'big' has a k-$K_{1,s}$-packing. Since the boundary given by the Lemma 4 does not depend on the main input, but only on the parameter and the problem in question. We can say that the reduced instance is a 'problem-kernel' and that the problem is in FPT .

Lemma 5. *(Kernelization Lemma) If a graph G is reduced and has $|V(G)| > k(s^3 + ks^2 + ks + 1)$, then it contains a k-$K_{1,s}$-packing.*

Proof. Assume in contradiction to the stated theorem that there exists a graph G of size $|V(G)| > k(s^3 + ks^2 + ks + 1)$, but where G has no k-$K_{1,s}$-packing.

Let $k' < k$ be the largest k' for which G is a YES-instance. By the Boundary Lemma 4 we know that $|V(G)| \leq k'(s^3 + k's^2 + k's + 1) < k(s^3 + ks^2 + ks + 1)$. This contradicts the assumption.   □

Thus for any k-$K_{1,s}$-packing we can prove a quadratic kernel. However, for the special case $s = 2$, we can improve on this. This is the topic of the next section.

4   The Special Case of P_3: A Linear Kernel

A 2-star can also be seen as a path with three vertices, denoted P_3. For this special case we can employ a different set of reduction rules to obtain a linear kernel for packing P_3's into a graph.

k-P_3-PACKING
INSTANCE: Graph $G = (V, E)$
PARAMETER: k
QUESTION: Are there k vertex disjoint instances of P_3 in G?

To improve on the quadratic kernel obtained in the previous section, we will make use of a series of reduction rules based on the ideas of crown decompositions [CFJ03].

Definition 2. *A* crown decomposition (H, C, R) *in a graph* $G = (V, E)$ *is a partition-ing of the vertices of the graph into three sets* H, C, *and* R *that have the following properties:*

1. H *(the head) is a separator in* G *such that there are no edges in* G *between vertices belonging to* C *and vertices belonging to* R.
2. $C = C_u \cup C_m$ *(the crown) is an independent set in* G.
3. $|C_m| = |H|$, *and there is a perfect matching between* C_m *and* H.

There are several recent papers that use crown decompositions of graphs to obtain good results in parameterized complexity [CFJ03, FHRST04, F03, ACFL04, PS04]. [CFJ03, ACFL04] solving VERTEX COVER can use the crown structure directly. The others have to either modify the input graph or create an auxiliary graph where a crown decomposition will lead to a reduction in the size of the input graph.

In this paper we instead modify the crown decomposition to fit our particular prob-lem. The first variation is *'double crown'*-decomposition[3] where each vertex in H has two vertices from C matched to it (as opposed to only one).

Definition 3. *A double crown decomposition* (H, C, R) *in a graph* $G = (V, E)$ *is a partitioning of the vertices of the graph into three sets* H, C, *and* R *that have the following properties:*

1. H *(the head) is a separator in* G *such that there are no edges in* G *between vertices belonging to* C *and vertices belonging to* R.
2. $C = C_u \cup C_m \cup C_{m2}$ *(the crown) is an independent set in* G.
3. $|C_m| = |H|$, $|C_{m2}| = |H|$ *and there is a perfect matching between* C_m *and* H, *and a perfect matching between* C_{m2} *and* H.

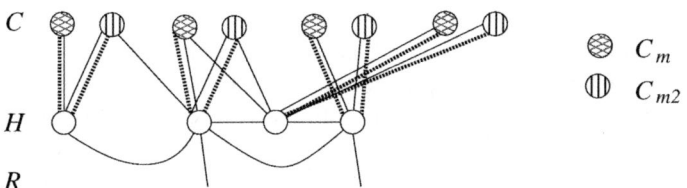

Fig. 1. Example of 'double crown'

Another variation of the crown is the *'fat crown'*-decomposition[4] where instead of independent vertices in C we have K_2's as shown in figure 2.

---

[3] The dashed lines in the figure indicate how each vertex in H is matched to two vertices in C.

[4] As in the case of the 'double crown', the dashed lines indicate the matching between H and C_m and the dashed ellipses show which K_2 the vertex in H is matched to.

Definition 4. *A fat crown decomposition* (H, C, R) *in a graph* $G = (V, E)$ *is a partitioning of the vertices of the graph into three sets* H, C *and* R *that have the following properties:*

1. H *(the head) is a separator in* G *such that there are no edges in* G *between vertices belonging to* C *and vertices belonging to* R.
2. $G[C]$ *is a forest where each component is isomorphic to* K_2
3. $|C| \geq |H|$, *and if we contract the edges in each* K_2 *there is a perfect matching between* C *and* H.

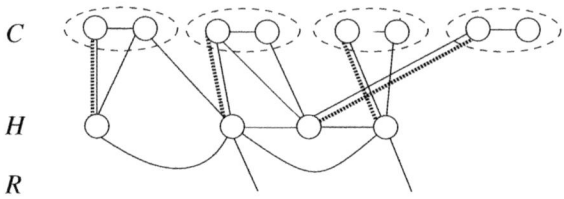

Fig. 2. Example of 'fat crown'

Using the 'crown', 'double crown' and 'fat crown' we can create powerful reduction rules.

Lemma 6. *A graph* $G = (V, E)$ *that admits a 'double crown'-decomposition* (H, C, R) *has a* k-P_3-*packing if and only if* $G \setminus (H \cup C)$ *has a* $(k - |H|)$-P_3-*packing.*

Proof. ($\Leftarrow$:) If $G \setminus (H \cup C)$ has a $(k - |H|)$-P_3-packing then it is obvious that G has a k-P_3-packing as $H \cup C$ has a $|H|$-P_3-packing ($v \in H$ and v's matched vertices from C_m and C_{m2} form a P_3).

($\Rightarrow$:) We want to prove that if G has a k-P_3-packing then $G \setminus (H \cup C)$ has a $(k - |H|)$-P_3-packing. Assume in contradiction that there exists a graph G' that has a crown-decomposition (H', C', R') that contradicts the lemma. This implies that $H' \cup C'$ participates in $x > |H'|$ P_3's. Since H' is a cutset, and C is an independent set in the graph, every P_3 in G that has vertices in $H' \cup C'$ must have at least one vertex from H'. Thus we can have at most $|H'|$ P_3's which is a contradiction. $\qquad\square$

Lemma 7. *A graph* $G = (V, E)$ *that admits a 'fat crown'-decomposition* (H, C, R) *has a* k-P_3-*packing if and only if* $G \setminus (H \cup C)$ *has a* $(k - |H|)$-P_3-*packing.*

The proof of Lemma 7 is analogue to Lemma 6, thus omitted.

To apply crown-decompositions we need to know when we can expect to find one. A very useful result in this regard can be deducted from [CFJ03, page 7], and [F03, page 8]. Fortunately, the results also apply to the variations of crown decomposition described here.

Lemma 8. *Any graph G with an independent set I, where $|I| \geq |N(I)|$, has a* crown *decomposition (H, C, R), where $H \subseteq N(I)$, that can be found in linear time, given I.*

Corollary 1. *Any graph G with a collection J of independent K_2s such that $|N(J)| \leq |J|$, has a* fat crown *decomposition (H, C, R), where $H \subseteq N(I)$, that can be found in linear time, given J.*

Proof. This follows from the previous Lemma. If we replace each K_2 with a single vertex, then by Lemma 8 this graph admits a crown-decomposition. We can reintroduce the K_2s to obtain a 'fat-crown'. □

Lemma 9. *Any graph G with an independent set I, where $|I| \geq 2 \cdot |N(I)|$, has a double crown decomposition (H, C, R), where $H \subseteq N(I)$, that can be found in linear time, given I.*

Proof. Let G be a graph with an independent set $I \subseteq V(G)$ such that $|N(I)| \leq |I|$. Create a graph with $G' = G$, but for every vertex $v \in N(I)$ add a copy v', such that $N(v) = N(v')$. By Lemma 8 G' has a crown-decomposition (H, C, R) such that $H \subseteq N_{G'}(I)$. We now claim that we can use this crown to construct a 'double crown' (H', C', R') in G.

First observe that $v \in H$ if and only if $v' \in H$. Assume in contradiction that $v' \in H$ but $v' \notin H$. v must be matched to some vertex u in C. Since $N(v) = N(v')$ we have that v' cannot be in C as it would contradict that C is an independent set. Also v' is not in R as that would contradict that H is a cut-set. Thus v' must be in H, contradicting the assumption.

With this observation the result follows easily as H consists of pairs of vertices, a vertex and its copy. Each pair v and v' in H is matched to two vertices u_1 and u_2. In G, let v be in H' and let it be matched to both u_1 and u_2. Do this for every pair in H. It is easy to see that this forms a double crown in G. □

We will now describe a polynomial time preprocessing algorithm that reduces the graph to a kernel of size at most $15k$.

```
Step 1. Compute an arbitrary maximal P₃-packing
        W. Let Q = V \ W.
Step 2. Let X be the components in G[Q]
        isomorphic to K₂. If |X| ≥ |N(X)| in G
        then reduce by Lemma 7.
Step 3. Let I be the isolated vertices I in G[Q].
        If |I| ≥ 2|N(I)| in G then reduce by Lemma
        6.
```

Lemma 10. *If $|V(G)| > 15k$ then the preprocessing algorithm will either find a k-P_3-packing or it will reduce G.*

Proof. Assume in contradiction to the stated lemma that $|V(G)| > 15k$, but that the algorithm produced neither a k-P_3-packing nor a reduction of G.

By the assumption the maximal packing W is of size $|W| < 3k$. Let $Q = V \setminus W$. Let Q_i be the vertices in Q that have degree i in the graph induced by Q.

Claim 5. $\forall i \geq 2, Q_i = \emptyset$

Proof. This is clear as otherwise W could not be maximal.   □

Claim 6. $|Q_1| \leq 6k$

Proof. Assume in contradiction that $|Q_1| > 6k$. This implies that number of K_2s X in Q is greater than $3k$, but then $|X| > |W|$. By Corollary 1 G has a 'fat crown' and should have been reduced in step 2 of the algorithm, contradicting that no reduction took place.   □

Claim 7. $|Q_0| \leq 6k$

Proof. Assume in contradiction that $|Q_0| > 6k$, but them $|Q_0|$ is more than $2|W|$ and by Lemma 9 G has a 'double crown' and by Lemma 6 should have been reduced in step 3 of the algorithm, contradicting that no reduction took place.   □

Thus the total size $|V(G)| = |W| + |Q_0| + |Q_1| + |Q_2| + \ldots \leq 3k + 6k + 6k + 0 = 15k$. This contradict the assumption that $|V(G)| > 15k$.

     □

Corollary 2. *Any instance* (G, k) *of* P_3-*packing can be reduced to a problem kernel of size* $\mathcal{O}(k)$.

Proof. This follows from the Lemma, as we can run the preprocessing algorithm until it fails to reduce G. By Lemma 10 the size is then at most $15k$.   □

5   Running Time

We will apply a straightforward brute-force algorithm on the kernels to find the optimal solution. In the case of P_3-packing we will select the center-vertices of the P_3s in a *brute force* manner. There are $\binom{15k}{k}$ ways to do this. By Stirling's formula this is approximately $2^{5.3k}$. With k-center vertices already selected the problem reduces to a bipartite problem where the question is if the left hand side each can have 2 neighbors assigned to it. This can easily be transformed to MAXIMUM BIPARTITE MATCHING by making 2 copies of each vertex on the left hand side. MAXIMUM BIPARTITE MATCHING can be solved in time $\mathcal{O}(\sqrt{|V|}|E|)$ [HK73]. Since we now have $15k + k$ vertices, and thus $\mathcal{O}(k^2)$ edges. We can solve each of these in time $\mathcal{O}(k^{2.5})$. Giving a total running time of $\mathcal{O}(2^{5.3k}k^{2.5})$, or $\mathcal{O}^*(2^{5.3k})$ when written in $\mathcal{O}^*$ notation.

Applying the same technique for the s-stars we will achieve $\mathcal{O}^*(2^{\mathcal{O}(k \log k)})$, asymptotically worse due to the quadratic kernel.

6   Conclusions and Further Research

Packing vertex-disjoint copies of a graph H into another graph G is NP-complete as long as H has more than two vertices [HK78]. We have analyzed within the framework of parameterized complexity a specific instance of this problem, the packing of vertex-disjoint stars with s leaves. We have proved that packing $K_{1,2}$s in a graph G, equivalently k-P_3-PACKING has a linear kernel.

Our algorithm for k-P_3-Packing runs in time $\mathcal{O}^*(2^{5.3k})$. This running time arises from reducing the problem to a kernel of size $15k$. We believe that this kernel can be further improved and thus the running time substantially decreased, however, it is already much better than $2^{\mathcal{O}(|H|k\log k + k|H|\log|H|)}$, the running time of the general algorithm in [FHRST04].

We have also proved that s-Star Packing ($K_{1,s}$-Packing) is in general fixed-parameter tractable with a quadratic kernel size. We proved a running time for this general case of $\mathcal{O}^*(2^{\mathcal{O}(k\log k)})$.

An open problem here is whether it is possible to obtain a linear sized kernel for s-Star packing for any $s \geq 3$ such as the one obtained in the case of k-P_3-PACKING.

There are several related problems that could be considered on the light of the techniques used in Section 3. The most obvious one is the following:

k-$K_{1,s}$-PACKING
INSTANCE: Graph $G = (V, E)$
PARAMETER: k
QUESTION: Are there k edge-disjoint instances of $K_{1,s}$ in G?

This problem is fixed-parameter tractable when $s = 2, 3$ using Robertson and Seymour's Graph Minor Theorem [RS99] since it can be easily proved that its NO-instances are closed under the minor operations. The issue here is that this method is non-constructive and carries a fast growing function $f(k)$. Possibly, applying similar arguments as those in Section 4 would lead to a much better running time.

Acknowledgements. We would like to thank Mike Fellows for all the inspiring conversations leading to the completion of this paper.

References

[ACFL04] F. Abu-Khzam, R. Collins, M. Fellows and M. Langston. Kernelization Algorithms for the Vertex Cover Problem: Theory and Experiments. *Proceedings ALENEX 2004*, Springer-Verlag, *Lecture Notes in Computer Science* (2004), to appear.

[AC99] G. Ausiello, P. Crescenzi, G. Gambosi, V. Kann, A. Marchetti-Spaccamela, M. Protasi. Complexity and Approximation *Springer Verlag* (1999).

[BM02] R. Bar-Yehuda, M. Halldrsson, J. Naor, H. Shachnai, I. Shapira. Scheduling Split Intervals. *Proceedings of the Thirteenth Annual ACM-SIAM Symposium on Discrete Algorithms*, pages 732-741 (2002).

[BK01] R. Bejar, B. Krishnamachari, C. Gomes, and B. Selman. Distributed constraint satisfaction in a wireless sensor tracking system. *Workshop on Distributed Constraint Reasoning, International Joint Conference on Artificial Intelligence*, 2001

[CFJ03] B. Chor, M. Fellows, D. Juedes. An Efficient FPT Algorithm for Saving k colors. *Manuscript* (2003).

[DFS00] R. Downey, M. Fellows, U. Stege. Parameterized Complexity: A Framework for Systematically Confronting Computational Intractability. *AMS-DIMACS Series in Discrete Mathematics and Theoretical Computer Science*, Volume 49, pages 49-99 (1999).

[F03] M. Fellows. Blow-Ups, Win/Win's, and Crown Rules: Some new Directions in FPT . *Proceedings 29th Workshop on Graph Theoretic Concepts in Computer Science* (2003).

[FHRST04] M.Fellows, P.Heggernes, F.Rosamond, C. Sloper, J.A.Telle, Exact algorithms for finding k disjoint triangles in an arbitrary graph, *To appear WG2004*

[FM+00] M.R. Fellows, C. McCartin, F. Rosamond, and U.Stege. Coordinatized Kernels and Catalytic Reductions: An Improved FPT Algorithm for Max Leaf Spanning Tree and Other Problems, *Foundations of Software Technology and Theoretical Computer Science*, (2000).

[HK73] J. Hopcroft and R. Karp. An n 5=2 Algorithm for Maximum Matchings in Bipartite Graphs. *SIAM Journal on Computing*, 2 pages 225–231 (1973).

[HK78] P. Hell and D. Kirkpatrick. On the complexity of a generalized matching problem. *Proceedings of 10th ACM Symposium on theory of computing*, pages 309–318 (1978).

[HK86] P. Hell and D. Kirkpatrick. Packings by complete bipartite graphs. *SIAM Journal of Algebraic Discrete Methods*, number 7, pages 199–209 (1986).

[HS89] C. Hurkens and A. Schrijver. On the size of systems of sets every t of which have an SDR, with application to worst case ratio of Heuristics for packing problems. *SIAM Journal of Discrete Mathematics*, number 2, pages 68–72 (1989).

[K94] V. Kann. Maximum bounded H-matching is MAX-SNP-complete. *Information Processing Letters*, number 49, pages 309–318 (1994).

[PS04] E.Prieto, C. Sloper. Creating Crown Structure — The case of Max Internal Spanning Tree. *Submitted*

[RS99] N. Robertson, PD. Seymour. *Graph Minors XX. Wagner's conjecture*, to appear.

Moving Policies in Cyclic Assembly-Line Scheduling*

Matthias Müller-Hannemann[1] and Karsten Weihe[2]

[1] Darmstadt University of Technology, Hochschulstraße 10, 64289 Darmstadt,
Germany
muellerh@algo.informatik.tu-darmstadt.de
[2] The University of Newcastle, Callaghan, NSW 2308, Australia
weihe@cs.newcastle.edu.au

Abstract. We consider an assembly line problem that occurs in various
kinds of production automation, for example, in the automated manufacturing of PC boards[3]. The assembly line has to process a (potentially
infinite) number of identical workpieces in a cyclic fashion. In contrast to
common variants of assembly–line scheduling, the forward steps may be
smaller than the distance of two stations. Therefore, each station may
process parts of several workpieces at the same time, and parts of a
workpiece may be processed by several stations at the same time. The
throughput rate is determined by the number of (cyclic) forward steps,
the offsets of the individual forward steps, and the distribution of jobs
over the stationary stages between the forward steps. Even for a given
number of forward steps and for given offsets of the forward steps, the
optimal assignment of the jobs to the stationary stages is at least weakly
$\mathcal{NP}$–hard.
We will base our algorithmic considerations on some quite conservative
assumptions, which are greatly fulfilled in various application scenarios,
including the one in our application: the number of jobs may be huge, but
the number of stations and the number of forward steps in an optimal
solution are small, the granularity of forward steps is quite coarse, and
the processing times of the individual items do not differ by several
orders of magnitude from each other. We will present an algorithm that is
polynomial and provably deviates from optimality to a negligible extent
(under these assumptions). This result may be viewed as an application
of fixed–parameter tractability to a variety of real–world settings.
Keywords: assembly line balancing, printed circuit boards, fast component mounters, stepping problem, cycle time minimization, fixed-parameter analysis

1 Introduction

Problem description A (potentially infinite) number of identical workpieces is to
be processed by an assembly line. Several workpieces are simultaneously in the

---

* This work has been partially supported by DFG grant MU 1482/2.
[3] In cooperation with Philips/Assembléon B.V., Eindhoven, the Netherlands.

R. Downey, M. Fellows, and F. Dehne (Eds.): IWPEC 2004, LNCS 3162, pp. 149–161, 2004.

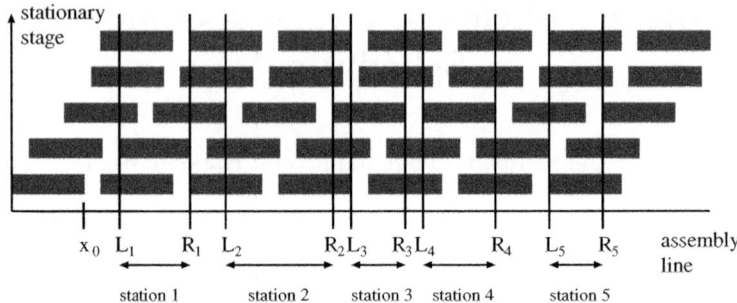

Fig. 1. An assembly line with five stations. In this example, the (cyclic) stepping scheme consists of four stationary stages (the fifth stage is identical with the first stage).

assembly line, and the offset of two workpieces is fixed. The stations (machines) of the assembly line must perform a set of specific jobs on each workpiece. Jobs have to be processed without preemption. Each job has a certain location on the workpiece. A job can only be performed if it is in the visibility region of the station on which it shall be executed. The visibility region of station i is an interval $[L_i \ldots R_i]$ of the assembly line. See Figure 1. The process decomposes into stationary stages, in which the assembly line does not move forward. After a stationary stage is finished, the band (and thus all workpieces) are moved forward by some common offset, and the next stationary stage commences.

In many application scenarios of flexible manufacturing the described model of an assembly line is the appropriate variant. For us, this problem arose as a subproblem in a cooperation with Philips/Assembléon B.V., Eindhoven, the Netherlands. Here the workpieces are printed boards. Each station holds a robot arm, which places various components on the boards. Jobs correspond to mounting steps: picking a component from a feeder; moving from the pick–up position to the position of this mounting job; mounting the component; moving back to the feeder for the next mounting job.

In the literature, typically, the following special case is considered: each workpiece moves from station to station, each station may access and process exactly one workpiece in a stationary stage (all its jobs on this workpiece in fact), and each workpiece can only be processed by one station at a time.

Motivated by our concrete application, the scenario of *this* paper is a bit different: the stations are very close together, and forward steps may be relatively small. Therefore, one workpiece may be processed by several stations in the same stationary stage, and one station may process several workpieces in the same stationary stage. See Figure 1 again. All workpieces must be processed by an identical schedule. In particular, each workpiece must exactly take over the position of its immediate predecessor after a certain number s of forward steps.

With respect to an initial start configuration, the movement of the assembly line is completely described by a *stepping scheme*, namely by the overall number

s of forward steps and the offsets of the individual forward steps. The stepping scheme determines the positions of the workpieces for each stationary stage. Due to the restricted visibility region of each station, a job can usually only be performed on a subset of the stationary stages.

Now the optimization problem is this: we have to determine (i) a stepping scheme and (ii) for each job an assignment to a stationary stage where the location of this job on the workpiece is visible to the station of this job. We call the stepping scheme together with such an assignment a (cyclic) *assembly-line schedule*. The objective is to minimize the *cycle time* of the schedule, that is, the time for one cycle of the process. This means the sum of the process times of all s stationary stages plus the time for moving the assembly line between two stationary stages (and between the last and first stationary stage of a cycle).

Previous work Scholl [Sch99] surveys general problems and approaches to the balancing and sequencing of assembly lines. Ayob et. al. [ACK02] compare different models and assembly machine technologies for surface mount placement machines. A recent overview on printed circuit board assembly problems has been given by Crama et al. [CKS02].

Assigning jobs to stationary stages seems to resemble batching problems on m machines which appear in several versions but different from the one considered in this paper [Bru97]. The most important differences to our problem are the definitions of the objective function and of the completion times of jobs. In the literature, the completion time of a job equals the completion time of the whole corresponding batch. Moreover, the assignment to machines is usually not given.

There is some engineering work on this kind of assembly-line problems [4]. However, we are not aware of any *mathematical* work which considers the problem of our paper (or variants thereof), although the problem might be quite natural and definitely occurs in practice in the form presented here.

Spirit of the paper: Our main research topic is the application of *fixed–parameter tractability* [DF99, AGN01, Fel02] to real–world optimization problems from Operations Research and the investigation of concrete problems and techniques. In this paper, we present a result of this project. Roughly speaking, a problem is fixed–parameter tractable with respect to a given set of parameters, if there is an algorithm for this problem whose run time is polynomial in the input size in case the values of these parameters are bounded. The literature on fixed–parameter tractability is mainly concerned with the analysis of "classical" problems such as those in [GJ79] with respect to (more or less) natural parameters. Here, we apply fixed–parameter tractability to real–world problems in a systematic, conscious effort.[5] The challenge in applied work is different because we are not free in the

---

[4] Internal reports of manufacturers like Assembléon.

[5] The underlying intuitive ideas have been significantly influencing applied work in virtually all application domains. Therefore, systematic approaches like ours might be valuable. However, we are not aware of any work of this kind, except that the research on "classical" problems includes some basic problems that are related to bioinformatics [BDF$^+$95].

choice of the parameters. In fact, the challenge is to find parameters that not only allow efficient algorithms but also turn out to have small values in typical instances from an application or from a variety of applications.

Our technique is different from the common techniques like, for example, *kernelization*[6] and search trees which are typically applied to derive fixed-parameter results. Here we will see that another approach is more promising for our optimization problem.

Specific contribution of the paper: Empirically, processing times of individual mounting jobs differ by not too much. The maximum processing time of a single job is typically very small in comparison with the cycle time.

In practice, the offsets of the forward moving steps cannot become arbitrarily small. More precisely, the size of an offset is bounded from below and is a multiple of some parameter Δ. For example, for printed circuit board machines as developed by Philips/Assembléon, the precision is about $\Delta \approx .5mm$.

The parameters that turn out to have bounded values are these: (1) the number m of stations, (2) the total number s of forward steps in an optimal solution, (3) the ratio q_1 of the offset of two workpieces divided by the granularity of forward steps, (4) the ratio q_2 of the maximal divided by the minimal processing times of jobs, and (5) $q_3 := s^2 \cdot m/2N$, where N denotes the total number of jobs. The first three parameters will bound the run time, and the last two parameters will bound the error. More specifically, the run time will be $\mathcal{O}((N + m^3 s^3) \cdot \min\{Ns^{q_1}, N^s\})$, and the relative error will be bounded by $q_2 \cdot q_3$. Note that the relative error vanishes as the number N of jobs increases. As a by–product, it will turn out that our algorithm is at least 2–approximative without assumptions on the parameters q_2 and q_3, and that it can be transformed to a PTAS. For a given relative error $\varepsilon > 0$, the run time of the PTAS algorithm is $\mathcal{O}((N + m^3 s^3) \cdot s^{ms/2\varepsilon})$.

Discussion: According to the "spirits" paragraph above, these five parameters were not chosen arbitrarily, but in fact they were chosen because quite often in practice the values of these five parameters might be *very* small. For example, in the application that motivated our work, m is in the order of $10 \ldots 20$, s is in the order of $3 \ldots 5$, q_1 is in the order of dozens or hundreds, q_2 is close to 1, and q_3 is orders of magnitude smaller than 1 because N is huge. Therefore, the relative error is negligible under the (quite realistic) assumptions that the processing times of the jobs do not differ by orders of magnitude from each other, and that the total number of N jobs is orders of magnitude larger than the number of forward steps.

In practice, $\min\{Ns^{q_1}, N^s\}$ might typically be assumed at the second term, N^s. However, the first term plays a role in theory: the fact that the run time is $\mathcal{O}((N + m^3 s^3) \cdot Ns^{q_1})$ means that the problem is fixed–parameter tractable in the sense of [DF99, AGN01, Fel02] if a relative error of $q_2 \cdot q_3$ is admitted.

---

[6] It can even be proved [DFS99] that every fixed–parameter result can be polynomially transformed into a kernelization result.

Remark: Additional precedence constraints among jobs are quite common in practice. Therefore, it is worth noting that our algorithm can cope with precedence constraints, and the asymptotic run time remains the same.

Organization: In Section 2, we reduce our problem to a certain weakly NP-hard core problem which has to be applied several times. We prove error bounds on a relaxed version of this core problem and derive a PTAS. In the second part of this section, we develop an efficient "core algorithm" for the relaxed version. Section 3 is devoted to the correctness proof of this core algorithm, whereas Section 4 achieves bounds on the number of core problems to be considered.

2   The Algorithm

In the following, jobs are denoted by J, J_1, J_2, etc. For a job J, $\ell[J]$ and $u[J]$ describe the intervals of stationary stages to which J could be feasibly assigned subject to the visibility constraints. The intervals $[\ell[J], u[J]]$ are to be interpreted as *wrap-around intervals*, that is, if $u < \ell$, the notation $[\ell, u]$ is a shorthand for $[\ell, u] = \{\ell, \ell + 1, \ldots, s, 1, \ldots, u - 1, u\}$.

2.1   Reduction to a Core Problem

The core problem requires $\mathcal{O}(N + m^3 s^3)$ time, and the algorithm for the core problem is applied $\mathcal{O}(\min\{N s^{q_1}, N^s\}))$ times. Both parts together give the total run time claimed above. The core problem differs from the original problem in two respects:

1. The stepping scheme is fixed. In other words, it remains to compute the assignments of jobs to stationary stages. (Recall that the stepping scheme comprises the number s of forward steps and the offsets of the individual forward steps.)
   Already this assignment problem of jobs to stationary stages can be shown to be weakly $\mathcal{NP}$-hard by a straightforward reduction from PARTITION (the decision version of which is well-known to be weakly $\mathcal{NP}$-complete [GJ79]).
2. Due to this hardness result, the problem is slightly relaxed in the following way: we do not necessarily assign each job to a single stationary stage. Instead, a stationary stage may be finished while a job is being processed, so this job is resumed at the beginning of the very next stationary stage. In a postprocessing at the end of the algorithm, each such job is then assigned to exactly one stationary stage to which a part of the job was assigned by the core algorithm.

Fixing the stepping scheme amounts to enumerating all possible stepping schemes. In Section 4 we will show that it suffices to consider $\mathcal{O}(\min\{N s^{q_1}, N^s\})$ stepping schemes as claimed above.

Linear job orderings: For each machine, we can define an *interval-induced partial order* on the jobs. In this partial order, job J_1 is a predecessor of job J_2 if and only if either $\ell[J_1] < \ell[J_2]$ or $\ell[J_1] = \ell[J_2]$ and $u[J_1] < u[J_2]$. Visibility conditions in an assembly line ensure that no interval of a job is strictly contained in that of another job. Hence, two jobs J_1, J_2 are incomparable if and only if $\ell[J_1] = \ell[J_2]$ and $u[J_1] = u[J_2]$.

Note that in the fractional model the precise order of two incomparable jobs is irrelevant for the overall cycle time. Hence, we may simply fix one linear extension of the interval-induced partial order for each machine. We call such an extension an *interval-consistent linear order*. For notational simplicity, we assume that jobs are numbered with respect to this order, i.e. $J_{i,1} \rightarrow J_{i,2} \rightarrow \cdots \rightarrow J_{i,k_i}$ on machine i.

Simplified instances: In the relaxed model, we may treat all jobs with the same interval of feasible stages as one larger job. Hence, a simplified instance will have at most $O(s^2)$ jobs on each machine, for a total of $O(ms^2)$ jobs over all machines.

Error bound: The relaxation introduces an error. The following lemma will show that this error can be reasonably bounded. As discussed above, $q_2 \cdot q_3$ may be assumed to be orders of magnitude smaller than 1.

Lemma 1. *The relative error is at most* $\min\{1, q_2 \cdot q_3\}$.

Proof. Obviously, the *absolute* error is bounded by $s \cdot p_{max}$, where p_{max} denotes the maximal job length. First we will show that the absolute error is even bounded by $s \cdot p_{max}/2$. This follows from the before-mentioned postprocessing to convert a relaxed solution to an integral one. Namely, we decide iteratively for each stage where to assign the splitted jobs. For stage i, let $split(i)$ denote the set of splitted jobs which start in stage i and are continued in stage $i + 1$. Let $p[J]$ be the length of a job and $p_i[J]$ be the amount of time that job J is executed within stage i. Define

$$T_1 := \max\{p_i[J] \mid J \in split(i)\} \text{ and } T_2 := \max\{p[J] - p_i[J] \mid J \in split(i)\}.$$

If $T_1 \geq T_2$, then we assign all jobs in $split(i)$ completely to stage i (which implies that the length of stage i increases by T_2). Otherwise, we shift for all jobs in $split(i)$ the portion which is executed in stage i to stage $i + 1$. In this case, the length of stage $i + 1$ increases by at most T_1. (The increase might be strictly shorter if a shifted job fills up idle time.)

Let CT_{frac} denote the optimal cycle time with respect to the relaxed model, and let CT_{app} be the cycle time of the approximation which we obtain by the above conversion routine. In each rounding step, the makespan of a stage can at most double. Hence, we immediately get

$$CT_{app} \leq 2 \cdot CT_{frac}.$$

Thus, we obtain at least a 2–approximation for the minimum cycle time, and it remains to show that the relative error is bounded by $q_2 \cdot q_3$. Denote by P the sum of processing times over all jobs, and by p_{max} and p_{min} the maximal and minimal processing time over all jobs, respectively. Clearly, we also get

$$CT_{app} \leq CT_{frac} + \frac{s}{2} p_{max}.$$

To bound the *relative* error, we bound the optimal cycle time from below by the ratio of the total sum of all job lengths divided by m. The total sum of all jobs is the product of N and the average job lengths. Replacing the average job length by the minimal job length eventually gives an upper bound of $q_2 \cdot q_3$ for the relative error as claimed in the introduction.

If the number of machines m and the number of stages s are both fixed constants (that is, they are not part of the input), then we can easily derive a polynomial approximation scheme.

We use the standard trick to divide jobs into long and small jobs. Denote by LB a lower bound for the cycle time CT. As a lower bound, we may simply take the maximum load of all machines or fractional solution value CT_{frac}. For a given $\varepsilon > 0$, we define that job J is a *long job* if $p[J] \geq \varepsilon \frac{2LB}{s}$, and J is a *small job*, otherwise.

Let $long_i$ be the number of long jobs on machine i. As $LB \geq \sum_{j=1}^{k_i} p[J_{i,j}] \geq long_i \frac{2\varepsilon LB}{s}$, we have at most $long_i \leq \frac{s}{2\varepsilon}$ long jobs on machine i, for a total of at most $\frac{ms}{2\varepsilon}$ on all machines.

Our approximation scheme works as follows. The idea is to fix the assignment to stages for all long jobs. For each possible assignment of long jobs to stages we iteratively solve the fractional relaxation and use the conversion routine to an integral assignment from above. The crucial observation is that our core routine can also solve the relaxed problem subject to these side constraints. The algorithm returns the best solution obtained in one of the iterations. We need at most $O(s^{\frac{ms}{2\varepsilon}})$ iterations, and each iteration runs in time $O(m^3 s^3 + N)$. Hence, as m and s are fixed, we get a polynomial algorithm.

Denote by $T_{max,i}$ the maximum processing time of a job which is only partially assigned to stage i. Since now only small jobs are splitted in the relaxed solution, we conclude for the obtained integral assignment

$$LB \leq CT_{frac} \leq CT \leq CT_{frac} + \sum_{i=1}^{s} T_{max,i}/2$$

$$\leq CT_{frac} + \frac{s}{2} \cdot \frac{2LB\varepsilon}{s}$$

$$\leq CT_{frac} + \varepsilon LB \leq CT_{frac}(1 + \varepsilon),$$

which proves the desired performance guarantee of the approximation scheme.

2.2   An Algorithm for the Core Problem

We are now going to present an $\mathcal{O}(N + m^3 s^3)$ time algorithm for the core problem.

Steps of the core algorithm: The algorithm consists of four steps. Step i computes an assignment $st_i[J]$ for every job J. More precisely, as we here allow jobs to be splitted over several consecutive stages, $st_i[J]$ denotes the stage when job J is started, whereas $st_{i,c}[J]$ refers to the stage when job J is completed.

Step 1: For each job we first choose the first stationary stage at which it can be scheduled subject to the visibility constraints, that is, we choose $st_1[J] := \ell[J]$. However, if a job can be realized at any stationary stage, we choose $st_1[J] := s$.

Step 2: Next we iterate over the stationary stages from 1 to $s - 1$ and do the following in the i–th iteration. Let $J[i]$ be the set of jobs such that $st_1[J] = i$. First we build an earliest start schedule $S[i]$ for the jobs in $J[i]$. Let $C_i[J]$ be the completion time of job J with respect to $S[i]$. Moreover, let $t[i]$ be the first moment in time where all those jobs of $J[i]$ are finished in $S[i]$ that have definitely to be finished in the i–th stationary stage (due to the visibility constraints). In other words,

$$t[i] := \max \left\{ C_i[J] \mid J \in J[i] \text{ and } u[J] = i \right\}.$$

Next we interrupt each job which remained active after $t[i]$ and shift the remaining portion to the following stationary stage $i + 1$. Similarly, each job which is started at or after $t[i]$ is shifted to stationary stage $i + 1$. This finishes the i–th iteration of the loop. The result after the last iteration is $st_2[\cdot]$.

Step 3: First we build an earliest start schedule for the stationary stage s. Then we iterate over all machines and shift jobs (or portions of them) assigned to the stationary stage s backwards (if possible) to avoid idle times in earlier steps. (Note that this will only be possible for jobs which can be assigned to every stationary stage.) The result is $st_3[\cdot]$.

Step 4: Denote by $C[i]$ the current makespan for the stationary stage i and by $C_M[i]$ the current makespan for station M in the stationary stage i. Denote by $C_i[J]$ the completion time of job J in the stationary stage i. As usual, the minimum over an empty set is defined to be $+\infty$. A station is said to be *critical* for the stationary stage i if the makespan of i is assumed at that station. Let $\mathcal{M}_{crit}[i]$ be the set of critical machines of the stationary stage i. Let ε_1 denote the maximal amount by which we can shift jobs from stationary stage s to stationary stage 1 until one more machine becomes critical for stationary stage s:

$$\varepsilon_1 := \min \left\{ C[s] - C_M[s] \mid C[s] > C_M[s] \text{ for } M \in \mathcal{M} \right\}.$$

Moreover, let ε_2 denote the maximal amount by which we can shift jobs from stationary stage s to stationary stage 1 without violating the right bounds for the assignment of any job:

$$\varepsilon_2 := \min\{C[s] - C_s[J] \mid st_3[J] = s \text{ and } u[J] = s\}.$$

For station M and stationary stage i let $last_M[i]$ be the last job which is executed on machine M within stationary stage i. For a job J on machine M denote by

$chain_f(J)$, called the *forward chain* of J, the inclusion-maximal chain of jobs $J \to J_1 \to \cdots \to J_k$ scheduled on machine M after J without any idle time. Let $last(chain_f(J)) = J_k$ be the last job within this chain. For each critical machine we consider the next available idle slot which we can fill. The value ε_3 denotes the minimum over all these slots:

$$\varepsilon_3 := \min\{C[j] - C_M[j] \mid j = st_{3,c}[last(chain_f(J))], J = last_M[s], M \in \mathcal{M}_{crit}[s]\}.$$

For each critical machine M of stationary stage s our goal is to shift the complete forward chain $chain_f(last_M[s])$ by some ε. Consider such a chain ch on machine M. If ch contains a stationary stage i and a job J with $C_M[i] = C[i]$, $last_M[i] = J$ and $u[J] = i$, then shifting is only possible if the beginning of stationary stage $i + 1$ is also shifted by ε. If, in turn, stationary stage $i + 1$ has a machine M' and a job J' with the property $J' = first_M[i + 1]$ and $\ell[J'] = i + 1$ then we also have to shift the forward chain of J'. Repeating this argument, we collect a set of chains $chset$ and a set of stationary stages $bset$ which all have to be shifted.

In this collection step, we may observe that there is a cyclic dependency, that is, there are two chains in $chset$ which end with the same job J. In such a case, the algorithm terminates immediately. We call this event *termination because of cyclic dependency*. Define

$$\varepsilon_4 := \min\{C[i] - C_i[J] \mid J \in chset, st_{3,c}[J] = i \text{ and } u[J] = i\},$$

and

$$\varepsilon_5 := \min\{S_i[J] \mid st_3[J] = i, i \in bset, \ell[J] = i \text{ and } J \notin chset\},$$

where $S_i[J]$ denotes the start time of J with respect to stationary stage i.

If $s \in bset$, redefine $\varepsilon_1 := \varepsilon_1/2$ and $\varepsilon_2 := \varepsilon_2/2$. After this potential redefinition, let $\varepsilon := \min\{\varepsilon_1, \varepsilon_2, \varepsilon_3, \varepsilon_4, \varepsilon_5\}$.

If $\varepsilon = 0$, the algorithm terminates. Otherwise, we shift from each critical machine of stationary stage s the amount of ε to stationary stage 1 and update all schedules appropriately. In particular, jobs of a stationary stage in $bset$ which do not belong to $chset$ will be shifted backwards if this is feasible and fills idle time. We repeat Step 4 until $\varepsilon = 0$. The result, $st_4[\cdot]$, is the output of the algorithm.

This concludes the description of our algorithm. See Figure 2 for a small example.

3   Correctness Proof

In Section 2.1, the correctness of the reduction was exhaustively discussed. Therefore, it remains to prove that the algorithm for the core problem is correct. The correctness proof is based on a couple of algorithmic invariants.

Invariant 1 *Throughout the algorithm, each job is legally assigned to stationary stages according to the definition of the core problem.*

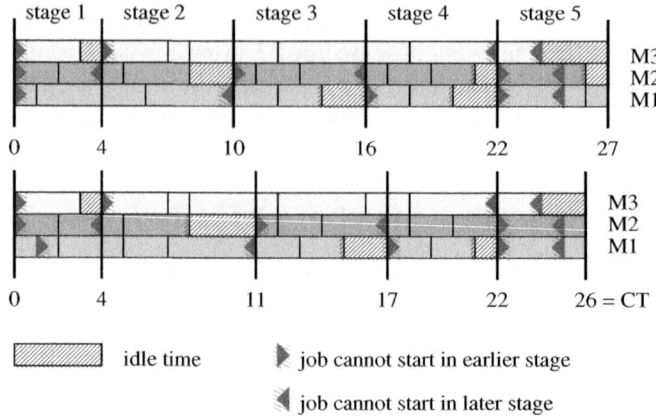

Fig. 2. A small example instance with $m = 3$ machines and $s = 5$ stationary stages. On top, we show a Gantt–chart of the situation after Step 3. Below, we show the situation after one iteration of Step 4. Note that this situation is already optimal.

Invariant 2 *A stationary stage $i \in \{1, 2, \ldots, s-1\}$ does not finish unless some job J has been completed such that $u[J] = i$.*

For a job J on machine M denote by $chain_b(J)$, called the *backward chain* of J, the inclusion-maximal chain of jobs $J_1 \to J_2 \to \cdots \to J_k \to J$ scheduled on machine M before J without any idle time. (If there is no idle time on machine M at all, the algorithm stops with an optimal solution.) Let $first(chain_b(J))$ be the first job of the backward chain $chain_b(J)$. Note that, by definition, $first(chain_b(J))$ is always the first job of some stationary stage.

Invariant 3 *After Step 3, the following holds throughout the rest of the algorithm: For a job J with $st_3[J] > \ell[J]$, we have $st_3[J_f] = \ell[J_f]$ where $J_f = first(chain_b(J))$.*

For stationary stage i, let $last(i)$ be a job which realizes its makespan and satisfies $u[last(i)] = i$. By Invariant 2, $last(i)$ always exists for $1 \le i \le s-1$. The backward chain of $last(i)$ starts with a job $J_f = first(chain_b(last(i)))$. By Invariant 3, we have $st_3[J_f] = \ell[J_f]$. Hence, the backward chain of $last(i)$ gives us a certificate that the difference between the completion time of stationary stage i and the start time of stationary stage $st_3[J_f]$ is as small as possible.

Certificates of optimality Clearly, the job assignment is optimal if there is a machine M without any idle time. A second type of certificate for optimality can be derived by combining backward chains from different machines: Let $ch_0, ch_1, \ldots, ch_k$ be chains of jobs where $ch_0 = chain_b(last(i_0))$ for some stationary stage i_0, $i_1 = first(chain_b(last(i_0))) - 1$ and $ch_1 = chain_b(last(i_1)), \ldots,$

$i_k = first(chain_b(last(i_k - 1))) - 1$ and $ch_k = chain_b(last(i_k)$. If the concatenation of these chains covers all index steps at least once then the job assignment is also optimal.

Lemma 2. *The proposed algorithm finds an optimal fractional job assignment for s stages in $O(m^3 s^3 + N)$ time, where N is the total number of jobs.*

Proof. By Invariant 1, the algorithm always yields a feasible fractional assignment. We first show that the algorithm terminates. At most $m - 1 = |\mathcal{M}| - 1$ times we can choose $\varepsilon = \varepsilon_1$ as each time we add at least one critical machine to $\mathcal{M}_{crit}[s]$ and $\mathcal{M}_{crit}[s]$ never decreases. Only once it can happen that $\varepsilon = \varepsilon_2$ as afterwards we have a certificate of optimality. At most $m(s - 1)$ times we can choose $\varepsilon = \varepsilon_3$ directly after another as each time one idle slot vanishes. As long as $\mathcal{M}_{crit}[s]$ remains unchanged, $\varepsilon = \varepsilon_4$ can be chosen at most once for each stage. Similarly, $\varepsilon = \varepsilon_5$ will occur at most $m \cdot s$ times before $\mathcal{M}_{crit}[s]$ must change. Thus, in total we have at most $O(m^2 s^2)$ executions of Step 4. Steps 1-3 each require $O(N)$ time, each execution of Step 4 can be done in $O(sm)$ time which yields the claimed time bound.

To show optimality at termination, we have to discuss the different possibilities for $\varepsilon = 0$. Note that we always have $\varepsilon_1 > 0$. Similarly, by definition $\varepsilon_4 > 0$ and $\varepsilon_5 > 0$. If $\varepsilon_3 = 0$, then there is a critical machine M without any idle time in the current schedule. So the job assignment is clearly optimal in this case.

It remains the case $\varepsilon_2 = 0$. Hence, there is a critical machine M with $u[last_M[s]] = s$. Now consider the backward chain of the job $J = last_M[s]$. Let $ch_0 = chain_b(J)$, $i_1 = first(chain_b(J) - 1)$ and $ch_1 = chain_b(last(i_1)$. Continue in this manner by defining $i_j = first(chain_b(last(i_j - 1))) - 1$ and $ch_j = chain_b(last(i_j)$ until the concatenation of these chains covers all stages at least once. Note that Invariant 2 guarantees that this process will not stop before all stages are covered. Hence, we have found a certificate for optimality. Finally, we obviously also have a certificate for optimality if the algorithm terminates because of a cyclic dependency.

4   Number of Subproblems to Be Considered

Consider a stationary stage. Such a *stage has position* x with respect to the reference point 0, if the left borders of the workpieces on the assembly line are at positions $x + k \cdot WP$ for integral k values. Here, WP denotes the *workpiece pitch*, that is, the distance between the left borders of two successive workpieces on the assembly line. For a stationary stage with position x, let $MT[x]$ be the set of jobs which can be executed at this position of the board by the corresponding machine. We say that stationary stage positions x_1, x_2 are *equivalent* if the sets $MT[x_1]$ and $MT[x_2]$ are identical. This equivalence relation partitions the interval $[0, WP]$ into at most $O(N)$ subintervals $I_1, I_2, \ldots, I_t$, $t \leq 2N$ (namely its equivalence classes), which can be determined by a straight-forward sweep-line algorithm in linear time. Just note that during the sweeping of the stage position x from $[0, WP]$ each job is added to and removed from $MT[x]$ exactly once. If we enumerate over every s-element subset of these intervals as candidates for the subroutine which solves the problem for a fixed stepping scheme to optimality, we will find the overall optimal solution. This requires at most $O(N^s)$ iterations.

This bound can be improved substantially, if we exploit that possible forward steps are quite granular (recall the discussion from the Introduction): the offset by which a workpiece is moved in a forward step must be a multiple of Δ. Each stationary stage position x can therefore be expressed as $x = x_0 + k \cdot \Delta$ for an appropriate integer $k \geq 0$ and a shift parameter x_0 with $0 \leq x_0 < \Delta$. The shift parameter x_0 is the offset of the start configuration to the reference point 0. Obviously, the number of stationary stages is bounded from above by the parameter $q_1 := \lceil WP/\Delta \rceil$.

Hence, for fixed parameter x_0 we have to choose s elements from at most $\lceil WP/\Delta \rceil$ positions (instead from $O(N)$ positions). For a given x_0, we say that we *hit* interval $I_j = [\ell_j, u_j]$ if there is an integer k such that $\ell_j \leq x_0 + k\Delta \leq u_j$. Now we can also define an equivalence relation for the parameter x_0 with respect to the intervals $I_1, I_2, \ldots, I_N$ for the N jobs. Namely, x_0 and $\tilde{x}_0$ are equivalent if the sets of intervals hit by them are identical. Again, we can use a sweep-line technique to determine all equivalence classes for the parameter x_0. There are at most $O(N)$ different classes for the following reason: if we slide the parameter x_0 from 0 to Δ each of the $t \leq N$ intervals I_j will be inserted to or removed from the set of hit intervals at most once. In summary, we have to consider at most $O(\min\{Ns^{q_1}, N^s\})$ different selections of stationary stage positions.

5   Conclusion

In this paper, we have considered a variant of an assembly-line scheduling problem of high practical importance.

The main idea of our methodology is to identify parameters that not only allow efficient algorithms but also turn out to have small values in practice. In our specific application, we used a collection of such parameters which enabled us to develop an efficient approach with provably near-optimal results. In the future, we will try to extend this kind of analysis to other applications form our cooperation partners.

References

[ACK02]   M. Ayob, P. Cowling, and G. Kendall, *Optimisation of surface mount placement machines*, Proceedings of IEEE International Conference on Industrial Technology, 2002, pp. 486–491.

[AGN01]   J. Alber, J. Gramm, and R. Niedermeier, *Faster exact algorithms for hard problems: a parameterized point of view*, Discrete Mathematics **229** (2001), 3–27.

[BDF+95] H. Bodlaender, R. Downey, M. Fellows, M. Hallett, and H. T. Wareham, *Parameterized complexity analysis in computational biology*, Computer Applications in the Biosciences **11** (1995), 49–57.

[Bru97]   P. Brucker, *Scheduling algorithms*, second revised and enlarged ed., Springer-Verlag, 1997.

[CKS02]   Y. Crama, J. van de Klundert, and F. C. R. Spieksma, *Production planning problems in printed circuit board assembly*, Discrete Applied Mathematics **123** (2002), 339–361.

[DF99]   R. G. Downey and M. R. Fellows, *Parameterized complexity*, Monographs in Computer Science, Springer-Verlag, 1999.

[DFS99]   R. G. Downey, M. R. Fellows, and U. Stege, *Parameterized complexity: A framework for systematically confronting computational intractability*, DIMACS Series in Discrete Mathematics and Theoretical Computer Science (R. L. Graham, J. N. Kratochvil, and F. S. Roberts, eds.), vol. 49, 1999, pp. 49–99.

[Fel02]   M. Fellows, *Parameterized complexity: The main ideas and connections to practical computing*, Electronic Notes in Theoretical Computer Science (J. Harland, ed.), vol. 61, Elsevier Science Publishers, 2002.

[GJ79]   M. R. Garey and D. S. Johnson, *Computers and intractability: A guide to the theory of NP-completeness*, W. H. Freeman, 1979.

[Sch99]   A. Scholl, *Balancing and sequencing of assembly lines*, 2nd ed., Physica-Verlag, Heidelberg, 1999.

A Structural View on Parameterizing Problems: Distance from Triviality[*]

Jiong Guo, Falk Hüffner, and Rolf Niedermeier

Wilhelm-Schickard-Institut für Informatik, Universität Tübingen, Sand 13, D-72076 Tübingen, Fed. Rep. of Germany
{guo,hueffner,niedermr}@informatik.uni-tuebingen.de

Abstract. Based on a series of known and new examples, we propose the generalized setting of "distance from triviality" measurement as a reasonable and prospective way of determining useful structural problem parameters in analyzing computationally hard problems. The underlying idea is to consider tractable special cases of generally hard problems and to introduce parameters that measure the distance from these special cases. In this paper we present several case studies of distance from triviality parameterizations (concerning CLIQUE, POWER DOMINATING SET, SET COVER, and LONGEST COMMON SUBSEQUENCE) that exhibit the versatility of this approach to develop important new views for computational complexity analysis.

1 Introduction

VERTEX COVER is one of the NP-complete problems that stood at the cradle of parameterized algorithm design and analysis [11]. Given an undirected graph with n vertices and a nonnegative integer k, the question is whether we can find a set of at most k graph vertices such that each graph edge has at least one of its endpoints in this set. The currently best fixed-parameter algorithms exactly solve VERTEX COVER in $O(1.3^k + kn)$ time [8,22]; that is, VERTEX COVER is fixed-parameter tractable when parameterized by k. A different way of parameterizing VERTEX COVER is to consider the structure of the input graph. If the given graph allows for a tree decomposition [26,4] of width w, then it is well-known that VERTEX COVER can be solved in $O(2^w \cdot n)$ time [30] independent of the size k of the cover set we are searching for. Hence, VERTEX COVER is also fixed-parameter tractable when parameterized by w. As a rule, most problems can be parameterized in various reasonable ways.[1] The example VERTEX COVER exhibits two fundamentally different ways of parameterization—"parameterizing by size" (i.e., the size of the vertex cover) and "parameterizing by structure" (i.e., the treewidth of the underlying graph). In this paper we propose to take

---

[*] Supported by the Deutsche Forschungsgemeinschaft (DFG), Emmy Noether research group PIAF (fixed-parameter algorithms), NI 369/4.
[1] For instance, Fellows [13] discusses how to parameterize the MAX LEAF SPANNING TREE problem in at least five different ways.

R. Downey, M. Fellows, and F. Dehne (Eds.): IWPEC 2004, LNCS 3162, pp. 162–173, 2004.

a broader, generalized view on parameterizing problems by structure, leading to a generic framework of new research questions in parameterized complexity analysis.

The leitmotif of parameterized complexity theory [11] is to gain a better understanding of problem hardness through a refined complexity analysis that uses a two-dimensional view on problems by means of parameterization. A natural way to do this is as follows. Consider a problem such as VERTEX COVER and find out what efficiently solvable special cases there are known. For instance, VERTEX COVER is trivially solvable on trees. Now, for example, consider the parameter d defined as the number of edges that have to be deleted from a graph to transform it into a tree. In this sense parameter d measures the "distance from triviality" and one may ask whether VERTEX COVER is fixed-parameter tractable when parameterized by d. In this simple example the answer is clearly "yes" because such a graph has treewidth bounded by $d + 1$ [3] and, thus, VERTEX COVER can be solved using the tree decomposition approach [30]. But in other cases this "distance from triviality" approach to parameterization often leads to interesting new research questions: For instance, in a recent work Hoffmann and Okamoto [19] describe a fixed-parameter algorithm for the TRAVELING SALESMAN PROBLEM in the two-dimensional Euclidean plane based on the following distance from triviality parameterization: Consider a set of n points in the Euclidean plane. Determine their convex hull. If all points lie on the hull, then this gives the shortest tour. Otherwise, Hoffmann and Okamoto show that the problem is solvable in $O(k! \cdot k \cdot n)$ time where k denotes the number of points inside the convex hull. Thus, the distance from triviality here is the number k of inner points.

In this paper we extend the distance from triviality concept to a broader setting and we discuss further examples for the fruitfulness of this new parameterization methodology. We present further recent examples from the literature concerning SATISFIABILITY [28] and GRAPH COLORING [7] that fit into our framework. In addition, we provide four new fixed-parameter tractability results using this framework for CLIQUE, POWER DOMINATING SET, SET COVER, and LONGEST COMMON SUBSEQUENCE. Given all these case studies, we hope to convince the reader that, in a sense, "parameterizing away from triviality" yields a generic framework for an extended parameterized complexity analysis to better understand computational (in)tractability. Further aspects of our scenario and its prospects for future research are discussed in the concluding section.

2   Preliminaries and Previous Work

Preliminaries. Parameterized complexity theory [11] offers a two-dimensional framework for studying the computational complexity mostly of NP-hard problems. A *parameterized language* (problem) L is a subset $L \subseteq \Sigma^* \times \Sigma^*$ for some finite alphabet Σ. For $(x, k) \in L$, by convention, the second component denotes the *parameter*. The two dimensions of parameterized complexity analysis are constituted by the input size $n := |(x, k)|$ and the parameter value k (usually a nonnegative integer). A parameterized language is *fixed-parameter tractable* if it can be determined in $f(k) \cdot n^{O(1)}$ time whether $(x, k) \in L$, where f is a computable function only depending on k. Since the parameter k represents some aspect(s) of the input or the solution, there usually are many meaningful ways to *parameterize* a problem. An important issue herein is whether a problem is fixed-parameter tractable with respect to a chosen parameter or not (i.e., W[1]-hard, see [11] for details), and, in case of fixed-parameter tractability, how small the usually exponential growth of the function f can be kept. Hence, investigating different parameterizations gives insight into what causes the computational (in)tractability of a problem and in which qualitative and quantitative sense this happens. Refer to [10,14,21] for recent surveys on parameterized complexity.

Previous Work. The aim of this paper is to stimulate research on the structural parameterization "distance from triviality." Clearly, one of the most sophisticated examples in this context is the notion of bounded treewidth developed by Robertson and Seymour [26]. Without going into details, we remark that the basic motivation for considering this concept can be derived from the fact that many NP-hard graph problems (such as VERTEX COVER) become easy (linear-time solvable) on trees. Treewidth then measures how tree-like a graph is, and if this parameter is small, then many otherwise hard graph problems can be solved efficiently (see [4] for a survey). In this sense treewidth measures the distance from the triviality "tree" and problems such as VERTEX COVER are fixed-parameter tractable with respect to this structural parameter [30].

Another prominent problem is GRAPH COLORING. Leizhen Cai [7] recently initiated a study of GRAPH COLORING which falls into our framework. For instance, considering split graphs (where GRAPH COLORING is solvable in polynomial time) he showed that GRAPH COLORING is fixed-parameter tractable with respect to parameter k on graphs that originate from split graphs when adding or deleting k edges. By way of contrast, it is W[1]-hard when deletion of k vertices leads to a split graph. Interestingly, the problem is much harder in case of bipartite graphs instead of split graphs: GRAPH COLORING becomes NP-complete for graphs that originate from bipartite graphs by adding three edges or if two vertex deletions are needed to make a graph bipartite. In summary, Cai states that "this new way of parameterizing problems adds a new dimension to the applicability of parameterized complexity theory" [7].[2]

Finally, to emphasize that not only graph problems fall into our framework we give an example with SATISFIABILITY. It is easy to observe that a boolean formula in conjunctive normal form which has a matching between variables and clauses that matches all clauses is always satisfiable. For a formula F, considered as a set of m clauses over n variables, define the *deficiency* as $\delta(F) := m - n$. The maximum deficiency is $\delta^*(F) := \max_{F' \subseteq F} \delta(F')$. Szeider shows that the satisfiability of a formula F can be decided in $O(2^{\delta^*(F)} \cdot n^3)$ time [28]. Note that

---

[2] Juedes et al. [20] show that coloring an n-vertex-graph with $n - k$ colors is fixed-parameter tractable with respect to k. Clearly, $k = 0$ is trivial. This parameterization, however, is not a structural one.

a formula F with $\delta^*(F) = 0$ has a matching as described above. Again, $\delta^*(F)$ is a structural parameter measuring the distance from triviality in our sense.

In the following sections we provide new case studies for the applicability of the distance from triviality concept in various contexts. Clearly, it is conceivable that several other examples from the literature will fit as examples into our concept.[3] An important point, however, is that all of the parameterizations discussed here have nothing to do with the solution itself (i.e., the value to be determined or optimized). Our parameterizations are structural ones.

3   Case Study Clique

The CLIQUE problem is defined as follows:

Input: A graph $G = (V, E)$ and a nonnegative integer s.
Question: Does G contain a *clique*, i.e., a complete subgraph, of size s?

CLIQUE is W[1]-complete with respect to the natural parameter s [11]. It is also hard to approximate to an arbitrary constant factor. Here we exhibit fixed-parameter tractability with respect to the distance from a trivial case.

Our trivial case is the class of *cluster graphs*: graphs which are a disjoint union of cliques. CLIQUE can be trivially solved in linear time on such graphs. We examine CLIQUE on graphs which are "almost" cluster graphs, namely, on graphs which are cluster graphs with k edges added. From a general result on graph modification problems by Leizhen Cai [6] it follows that finding the added k edges is fixed-parameter tractable with respect to k. Improved algorithms for this problem (which is known as CLUSTER DELETION) were given by Gramm et al. [15,16], providing an algorithm running in $O(1.53^k + |V|^3)$ time.

It remains to show how to solve CLIQUE for the "almost cluster graph" G after identifying the k added edges and the corresponding cluster graph G'. If the largest clique in G is not one which is already contained in G', then each of its vertices must have gained in degree by at least one compared to G'. This means it can only be formed by a subset of the up to $2k$ vertices "touched" by the added edges. Hence, we solve CLIQUE for the subgraph of G induced by the up to $2k$ vertices which are endpoints of the added edges. This step can be done for example by using Robson's algorithm for INDEPENDENT SET [27] on the complement graph in $O(1.22^{2k}) = O(1.49^k)$ time, which is dominated by the above time bound for the CLUSTER DELETION subproblem. The largest clique for G is simply the larger of the clique found this way and the largest clique in G'. We obtain the following theorem:

Theorem 1. CLIQUE *for a graph* $G = (V, E)$ *which is a cluster graph with k edges added can be solved in* $O(1.53^k + |V|^3)$ *time.*

---

[3] For instance, Nishimura et al. [23] developed algorithms for recognizing general classes of graphs generated by a base graph class by adding at most k vertices. Their fixed-parameter tractability studies are closely related to our methodology.

4   Case Study Power Dominating Set

Domination in graphs is among the most important problems in combinatorial optimization. We consider here the POWER DOMINATING SET problem [18], which is motivated from applications in electric networks. The task is to place monitoring devices (so-called *PMUs*) at vertices such that all edges and vertices are *observed*. The rules for observation are:

1. A PMU in a vertex v observes v and all incident edges and neighbors of v.
2. Any vertex that is incident to an observed edge is observed.
3. Any edge joining two observed vertices is observed.
4. If a vertex is incident to a total of $i > 1$ edges and if $i - 1$ of these edges are observed, then all i edges are observed. This rule is reminiscent of Kirchhoff's current law from electrical engineering.

We can now formulate the POWER DOMINATING SET problem:

Input: A graph $G = (V, E)$ and a nonnegative integer k.
Question: Does G have a *power dominating set* of size at most k, that is, a subset $M \subseteq V$ of vertices such that by placing a PMU in every $v \in M$, all vertices in V are observed?

POWER DOMINATING SET is NP-complete [18]. There is an algorithm known which solves POWER DOMINATING SET in linear time on trees [18]. Since we use this algorithm as a building block for our result, we briefly sketch how it proceeds. This algorithm works bottom-up from the leaves and places a PMU in every vertex which has at least two unobserved children. Then it updates observation according to the four observation rules and prunes completely observed subtrees, since they no longer affect observability of other vertices.

Our goal is now to find an efficient algorithm for input graphs that are "nearly" trees. More precisely, we aim for a fixed-parameter algorithm for graphs which are trees with k edges added.

Note that a tree with k edges added has treewidth bounded by $k+1$ [3]. While DOMINATING SET is fixed-parameter tractable with respect to the parameter treewidth [2], no such result is currently known for POWER DOMINATING SET. This motivates our subsequent result.

As a first step we present a simple algorithm with quadratic running time for the case of one single added edge.

Lemma 1. POWER DOMINATING SET *for a graph* $G = (V, E)$ *with* $n := |V|$ *which is a tree with one edge added can be solved in* $O(n^2)$ *time.*

Proof. Graph G contains exactly one cycle and a collection of trees T_i touching the cycle at their roots.

We use the above mentioned linear time algorithm to find an optimal solution for each T_i. When it reaches the root r_i, several cases are possible:

– The root r_i needs to be in M, and we can remove it. This breaks the cycle, and we can solve the remaining instance in linear time.

- The root r_i is not in M, but already observed. Then all children of r_i in T_i except for at most one are observed, or we would need to take r_i into M. Then, we can remove T_i except for r_i and except for the unobserved child, if it exists. The root r_i remains as an observed degree-2 or degree-3 vertex on the cycle.
- The root r_i still needs to be observed. This is only possible if it has exactly one child in T_i which is unobserved since otherwise r_i either would be in M, or be observed. As in the previous case, we keep r_i and the unobserved child, and the rest of T_i can again be removed.

If after these data reductions two observed vertices are adjacent on the cycle, their connecting edge becomes observed, and we can break the cycle. Otherwise, we call it a *reduced cycle*.

At least one vertex on the reduced cycle has to be added to M. We simply try each vertex. After each choice, the rest of the cycle decomposes into a tree after pruning observed edges, and can be handled in linear time. From all possible initial choices, we keep the one leading to a minimal M, which then is an optimal choice for the initial problem. Since there are $O(n)$ vertices and edges on the cycle, this takes $O(n^2)$ time. $\qquad\square$

We note without proof that Lemma 1 can be improved to linear time by examining a fixed-size segment of the cycle. In each possible case we can determine at least one vertex in the segment which has to be taken into M.

Lemma 1 is applicable whenever each vertex is part of at most one cycle. We now generalize this and Haynes et al.'s [18] result.

Theorem 2. POWER DOMINATING SET *for a graph which is a tree with k edges added is fixed-parameter tractable with respect to k.*

Proof. We first treat all trees which are attached in single points to cycles as in the proof of Lemma 1. What remains are degree-2 vertices, degree-3 vertices with a degree-1 neighbor, and other vertices of degree 3 or greater, the *joints*. For a vertex v, let $\deg v$ denote its degree, that is, the number of its adjacent vertices. We branch into several cases for each joint:

- The joint v is in M. We can prune it and its incident edges.
- The joint v is not in M. Note that the only effect v can still have is that a neighbor of v becomes observed from application of observation rule 4 ("Kirchhoff's current law") applied to v. We branch further into $\deg v \cdot (\deg v - 1)$ cases for each pair (w_1, w_2) of neighbors of v with $w_1 \neq w_2$. In each branch, we omit the edges between v and all neighbors of v except w_1 and w_2. Clearly any solution of such an instance provides a solution for the unmodified instance. Furthermore, it is not too hard to show that if the unmodified instance has a solution of size s, then on at least one branch we will also find a solution of size s. To see this, consider a solution M for the unmodified problem. Vertex v is observed; this can only be either because a neighbor w_1 of v was put into M, or because there is a neighbor w_1 of v

such that the edge $\{w_1, v\}$ became observed from observation rule 4. Furthermore, as mentioned, there can be at most one vertex w_2 which becomes observed by observation rule 4 applied to v. Then M is also a valid solution for the branch corresponding to the pair (w_1, w_2).

In each of the less than $(\deg v)^2$ branches we can eliminate the joint. If we branch for all joints in parallel, we end up with an instance where every connected component is a tree or a cycle with attached degree-1 vertices, which can be solved in linear time. The number of cases to distinguish is $\prod_{v \text{ is joint}} (\deg v)^2$. Since there are at most $2k$ joints, each of maximum degree k, the total running time is roughly bounded by $O(n \cdot 2^{4k \log k})$, confirming fixed-parameter tractability. □

5 Case Study Tree-like Weighted Set Cover

SET COVER is one of the most prominent NP-complete problems. Given a base set $S = \{s_1, s_2, \ldots, s_n\}$ and a collection C of subsets of S, $C = \{c_1, c_2, \ldots, c_m\}$, $c_i \subseteq S$ for $1 \leq i \leq m$, and $\bigcup_{1 \leq i \leq m} c_i = S$, the task is to find a subset C' of C with minimal cardinality which covers all elements in S, i.e., $\bigcup_{c \in C'} c = S$. Assigning weights to the subsets and minimizing the total weight of the collection C' instead of its cardinality, one naturally obtains the WEIGHTED SET COVER problem. We call C' the *minimum set cover* of S resp. the *minimum weight set cover*. We define the *occurrence* of an element $s \in S$ in C as the number of subsets in C which contain s. SET COVER remains NP-complete even if the occurrence of each element is bounded by 2 [24].

Definition 1 (Tree-like subset collection).
Given a base set $S = \{s_1, s_2, \ldots, s_n\}$ and a collection C of subsets of S, $C = \{c_1, c_2, \ldots, c_m\}$. We say that C is a tree-like subset collection of S if we can organize the subsets in C in an unrooted tree T such that every subset one-to-one corresponds to a node of T and, for each element $s_j \in S$, $1 \leq j \leq n$, all nodes in T corresponding to the subsets containing s_j induce a subtree of T.

We call T the *underlying subset tree* and the property of T that, for each $s \in S$, the nodes containing s induce a subtree of T, is called the "*consistency property*" of T. Observe that the consistency property is also of central importance in Robertson and Seymour's famous notion of tree decompositions of graphs [26,4]. By results of Tarjan and Yannakakis [29], we can test whether a subset collection is a tree-like subset collection and, if so, we can construct a subset tree for it in linear time. Therefore, in the following we always assume that the subset collection is given in form of a subset tree. For convenience, we denote the nodes of the subset tree by their corresponding subsets.

Here, we consider the TREE-LIKE WEIGHTED SET COVER (TWSC) problem with bounded occurrence which is defined as follows:

TREE-LIKE WEIGHTED SET COVER WITH BOUNDED OCCURRENCE:
Input: Given a base set $S = \{s_1, s_2, \ldots, s_n\}$ and a tree-like collection C of subsets of S, $C = \{c_1, c_2, \ldots, c_m\}$. Each element of S can be in at

most d subsets for a fixed $d \geq 1$. Each subset in C has a positive real weight $w(c_i) > 0$ for $1 \leq i \leq m$. The weight of a subset collection is the sum of the weights of all subsets in it.

Task: Find $C' \subseteq C$ with minimum weight which covers all elements in S, i.e., $\bigcup_{c \in C'} c = S$.

TWSC with bounded occurrence $d \geq 3$ is NP-complete even if the underlying subset tree is a star [17]. However, it can be solved in $O(m^2 n)$ time if the underlying subset tree is a path [17]. Now our goal is, based on the "trivial" path-like case, to give a fixed-parameter algorithm for TWSC with bounded occurrence where the number of leaves of the subset tree functions as the distance parameter from the path-like case.

The fixed-parameter algorithm. Given a subset tree T with k leaves, the following observations are easy to prove.

Observation 1. The maximum degree of the nodes of T is upperbounded by k.

Observation 2. The number of tree nodes with more than 2 neighbors is upperbounded by k.

Observation 3. For each $c \in C$, the number of subsets $c' \in C$ which share some elements with c, i.e., $c \cap c' \neq \emptyset$, is upperbounded by $d \cdot k$. These subsets and c induce a subtree of T.

For a subset $c \in C$, let $\deg c$ denote the degree of its corresponding node in T. The basic idea of the algorithm is to divide the tree-like instance into several "independent" path-like instances and to solve these instances separately by using the $O(m^2 n)$-time algorithm. Instances are independent if they share no common elements with each other. For each tree node c with $\deg c \geq 3$, we construct a set $H_c := \{c' \mid c \cap c' \neq \emptyset\}$. Note that $|H_c| \leq d \cdot k$ by Observation 3. To cover all elements of c, we have to add some subsets from H_c into the set cover. We delete the subsets added into the set cover and all their adjacent edges from T. Furthermore, we delete elements of S which are already covered from all remaining subsets. Observe that if c is in the set cover, we retain several subtrees of T after deleting c; otherwise, c is now an empty subset. By deleting all empty subsets, the subset tree T is divided into several subtrees. Due to the consistency property of T, all these resulting subtrees are independent. Since the possible combinations of the subsets from H_c which cover all elements of c are upperbounded by 2^{dk}, we can have up to 2^{dk} new instances by dividing T at c. By processing all nodes c with $\deg c \geq 3$ in the same way, there are $O(2^{dk^2})$ new instances each of which consists of $O(m)$ independent path-like instances. Then the minimum set cover for one of these new instances is the union of the optimal solutions for the path-like instances with the subsets already added into the set cover while processing the nodes with degree at least 3. In summary, we get the following theorem:

Theorem 3. *TWSC with occurrence bounded by d can be solved in $O(2^{dk^2} \cdot m^2 n)$ time, where k denotes the number of the leaves of the subset tree.*

Note that while results from [17] only cover cases with bounded subset size, we impose no such restriction here. However, here we need the bounded occurrence restriction.

6   Case Study Longest Common Subsequence

In this section we deal with the LONGEST COMMON SUBSEQUENCE problem, an important problem in theoretical computer science and computational biology.

> LONGEST COMMON SUBSEQUENCE (LCS):
> **Input:** Given a set of k strings $X_1, X_2, \ldots, X_k$ over an alphabet Σ and a positive integer m.
> **Question:** Is there a string $X \in \Sigma^*$ of length at least m that is a subsequence of X_i for $i = 1, \ldots, k$?

LCS is NP-complete even if $|\Sigma| = 2$. Concerning the parameterized complexity of LCS with unbounded alphabet size, Bodlaender et al. [5] showed that LCS is W[t]-hard for $t \geq 1$ with k as parameter, W[2]-hard with m as parameter, and W[1]-hard with k and m as parameters. With a fixed alphabet size, LCS is trivially fixed-parameter tractable with m as parameter, but W[1]-hard with k as parameter [25].

Let n denote the maximum length of the input strings and s_i^a denote the number of occurrences of letter $a \in \Sigma$ in X_i. We consider a new parameterization of LCS with k and $s := \max_{a \in \Sigma} \max_{1 \leq i \leq k} s_i^a$ as parameters. To begin with, we show that the case $s = 1$ of this parameterization, where every letter occurs in each string only once, is solvable in polynomial time.

Without loss of generality, we assume that all input strings have the same length n, $\Sigma = \{1, 2, \ldots, n\}$, and $X_1 = 1\,2\,3\cdots n$. Then the strings $X_2, X_3, \ldots, X_k$ are permutations of X_1. We construct a directed graph G with $n \times k$ vertices; each vertex represents a position in a string. A directed edge is added from $v_{i,j}$ to $v_{i+1,l}$ for $1 \leq i < k$ iff the letters in position j of X_i and in position l of X_{i+1} are the same. It is easy to observe that a longest common subsequence one-to-one corresponds to a maximum set of directed paths in G which do not cross each other. Two paths P and P' cross each other if there is an edge $(v_{i,j}, v_{i+1,l})$ in P and an edge $(v_{i,j'}, v_{i+1,l'})$ in P' with $j \leq j'$ and $l \geq l'$. In order to find a maximum set of the noncrossing paths in G, we construct a "path-compatibility" graph $PC(G)$ from G: For each directed path in G, we create a vertex P_a in $PC(G)$ where a is the position in X_1 where the path starts. We add a directed edge (P_a, P_b) from P_a to P_b if P_a does not cross P_b and $a < b$. Thus, $PC(G)$ is an acyclic directed graph and a maximum set of noncrossing paths of G one-to-one corresponds to a longest path in $PC(G)$. By using depth-first search for each vertex in $PC(G)$ with in-degree of zero, we can easily find such a longest path.

Concerning the running time to solve LCS on such an instance, we note that graphs G and $PC(G)$ can be constructed in $O(k \cdot n)$ and $O(k \cdot n^2)$ time, respectively. Finding a longest path in $PC(G)$ can be done in $O(n)$ time. Summarizing, the running time for solving LCS on these instances is $O(k \cdot n^2)$.

The fixed-parameter algorithm. Given strings $X_1, X_2, \ldots, X_k$, we construct a graph G with $n \cdot k$ vertices as described above. However, a vertex $v_{i,l}$ with $1 \leq i < k$ has to be connected to all vertices $v_{i+1,h}, 1 \leq h \leq n$, where X_{i+1} has the same letter in position h as X_i in position l. Graph G can be constructed in $O(k \cdot n^2)$ time. Then we construct the path-compatibility graph $PC(G)$ from G. Since the number of paths in G can be up to $\sum_{a \in \Sigma} \prod_{1 \leq i \leq k} s_i^a = O(n \cdot s^k)$, we have $O(n \cdot s^k)$ vertices in $PC(G)$. Each vertex $P_{j_1, j_2, \ldots, j_k}$ represents a path in G, and the indices $j_1, j_2, \ldots, j_k$ denote the positions in the k strings where this path passes. Furthermore, a directed edge from $P_{j_1, j_2, \ldots, j_k}$ to $P_{j'_1, j'_2, \ldots, j'_k}$ is added into $PC(G)$ if the corresponding paths do not cross each other, are vertex-disjoint, and $j_1 < j'_1$. It is easy to verify that by finding a longest path in this acyclic directed graph, we get a longest common subsequence of the input instance. The construction of the edges in $PC(G)$ takes $O(k)$ time per vertex pair, and we obtain an algorithm running in $O(k \cdot (n \cdot s^k)^2) = O(2^{2k \log s} \cdot k \cdot n^2)$ time. This yields the following result.

Theorem 4. LONGEST COMMON SUBSEQUENCE *can be solved in* $O(2^{2k \log s} \cdot k \cdot n^2)$ *time, where s denotes the maximum occurrence number of a letter in an input string.*

7   Concluding Discussion

The art of parameterizing problems is of key importance to better understand and cope with computational intractability. In this work we proposed a natural way of parameterizing problems—the parameter measures some distance from triviality. The approach consists of two fundamental steps. Assume that we study a hard problem X.

1. Determine efficiently solvable special cases of X (e.g., in case of graph problems, the restriction to special graph classes)—the triviality.
2. Identify useful distance measures from the triviality (e.g., in case of trees and graphs the treewidth of a graph)—the (structural) parameter.

As to step 2, observe that various distance measures are possible such as edge deletions or vertex deletions in case of graphs. It is important, however, that we can efficiently determine the distance of a given input instance to the chosen triviality with respect to the parameter considered. For instance, it is "easy" to determine the distance of a graph to being acyclic with respect to edge deletion (this leads to the polynomial-time solvable FEEDBACK EDGE SET problem) whereas it is hard to determine the distance of a graph to being bipartite with respect to edge deletion (this leads to the NP-hard GRAPH BIPARTIZATION problem). However, in case we are explicitly given the "operations" that transform the given object into a trivial one (e.g., the edges to be deleted to make a graph bipartite), the question for the parameterized complexity of the underlying problem with respect to the distance parameter might still be of interest. In the new case studies presented in this paper the distance measurement for LONGEST

COMMON SUBSEQUENCE was easy to determine whereas in the CLIQUE case the measurement led to an NP-hard but fixed-parameter tractable problem.

We do not claim that all the parameterizations we considered generally lead to small parameter values. This was not the central point, which, by way of contrast, was to extend the range of feasible special cases of otherwise computationally hard problems. As pointed out by an anonymous referee, it would be interesting to study more drastic distance measures such as considering relative distances—for instance, what if 1 % of all edges may be edited in a graph.

It is worth emphasizing that the tractable trivial case may refer to polynomial-time solvability as well as fixed-parameter tractability.[4] An example for the latter case is DOMINATING SET on planar graphs which is fixed-parameter tractable [1,2]. These results were extended to graphs of bounded genus [12,9], genus here measuring the distance from the "trivial case" (because settled) planar graphs. Moreover, the proposed framework might even be of interest in the approximation algorithms context where triviality then might mean good polynomial-time approximability (e.g., constant factor or approximation scheme). In summary, we strongly believe that distance from triviality parameterization leads to a wide range of prospective research opportunities.

References

1. J. Alber, H. L. Bodlaender, H. Fernau, T. Kloks, and R. Niedermeier. Fixed parameter algorithms for Dominating Set and related problems on planar graphs. *Algorithmica*, 33(4):461–493, 2002.
2. J. Alber, H. Fan, M. R. Fellows, H. Fernau, R. Niedermeier, F. Rosamond, and U. Stege. Refined search tree technique for Dominating Set on planar graphs. In *Proc. 26th MFCS*, volume 2136 of *LNCS*, pages 111–122. Springer, 2001.
3. H. L. Bodlaender. Classes of graphs with bounded treewidth. Technical Report RUU-CS-86-22, Dept. of Computer Sci., Utrecht University, 1986.
4. H. L. Bodlaender. Treewidth: Algorithmic techniques and results. In *Proc. 22nd MFCS*, volume 1295 of *LNCS*, pages 19–36. Springer, 1997.
5. H. L. Bodlaender, R. G. Downey, M. R. Fellows, and H. T. Wareham. The parameterized complexity of the longest common subsequence problem. *Theoretical Computer Science*, 147:31–54, 1995.
6. L. Cai. Fixed-parameter tractability of graph modification problems for hereditary properties. *Information Processing Letters*, 58:171–176, 1996.
7. L. Cai. Parameterized complexity of Vertex Colouring. *Discrete Applied Mathematics*, 127(1):415–429, 2003.
8. J. Chen, I. A. Kanj, and W. Jia. Vertex Cover: Further observations and further improvements. *J. Algorithms*, 41:280–301, 2001.
9. E. D. Demaine, F. V. Fomin, M. T. Hajiaghayi, and D. M. Thilikos. Subexponential parameterized algorithms on graphs of bounded-genus and H-minor-free graphs. In *Proc. 15th SODA*, pages 830–839. SIAM, 2004.
10. R. G. Downey. Parameterized complexity for the skeptic. In *Proc. 18th IEEE Annual Conference on Computational Complexity*, pages 147–169, 2003.
11. R. G. Downey and M. R. Fellows. *Parameterized Complexity*. Springer, 1999.

---

[4] The latter being of particular interest when attacking W[1]-hard problems.

12. J. Ellis, H. Fan, and M. R. Fellows. The Dominating Set problem is fixed parameter tractable for graphs of bounded genus. In *Proc. 8th SWAT*, volume 2368 of *LNCS*, pages 180–189. Springer, 2002.
13. M. R. Fellows. Blow-ups, win/win's, and crown rules: Some new directions in FPT. In *Proc. 29th WG*, volume 2880 of *LNCS*, pages 1–12. Springer, 2003.
14. M. R. Fellows. New directions and new challenges in algorithm design and complexity, parameterized. In *Proc. 8th WADS*, volume 2748 of *LNCS*, pages 505–520. Springer, 2003.
15. J. Gramm, J. Guo, F. Hüffner, and R. Niedermeier. Graph-modeled data clustering: Fixed-parameter algorithms for clique generation. In *Proc. 5th CIAC*, volume 2653 of *LNCS*, pages 108–119. Springer, 2003. To appear in *Theory of Computing Systems*.
16. J. Gramm, J. Guo, F. Hüffner, and R. Niedermeier. Automated generation of search tree algorithms for hard graph modification problems. *Algorithmica*, 39(4):321–347, 2004.
17. J. Guo and R. Niedermeier. Exact algorithms and applications for Tree-like Weighted Set Cover. Manuscript, June 2004.
18. T. W. Haynes, S. M. Hedetniemi, S. T. Hedetniemi, and M. A. Henning. Domination in graphs applied to electric power networks. *SIAM J. Discrete Math.*, 15(4):519–529, 2002.
19. M. Hoffmann and Y. Okamoto. The traveling salesman problem with few inner points. In *Proc. 10th COCOON*, volume 3106 of *LNCS*. Springer, 2004.
20. D. Juedes, B. Chor, and M. R. Fellows. Linear kernels in linear time, or How to save k colors in $O(n^2)$ steps. In *Proc. 30th WG*, LNCS. Springer, 2004. To appear.
21. R. Niedermeier. Ubiquitous parameterization—invitation to fixed-parameter algorithms. In *Proc. 29th MFCS*, LNCS. Springer, 2004. To appear.
22. R. Niedermeier and P. Rossmanith. On efficient fixed-parameter algorithms for Weighted Vertex Cover. *J. Algorithms*, 47(2):63–77, 2003.
23. N. Nishimura, P. Ragde, and D. M. Thilikos. Fast fixed-parameter tractable algorithms for nontrivial generalizations of Vertex Cover. In *Proc. 7th WADS*, volume 2125 of *LNCS*, pages 75–86. Springer, 2001. To appear in *Discrete Applied Mathematics*.
24. C. H. Papadimitriou and M. Yannakakis. Optimization, approximation, and complexity classes. *J. Comput. Syst. Sci.*, 43:425–440, 1991.
25. K. Pietrzak. On the parameterized complexity of the fixed alphabet Shortest Common Supersequence and Longest Common Subsequence problems. *J. Comput. Syst. Sci.*, 67(4):757–771, 2003.
26. N. Robertson and P. D. Seymour. Graph minors. II: Algorithmic aspects of treewidth. *J. Algorithms*, 7:309–322, 1986.
27. J. M. Robson. Algorithms for maximum independent sets. *J. Algorithms*, 7:425–440, 1986.
28. S. Szeider. Minimal unsatisfiable formulas with bounded clause-variable difference are fixed-parameter tractable. In *Proc. 9th COCOON*, volume 2697 of *LNCS*, pages 548–558. Springer, 2003.
29. R. E. Tarjan and M. Yannakakis. Simple linear-time algorithms to test chordality of graphs, test acyclicity of hypergraphs, and selectively reduce acyclic hypergraphs. *SIAM J. Comput.*, 13(3):566–579, 1984.
30. J. A. Telle and A. Proskurowski. Practical algorithms on partial k-trees with an application to domination-like problems. In *Proc. 3rd WADS*, volume 709 of *LNCS*, pages 610–621. Springer, 1993.

Perfect Path Phylogeny Haplotyping
with Missing Data Is Fixed-Parameter Tractable

Jens Gramm⋆, Till Nierhoff⋆, and Till Tantau⋆

International Computer Science Institute
1947 Center Street, Suite 600, Berkeley, CA 94704.
{gramm,nierhoff,tantau}@icsi.berkeley.edu

Abstract. Haplotyping via perfect phylogeny is a method for retrieving haplotypes from genotypes. Fast algorithms are known for computing perfect phylogenies from complete and error-free input instances—these instances can be organized as a genotype matrix whose rows are the genotypes and whose columns are the single nucleotide polymorphisms under consideration. Unfortunately, in the more realistic setting of missing entries in the genotype matrix, even restricted forms of the perfect phylogeny haplotyping problem become NP-hard. We show that haplotyping via perfect phylogeny with missing data becomes computationally tractable when imposing additional biologically motivated constraints. Firstly, we focus on asking for perfect phylogenies that are paths, which is motivated by the discovery that yin-yang haplotypes span large parts of the human genome. A yin-yang haplotype implies that every corresponding perfect phylogeny *has* to be a path. Secondly, we assume that the number of missing entries in every column of the input genotype matrix is bounded. We show that the perfect path phylogeny haplotyping problem is fixed-parameter tractable when we consider the maximum number of missing entries per column of the genotype matrix as parameter. The restrictions we impose are met by a majority of the problem instances encountered in publicly available human genome data.

1   Introduction

1.1   Haplotype Inference from Genotypes Via Perfect Phylogeny

Single nucleotide polymorphisms (SNPs) are differences in a single base, across the population, within an otherwise conserved genomic sequence. The sequence of SNP states in contiguous SNP positions along a chromosomal region is called a *haplotype*. The knowledge of the haplotypes in the human genome is of particular importance since they are believed to be often linked to medical conditions. However, current technologies suitable for large-scale SNP detection in the human genome—which contains two versions of each chromosome—do not obtain the haplotypes but only the *genotype* information: The genotype specifies, for every SNP position, the two states at this site in the two chromosomes. The genotype contains information only on the combination of SNP states at a given site, but it does not tell us which of the states belongs to which

---

⋆ The authors were supported through DAAD (German academic exchange service) postdoc
  fellowships.

chromosome. It is an important goal to develop efficient methods for inferring haplotypes from genotypes.

Haplotyping *via perfect phylogeny* is a method for haplotype inference where it is
assumed that the (unknown) haplotypes underlying the (observed) genotype data can
be arranged in a genetic tree in which each haplotype results from an ancestor haplotype via mutations. The perfect phylogeny approach is popular due to its applicability
to real haplotype inference problems and its theoretical elegance. It was introduced
by Gusfield [6] and received considerable attention which resulted, among others, in
quadratic-time algorithms for the case of complete and error-free input data [1, 2]. In
the special case where perfect *path* phylogenies are sought, even a linear time algorithm
is known [5].

The main hurdle for current haplotype inference methods is missing input data.
Real genotype data usually contain a small fraction of missing data caused by technical
problems in the process of genotype detection. In the presence of missing data, haplotyping via perfect phylogeny is NP-hard [8]. This is even true for the restricted case of
path phylogenies [5]. In an effort to solve the problem efficiently for restricted cases,
Halperin and Karp [7] show that perfect phylogeny haplotyping with missing data is
tractable if the input satisfies the 'rich-data hypothesis'. This hypothesis requires, intuitively, that the data contains enough information to 'locally' infer the haplotypes at any
two SNP sites. In this paper, we take a different approach which is independent of the
rich-data hypothesis. We show that perfect phylogeny haplotyping is fixed-parameter
tractable when restricted to path phylogenies and when taking the maximum number of
missing entries at a particular SNP site as parameter. Notably, experiments on publicly
available genotype data show a significant fraction of the data that allows perfect path
phylogenies but fails to satisfy the rich data hypothesis.

1.2   Computational Problems

When stripped of the biological context, the haplotype inference problem is a purely
combinatorial problem, which we describe in the following. In the combinatorial setting
a *haplotype* is a binary string. Each position of the string corresponds to a SNP site.
When we observe a certain base at the SNP site, the string contains a 0-entry at the
corresponding position; if we observe a certain other base, the string contains a 1-entry.
In rare situations one may observe three or even four different bases at a specific SNP
site, but these cases are not modeled. A *haplotype matrix* is a binary matrix whose rows
are haplotypes.

A *genotype* is a string over the alphabet $\{0, 1, 2\}$. A 2-entry corresponds to a heterozygous SNP site, meaning that the haplotypes underlying the genotype do not agree
at that site. The genotype *resulting* from two haplotypes has a 0-entry or a 1-entry at
all positions where both haplotypes agree on a 0-entry or 1-entry, respectively. It has a
2-entry at all positions where they disagree. We say that a genotype matrix A *admits a
perfect phylogeny* if there exists a rooted tree T, called *perfect phylogeny*, such that:

1. Each column of A labels exactly one edge of T.
2. Every edge of T is labeled by at least one column of A.
3. For each row r of A there are two nodes in T (possibly identical) labeled r' and r''.
   The labels r' and r'' are called *haplotype labels*.

A's column	1	2	3	4	5	6	7
row a	**0**	**1**	**2**	**2**	**0**	**0**	**2**
row b	0	1	1	0	2	2	0
row c	0	1	0	2	0	0	2
row d	2	2	0	0	0	0	0

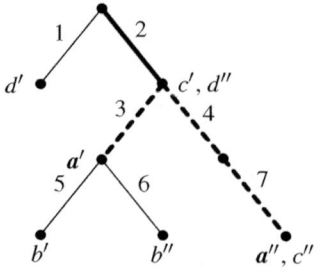

Fig. 1. Example of a genotype matrix A and a corresponding perfect phylogeny T. The edges of T correspond to the columns of A. Each genotype row r in A results from two haplotypes r' and r'', which label vertices of the tree. The paths induced by the first row a of A are shown in bold: The path corresponding to the 1-entries in row a leads from the root to an inner vertex of the tree (solid), the path corresponding to the 2-entries is rooted at this inner vertex (dashed). The latter path connects the two haplotype vertices a' and a''.

4. For every row r of A the set of columns with value 2 in this row forms a path p in T between r' and r''. The set of columns with value 1 in this row forms a path from T's root to the top-most node on the path p.

An example is depicted in Figure 1. A detailed biological justification for considering perfect phylogenies and for requiring the above properties for the tree T can be found in [5, 6].

Perfect phylogenies with the above properties are called *directed* since the ancestral state of every SNP site is assumed to be 0 or, equivalently, the root corresponds to the all-0 haplotype. In the *undirected* case the ancestral state of every site can be arbitrary (0 or 1).

The two versions of the haplotype inference problem via perfect phylogenies mentioned in the previous section are:

Problem 1.1 (Perfect Phylogeny Haplotyping Problem, {0, 1, 2}-PPH).
Input: A genotype {0, 1, 2}-matrix A.
Question: Does A admit a perfect phylogeny?

Problem 1.2 (Perfect Path Phylogeny Haplotyping Problem, {0, 1, 2}-PPPH).
Input: A genotype {0, 1, 2}-matrix A.
Question: Does A admit a perfect phylogeny that is a path?

It is known that these problems can be solved efficiently, but become computationally intractable if missing data is introduced [5]. Missing entries are denoted by ?-entries in the input matrix. Given such a genotype matrix with ?-entries, we say that it is *pp-realizable* (*ppp-realizable*) if we can replace the question marks with values from {0, 1, 2} such that the resulting genotype matrix admits a perfect (path) phylogeny. This leads to the problem {0, 1, 2, ?}-PPPH, whose fixed-parameter tractability is the focus of this paper:

Problem 1.3 (Perfect Path Phylogeny Haplotyping with Missing Entries,
{0, 1, 2, ?}-PPPH).
Input: A genotype $\{0, 1, 2, ?\}$-matrix A.
Question: Is A ppp-realizable?

1.3 Biological Relevance of Our Fixed-Parameter Tractability Result

The general perfect phylogeny haplotyping problem with missing entries is NP-complete. In order to solve it, one is thus forced to incorporate additional, biologically motivated constraints on the problem. To this end, firstly, we focus on *path* phylogenies. Secondly, we assume that for each SNP site only a small fraction of genotypes lack information about the state at this site. Thirdly, we focus on *directed* perfect phylogenies.

The first assumption is motivated by the recent discovery [10] that about 75 percent of the human genome can be covered by *yin-yang haplotypes*, i.e., long-region pairs of haplotypes that are heterozygous at every SNP site. In the model of perfect phylogeny haplotyping, the presence of a yin-yang haplotype pair implies that a perfect phylogeny of the haplotypes necessarily has to be a path phylogeny. These findings suggest that an efficient algorithm dealing with path phylogenies can already be applied to a large portion of the genome, leaving the remaining parts for heuristic or more time-intensive postprocessing.

The second assumption is motivated by the expectation that missing data entries should occur only rarely and randomly and should thus not amass at a specific SNP site. Furthermore, a SNP site for which most entries are missing is useless for classification purposes: filling up the numerous missing entries is pure guesswork.

The third assumption is a reasonable one to make since, due to the sparse distribution of missing entries, we can usually identify the labeling of one node that necessarily has to occur in a perfect phylogeny; this can be done using a technique described by Eskin *et al.* [2] (see there for details). Since within each column of the input matrix we are free to exchange 0 and 1 labels, we can choose the labels such that this particular node carries a 0-entry at every position, which makes the perfect phylogeny directed.

We have evaluated all three of our assumptions by considering publicly available genotype data provided by Gabriel *et al.* [3], available from `http://www.broad.mit.edu/mpg/hapmap/hapstruc.html`. The data consist of sets for four populations. Table 1 gives an overview on the distribution of ?-entries in the data. We see that the maximum number of question marks is approximately one fifth of the total column size but also that the average number of question marks per column is much smaller. We can also see from Table 1 that in all except two of the 248 investigated datasets we could identify one of the haplotypes, using the reduction given by Eskin *et al.* [2], thereby making the datasets suitable for our algorithm, which searches for *directed* perfect path phylogenies.

Table 2 treats the assumption of finding perfect phylogenies that are path phylogenies. For a given window length l, we computed for every set of l consecutive columns in the input genotype matrix whether these columns give rise to a perfect phylogeny and whether they give rise to a perfect path phylogeny. For our statistic we omitted rows within a window that contain missing entries. It turns out that, on average, approximately two third of the encountered perfect phylogenies are in fact path phylogenies.

Population	A	B	C	D
Number of individuals (rows per genotype matrix)	93	50	42	96
Number of genotype matrices	62	62	62	62
Average maximum number of ?'s per column	21.7	10.5	8.3	21.3
Average number of ?'s per column	6.2	2.7	1.6	6.8
Percentage of directed genotype matrices	97%	100%	100%	100%

Table 1. Statistics on the distribution of ?-entries in the human genotype data provided by Gabriel *et al.* [3]. For data from four populations, the table first lists the number of individuals per population whose genotypes were determined. For each individual, the genotypes were determined in 62 different regions of the genome. The table next lists, averaged over the genotype matrices for each population, the maximum and the average number of ?-entries per column. Finally, it lists the percentage of genotype matrices that can be directed.

Population	Window Size	Genotype Matrices	PP's	PPP's	RD_{pp}	RD_{ppp}
	5	3029	51%	40%	32%	29%
A	8	2843	29%	19%	9%	7%
	10	2721	20%	13%	3%	1%
	5	2900	35%	25%	26%	21%
B	8	2720	14%	9%	7%	4%
	10	2600	8%	5%	3%	3%
	5	3001	59%	51%	36%	30%
C	8	2821	37%	28%	15%	16%
	10	2701	27%	20%	9%	9%
	5	2819	36%	26%	43%	39%
D	8	2641	14%	8%	19%	16%
	10	2525	8%	5%	7%	6%

Table 2. Statistics on the frequency of perfect path phylogenies in genotype data provided by Gabriel *et al.* [3]. For each population we slid a window of varying length over the genotype matrices, resulting in the indicated number of genotype matrices that have as many columns as the window size. For these 'narrow' genotype matrices we list the percentage that admits a perfect phylogeny (PP) and a perfect path phylogeny (PPP). We also indicate what percentage of those genotype matrices admitting a perfect phylogeny (RD_{pp}) or perfect path phylogeny (RD_{ppp}), respectively, also satisfy the rich-data hypothesis.

Notably, less than one third of those genotype matrices admitting a perfect (path) phylogeny satisfy the rich-data hypothesis and, thus, meet the preconditions of the linear-time algorithm given by Halperin and Karp [7].

1.4   Our Contribution

In the present paper we show that the haplotype inference problem via perfect phylogeny becomes tractable if we impose the restrictions discussed before: restricting the perfect phylogenies to paths, restricting the number of missing entries per column to a small number, and focusing on directed perfect phylogenies. Due to lack of space, proofs are only given in the full version of this paper.

Theorem 1.4. $\{0,1,2,?\}$-PPPH *is fixed-parameter tractable, where the parameter is the maximum number of question marks per column.*

Recall that $\{0,1,2,?\}$-PPPH itself is NP-complete (even if we furthermore require that no 1-entries are present in the genotype matrix) [5]. The loophole, which is exploited in Theorem 1.4, is that the genotype matrices used in the completeness proof are unrealistic insofar as they contain huge amounts of missing data—much more than any real dataset contains.

The algorithm used in the proof of Theorem 1.4 is composed of a powerful preprocessing and a subsequent exponential-time dynamic programming strategy. The preprocessing relies on graph-theoretic arguments and the dynamic programming makes use of concepts from order theory, see for example [9] for an introduction.

Different authors, see for example [6, 4], have shown that there are numerous connections between perfect phylogeny haplotyping problems and more traditional graph realization problems. In [5] it is shown that $\{0,2,?\}$-PPPH is equivalent to a special version of the *interval hypergraph sandwich problem*, which is defined as follows: A *hypergraph* is a pair $H = (V,E)$ consisting of a vertex set V and a set E of subsets of V. The elements of E are called *hyperedges*. For a *hypergraph sandwich problem* we are as input given two hypergraphs $H^1 = (V,E^1)$ and $H^2 = (V,E^2)$ such that $E^1 = \{e_1^1,\ldots,e_m^1\}$ and $E^2 = \{e_1^2,\ldots,e_m^2\}$ with $e_i^1 \subseteq e_i^2$ for all $i \in \{1,\ldots,m\}$. The goal is to find a hypergraph $H = (V,E)$ that is 'sandwiched' between H^1 and H^2, that is, $E = \{e_1,\ldots,e_m\}$ with $e_i^1 \subseteq e_i \subseteq e_i^2$ for all $i \in \{1,\ldots,m\}$.

Problem 1.5 (Interval Hypergraph Sandwich Problem).
Input: Two hypergraphs $H^1 = (V,E^1)$, $H^2 = (V,E^2)$.
Question: Is there a hypergraph $H = (V,E)$ sandwiched between H^1 and H^2 and a linear ordering of V such that each hyperedge $e \in E$ is an interval?

For the special case of the *intersecting interval hypergraph sandwich problem* all hyperedges in E are required to share a common vertex. Even this restricted problem is NP-complete since it is shown in [5] to be equivalent to $\{0,2,?\}$-PPPH. Therefore, our fixed-parameter tractability result for $\{0,1,2,?\}$-PPPH allows us to state the following:

Theorem 1.6. *The intersecting interval hypergraph sandwich problem is fixed-parameter tractable, where the parameter is the maximum, taken over all vertices v, of edge pairs (e^1,e^2) with $v \in e^2 - e^1$.*

2   Fixed-Parameter Algorithm for $\{0,1,2,?\}$-PPPH

In this section we present the fixed-parameter algorithm that proves Theorem 1.4. The input of the algorithm is a $\{0,1,2,?\}$-matrix A of dimension $n \times m$ in which there are at most k question marks per column.

2.1 Structure of the Algorithm

The algorithm proceeds in two phases. In the first phase, which we call the *preprocessing phase*, the input is first simplified by collapsing multiple columns to one column if we can safely do so. In the second phase dynamic programming is used to compute a completion of the question mark entries such that a perfect path phylogeny results.

The core idea of the preprocessing phase is the following: Suppose several columns become identical by resolving some ?-entries and we replace these columns with that one 'consensus' column. Clearly, if we can find a perfect path phylogeny for this new matrix, we can also find one for the original one. However, the reverse implication is also true if the number of columns that formed the consensus was large enough.

Gusfield [6] introduced the *leaf count*, a 'weight' of a column that depends on the number of 1's and 2's in this column. The main benefit of the preprocessing is that there will be a uniform bound, depending only on k, on the number of columns that have the same leaf count. The edges, corresponding to columns, on a path from a leaf to the root in a perfect phylogeny have to carry increasing leaf counts. The core idea of the dynamic programming used in the second phase is to process the columns in order of ascending leaf counts, building a perfect path phylogeny from bottom to top. For each column, we consider all possible ways to resolve the ?-entries in it. For each possible resolution, we check whether a perfect path phylogeny exists that accommodates the resolution and, if so, construct it. The crucial observation, which makes dynamic programming feasible, is the fact that the desired perfect path phylogeny for columns up to a certain leaf count can be constructed from the information we stored for columns of a constant-size range of leaf counts preceding the current leaf count.

In the following, we describe the two algorithm phases in more detail. The correctness proofs can be found in the full version of the paper.

2.2 Preprocessing Phase

Terminology. We start with some technical terminology that will be used in the description of the preprocessing phase.

An *antichain* in a partially ordered set (poset) P is a set of incomparable elements. The *width* of a poset is the size of its largest antichain. An antichain is *maximal* if none of its proper supersets is also an antichain. A maximal antichain is *highest of cardinality i* if it has cardinality i and no element of any other maximal antichain of cardinality i strictly dominates any of its elements. We write $\mathrm{hma}_i(P)$ for the set containing the highest maximal antichain of cardinality i, if it exists, or for the empty set, if not. Let $\mathrm{hma}(P) := \bigcup_{i=1}^{\infty} \mathrm{hma}_i(P)$. In the following, $\mathrm{hma}_i(P)$ will typically be empty for $i \geq 3$, but it will sometimes be useful to allow for these sets to be nonempty in order to 'catch errors'.

We define a relation $\succ$ on the set $\{0, 1, 2\}$ by $1 \succ 2 \succ 0$. It is extended to $\{0, 1, 2\}$-columns by setting $c \succeq c'$ if $c[i] \succeq c'[i]$ for all rows i. This extended relation is a poset relation. The *leaf count* $\ell(c)$ of a column c is twice the number of 1-entries plus the number of 2-entries (?- and 0-entries are not counted). For a matrix A let A_i denote the set of columns that have leaf count exactly i. Note that if $c \succ c'$ then $\ell(c) > \ell(c')$. Thus the leaf count is a linear extension of the partial order $\succeq$.

A *(partial) resolution* of a column c with ?-entries is a column c' obtained by replacing (some of) the ?-entries with 0, 1, or 2. A *resolution of a set C* of columns with ?-entries is a set containing one resolution of each $c \in C$. Let res$C$ denote the set of all resolutions of a set C of columns and let pres$C$ denote the set of all partial resolutions. Since every column has at most k question mark entries, every column has at most 3^k resolutions. Thus $|\text{res}\,C| \leq 3^{k|C|}$.

A *consensus* for a set C of columns is a column that is a partial resolution of all $c \in C$. Note that two columns c and c' have a consensus iff $c[i] \neq c'[i]$ implies $c[i] = ?$ or $c'[i] = ?$ for all rows i. A consensus can contain ?-entries only in rows in which all columns in C have a ?-entry. For an incomplete genotype matrix A and a column c (not necessarily in A), the *dimension* of c in A is the size of largest subset C of columns of A such that c is a consensus of C.

Our final technical tool is the perfect path phylogeny property (ppp-property), which is defined as follows: Let C be a set of $\{0, 1, 2\}$-columns We say that C has the *ppp-property* if the following two conditions hold: First, the width of $(C, \succeq)$ is at most two. Second, there are two (possibly empty) chains $(C_1, \succeq)$ and $(C_2, \succeq)$ covering $(C, \succeq)$ such that for each row the following holds: If some column in one of the two chains has a 1-entry in that row, all columns in the other chain must have a 0-entry in that row. The name of the ppp-property is justified by the following lemma, see [5].

Lemma 2.1. *A $\{0, 1, 2\}$-matrix A admits a perfect phylogeny that is a path iff the set of columns of A has the ppp-property.*

Algorithm. The objective of the preprocessing is to ensure that, during the actual algorithm, there will only be a fixed number of columns that have any given leaf count. Normally, if a genotype matrix without missing data and without duplicate columns has more than two columns of the same leaf count, the matrix does not permit a perfect phylogeny that is a path. The reason is that different columns with the same leaf count are always incomparable with respect to the partial order $\succ$ and thus cannot lie on the same path from the root to a leaf of a perfect phylogeny—and if the perfect phylogeny is a path, there are only two paths starting at the root.

In the presence of missing data there can be an arbitrary number of different columns that have the same leaf count, but still the genotype matrix allows a perfect path phylogeny. The reason is that we might be able to 'collapse' all the different columns into one column by filling up the missing entries appropriately.

The idea of the preprocessing step is to perform such collapsing, but we cannot just 'collapse everything that can be collapsed' as it may be necessary *not* to collapse columns in order to realize the matrix. We show that it is 'safe' to collapse columns in the input matrix if their number is large. From here on, we denote $\kappa(i) := 6^i \cdot i!$.

Lemma 2.2. *Let A be a genotype matrix with at most k question mark entries per column. Let C be a set of columns in A and let $\ell \geq 1$ be minimal such that the columns in C have a consensus c containing $k - \ell$ questions mark entries. If $|C| > \kappa(\ell)$, then the following holds: The matrix B obtained from A by replacing all columns in C by c is ppp-realizable iff B is realizable.*

To give a flavor of the proof, which is omitted due to lack of space, consider the case $\ell = 1$ and consider seven columns in A that have a consensus c. We are done if there is a ppp-realization of A that contains the consensus c. So suppose that this is not the case and that the realizations of the seven columns of A *differ* from the consensus. Since $\ell = 1$, for each of the seven columns there is exactly one row where the column differs from the consensus, namely by having a ?-entry in that *disputed* position.

We extract the rows containing disputed positions. Since any two of the seven columns are different, every disputed position must be in a different row. Thus we extract at least seven rows. For any seven rows, at least three rows must have a 0-entry, a 1-entry, or a 2-entry in the consensus. We extract these three rows and sort them according to the row that contains the ?-entry, resulting in a matrix that looks this: $\begin{pmatrix} ? & x & x \\ x & ? & x \\ x & x & ? \end{pmatrix}$ with $x \in \{0,1,2\}$. This matrix, and hence the original one, is not ppp-realizable, unless one of the ?-entries is resolved by x.

The rest of the lemma is proved is proved by induction on ℓ. For the inductive step, we use graph-theoretic arguments on a graph defined on the columns of the genotype matrix A and combine them with a contradictory argument similar to the one just given.

While Lemma 2.2 allows us to perform collapses whenever numerous columns have a common consensus, it is not clear how we can *find* these columns efficiently. Given a genotype matrix A in which every column has at most k question marks, we call a set C of columns from A a *collapsible* set if it meets the requirements of Lemma 2.2, i.e., if there exists a consensus c with exactly ℓ question mark entries and if $|C| > \kappa(\ell)$. The following lemma states that a *collapsible* set of columns can be found efficiently.

Lemma 2.3. *Given an $n \times m$ genotype matrix A with at most k question mark entries per column, we can, in time $O(4^k m^2 n)$ and space $O(mn)$, find a collapsible set C of columns with its corresponding consensus c or determine that no such set exists.*

In the preprocessing step we remove duplicate columns and then invoke the algorithm from Lemma 2.3 as often as possible. As long as the algorithm outputs some set C and a consensus c, we collapse the columns in C, remove duplicates, and repeat. The resulting preprocessing is summarized in Figure 2.

The essential property of the preprocessed matrix is that Lemma 2.2 'does not apply' to it. Thus the dimension of any column c in it is at most $\kappa(k) = 6^k \cdot k!$. This allows us to sum up the effect of the preprocessing phase in the following theorem, which will be crucial for keeping the dynamic programming tables of our main algorithm 'small'. From here on, we use $\lambda(i) := (4i+2)\kappa(i)$.

Theorem 2.4. *Let B result from A by the preprocessing algorithm in Figure 2. Then B is ppp-realizable iff A is. Furthermore, if more than $\lambda(k)$ columns in B have the same leaf count, then both A and B are not ppp-realizable.*

Running time. By Lemma 2.3, finding a set of columns that can be collapsed (realized inside the repeat-loop in Figure 2) can be done in time $O(4^k m^2 n)$. Since we can perform at most m collapsing steps before running out of columns, the preprocessing algorithm will stop after at most $O(4^k m^3 n)$ steps.

Input: Genotype matrix A with at most k question mark entries per column.
Output: Genotype matrix B in which every column has dimension at most $\kappa(k)$ and
    which is ppp-realizable iff A is ppp-realizable.

$B \leftarrow A$
repeat
   for $\ell \leftarrow 1$ **to** k **do**
      foreach column b of B **do**
         foreach $p \in \mathrm{pres}\{b\}$ **do**
            if p has $k - \ell$ many ?-entries **then**
               $C \leftarrow \{c \mid c$ is a column of B and a resolution of $p\}$
               if $|C| > \kappa(l)$ **then**
                  $B \leftarrow B$ with all columns in C replaced by p
                  break for loop
until B remains unchanged
return B

Fig. 2. The preprocessing algorithm

2.3   Dynamic Programming Phase

Terminology. In the following, we introduce terminology for the description of the
dynamic programming table. The input for the dynamic programming phase is a preprocessed $n \times m$ matrix B. Observe that the leaf counts of the columns of B range between 0 and some number $L \leq 2n$. During the execution of our algorithm, for increasing
numbers $i \in \{0,\ldots,L\}$ we consider the set B_i of columns of B that have leaf count i.
By Theorem 2.4, if the size of B_i is larger than $\lambda(k)$, we can output that no realizable
resolution exists.

   In the following, we define sets R_i that, intuitively, contain all ppp-realizable candidate resolutions for the columns whose leaf count 'could become' i. Consider the
union $B_{[i-2k,i]}$ of the sets $B_{i-2k}, B_{i-2k+1}, \ldots, B_i$ (we assume that $B_j := \emptyset$ for negative j). Thus, $B_{[i-2k,i]}$ contains all columns whose leaf count, after resolution of the
question marks, could become i. With Theorem 2.4, the cardinality of $B_{[i-2k,i]}$ is at
most $(2k+1)\lambda(k)$. We require that a set $R \in \mathrm{res}\, B_{[i-2k,i]}$ is in R_i if $(R,\succeq)$ has the ppp-
property:

$$R_i := \{R \mid R \in \mathrm{res}\, B_{[i-2k,i]}, R \text{ is ppp}\}.$$

We can bound the size of R_i by bounding the size of $\mathrm{res}\, B_{[i-2k,i]}$: each question mark
in a column in $B_{[i-2k,i]}$ can be resolved in three different ways and there are at most
$k(2k+1) \cdot \lambda(k)$ question marks altogether in $B_{[i-2k,i]}$. Thus, $|R_i| \leq 3^{k(2k+1)\lambda(k)}$.

   We next extend our notion of candidate resolutions which are so far defined for a
set of columns having a limited range of leaf counts. We introduce sequences of these
candidate resolutions covering all columns having leaf counts up to i. Let us call a
sequence $(R_0,\ldots,R_i) \in R_0 \times \cdots \times R_i$ *consistent* if every column of B is resolved in
exactly the same way in all R_j in which it occurs. Let

Input: Genotype matrix A with at most k question mark entries per column.
Output: Statement whether A admits a perfect path phylogeny.

$B \leftarrow preprocess\, A$
for $i \leftarrow 0$ **to** L **do**
    if $|B_i| > \lambda(k)$ **then output** 'A has no realizable resolution'; **stop**
$j \leftarrow \min\{j \mid B_j \neq \emptyset\}$
foreach $R_j \in R_j$ **do**
    if R_j is ppp **then** $H(R_j) \leftarrow \{\text{hma}(R_j)\}$
for $i \leftarrow j+1$ **to** L **do**
    if $B_{[i-2k,i]} \neq \emptyset$ **then**
        $j \leftarrow \max\{j \mid j < i \text{ and } B_{[j-2k,j]} \text{ is not empty}\}$
        foreach $R_i \in R_i$ **do**
        $H(R_i) = \{\text{hma}(H_j \cup [R_i]^{=i}) \mid R_j \in R_j, (R_j, R_i) \text{ is consistent},$
                                $H_j \in H(R_j), H_j \cup [R_i]^{\geq i} \text{ is ppp}\}$
if $H(R_L) \neq \emptyset$ **then output** 'A admits a perfect path phylogeny'
                    else output 'A does not admit a perfect path phylogeny'

Fig. 3. Main algorithm. Using standard dynamic programming techniques, the above algorithm can be modified to output a realizable resolution of A instead of just deciding whether such a resolution exists.

$$R_{\leq i} := \{(R_0, \ldots, R_i) \in R_0 \times \cdots \times R_i \mid (R_0, \ldots, R_i) \text{ consistent}, R_0 \cup \cdots \cup R_i \text{ is ppp}\}.$$

Intuitively, each $(R_0, \ldots, R_i) \in R_{\leq i}$ (which does not contain ?-entries) is a possible 'consistent' realization of columns in $B_0, \ldots, B_i$ (which may contain ?-entries). The columns in $R_0 \cup \cdots \cup R_i$, however, may have leaf counts larger than i, namely up to $i + 2k$. Being interested in 'filtering out' those columns that actually have a certain leaf count, for a set R of columns we define the set $[R]^{=j}$ to contain the those columns in R that have leaf count j. In the same way we define $[R]^{\leq j}$ and $[R]^{>j}$. Thus, we use subscripts to refer to the leaf count *before* replacing question marks, and superscripts to refer to the leaf count *after* the resolution.

Our final technical definition concerns the highest maximal antichains of the column sets. For $R_i \in R_i$ we define:

$$H(R_i) := \{\text{hma}([R_0 \cup \cdots \cup R_i]^{\leq i}) \mid (R_0, \ldots, R_i) \in R_{\leq i}\}.$$

Algorithm. The algorithm uses dynamic programming. The employed dynamic programming table has one column for every leaf count ranging from 0 to L. The column corresponding to leaf count i has one entry for every $R_i \in R_i$. The table entry corresponding to $R_i \in R_i$ is defined to contain $H(R_i)$. The dynamic programming step computes $H(R_i)$ based only on $H_j(R_j)$ for all $R_j \in R_j$, where $j < i$ is maximal such that $B_{[j-2k,j]}$ is not empty. Figure 3 shows pseudo-code for solving $\{0,1,2,?\}$-PPPH. Proving the correctness of this algorithm is deferred to the full version of the paper.

Running time. The algorithm iterates $L \leq 2n$ times. In each iteration it computes $H(R_i)$ for each $R_i \in R_i$. For computing $H(R_i)$, we consider each R_j and, for each of these, each $H_j \in H(R_j)$. We know that both $|R_j|$ and $|R_i|$ are bounded by $3^{k(2k+1)\lambda(k)}$ (see definition of R_i). The size of H_j can be bounded by $3^{k(2k+1)\lambda(k)} \cdot 2n$. The test whether (R_j, R_i) is consistent and the test whether $H_j \cup R_i$ has the ppp-property can both be done in time $O(|R_i|)$. In summary, the running time is $O\left(4^k m^3 n + 3^{O(k^3 \cdot 6^k \cdot k!)} \cdot n^2\right)$, including the preprocessing.

3 Conclusion and Open Problems

We have shown that the haplotype inference problem via perfect phylogeny in the presence of missing data becomes feasible if the input data meets two conditions: the number of missing entries per SNP site is small and the perfect phylogeny is a path. An analysis of publicly available data shows that these requirements are often met by genomic data available for humans.

While, admittedly, the factor in the running time of our algorithm that depends on k grows quickly with k, the estimate that we give is a worst-case estimate that might be far too pessimistic in real instances.

Two main open questions remain concerning the presented problem parameterization, namely by the maximum number of missing SNP states per SNP site. First, is the undirected version of the perfect path phylogeny problem fixed-parameter tractable? Second, and more importantly, is the perfect (possibly non-path) phylogeny case fixed-parameter tractable?

References

[1] V. Bafna, D. Gusfield, G. Lancia, and S. Yooseph. Haplotyping as perfect phylogeny: A direct approach. *Journal of Computational Biology*, 10(3–4):323–340, 2003.

[2] E. Eskin, E. Halperin, and R. M. Karp. Efficient reconstruction of haplotype structure via perfect phylogeny. *Journal of Bioinformatics and Computational Biology*, 1(1):1–20, 2003.

[3] S. B. Gabriel, S. F. Schaffner, H. Nguyen, J. M. Moore, J. Roy, B. Blumenstiel, J. Higgins, M. DeFelice, A. Lochner, M. Faggart, S. N. Liu-Cordero, C. Rotimi, A. Adeyemo, R. Cooper, R. Ward, E. S. Lander, M. J. Daly, and D. Altshuler. Structure of halpotype blocks in the human genome. *Science*, 296:2225–2229, 2002.

[4] M. C. Golumbic and A. Wassermann. Complexity and algorithms for graph and hypergraph sandwich problems. *Graphs and Combinatorics*, 14:223–9, 1998.

[5] J. Gramm, T. Nierhoff, R. Sharan, and T. Tantau. On the complexity of haplotyping via perfect phylogeny. In *Proceedings of the 2nd RECOMB Satellite Workshop on Computational Methods for SNPs and Haplotypes*, LNBI. Springer, 2004. To appear.

[6] D. Gusfield. Haplotyping as perfect phylogeny: Conceptual framework and efficient solutions. In *Proceedings of the 6th RECOMB*, pages 166–75. ACM Press, 2002.

[7] E. Halperin and R. M. Karp. Perfect phylogeny and haplotype assignment. In *Proceedings of the 8th RECOMB*, pages 10–19. ACM Press, 2004.

[8] Gad Kimmel and Ron Shamir. The incomplete perfect phylogeny haplotype problem. In *Proceedings of the 2nd RECOMB Satellite Workshop on Computational Methods for SNPs and Haplotypes*, LNBI. Springer, 2004. To appear.

[9] W. T. Trotter. *Combinatorics and Partially Ordered Sets: Dimension Theory*. The Johns Hopkins University Press, Baltimore, 1992.

[10] J. Zhang, W. L. Rowe, A. G. Clark, and K. H. Buetow. Genomewide distribution of high-frequency, completely mismatching SNP haplotype pairs observed to be common across human populations. *American Journal of Human Genetics*, 73(5):1073–81, 2003.

Simplifying the Weft Hierarchy

Jonathan F. Buss and Tarique Islam⋆

School of Computer Science, University of Waterloo
{jfbuss,tmislam}@uwaterloo.ca

Abstract. We give simple, self-contained proofs of the basic hardness results for the classes $W[t]$ of the weft hierarchy. We extend these proofs to higher levels of the hierarchy and illuminate the distinctions among its classes. The anti-monotone collapse at $W[1, s]$ and the normalization of weft-t formulas arise as by-products of the proofs.

Introduction

The theory of fixed-parameter hardness and the W-hierarchy has been obscured by technicalities of the definitions of the classes of the W-hierarchy and by the difficulty of the known proofs of the basic completeness results. We present a significantly simpler formulation and proofs of basic completeness results for $W[1]$, which eliminate many of the technicalities. The new proofs closely follow the development of the theory of NP-completeness, greatly simplifying the complex manipulations of circuits and eliminating the ancillary graph-theoretic problems of the original proofs.

Starting from Chen, Flum and Grohe's [4] characterization of $W[t]$, we present self-contained and relatively straightforward proofs that

1. WEIGHTED c-SAT $\in W[1]$ for all constants $c \geq 2$,
2. SHORT NTM ACCEPTANCE is hard for $W[1]$, and
3. SHORT NTM ACCEPTANCE is reducible to WEIGHTED 2-SAT via a reduction that produces only anti-monotone formulas.

Therefore, an initial treatment of $W[1]$-completeness can follow entirely classical lines, closely mimicking the standard proofs that SAT is NP-complete:

1. SAT $\in NP$,
2. A Turing machine can simulate a log-cost random-access machine in polynomial time, and
3. CNFSAT is NP-hard.

Along the way, we obtain the restriction to anti-monotone 2-SAT formulas as a simple observation.

The full impact of the new formulation arises higher in the W-hierarchy. We present new proofs that

---

⋆ Both authors supported in part by a grant from the Natural Sciences and Engineering Research Council (NSERC) of Canada.

R. Downey, M. Fellows, and F. Dehne (Eds.): IWPEC 2004, LNCS 3162, pp. 187–199, 2004.

1. WEIGHTED WEFT-t-DEPTH-d SAT $\in W[t]$ for all constants $d \geq t \geq 2$, and

2. WEIGHTED t-NORMALIZED SAT is hard for $W[t]$.

Our new proofs, unlike the former ones, are straightforward extensions of the $W[1]$ case. Moreover, the full Normalization Theorem and the Monotone and Antimonotone Collapse Theorems arise as simple corollaries of the proofs.

Finally, we abstract the Chen-Flum-Grohe characterization to a more general setting, comparable to but different than the "guess-then-check" classes of Cai and Chen [1]. The new viewpoint illuminates the apparent disconnection between alternation and space in parameterized complexity [4].

1   Background

1.1   Parameterized Problems and the Weft of Circuits

We briefly review the definitions of parameterized problems and the weft of a circuit; for further history and discussion, see Downey and Fellows [6].

Definition 1. *A* parameterized problem *is a set of pairs $Q \subseteq \Sigma^* \times \Gamma^*$ over finite alphabets Σ and Γ. For each $(x, k) \in Q$, x is taken as input and k is taken as the parameter. A parameterized problem is* fixed-parameter tractable *if there is an algorithm that can determine whether $(x, k) \in Q$ using time $f(k)p(|x|)$ for some computable function f and polynomial p.*

Definition 2. *An* FPT-reduction *from a parameterized problem $Q \subseteq \Sigma^* \times \Gamma^*$ to a parameterized problem $Q' \subseteq (\Sigma')^* \times (\Gamma')^*$ is a mapping $R: \Sigma^* \times \Gamma^* \mapsto (\Sigma')^* \times (\Gamma')^*$ such that*

1. *For all $(x, k) \in \Sigma^* \times \Gamma^*$, $(x, k) \in Q \Leftrightarrow R(x, k) \in Q'$,*
2. *There exists a computable function $g: \mathbb{N} \mapsto \mathbb{N}$ such that for all $(x, k) \in \Sigma^* \times \Gamma^*$, where $R(x, k) = (x', k')$, we have $k' \leq g(k)$, and*
3. *There exist a computable function $f: \mathbb{N} \mapsto \mathbb{N}$ and a polynomial p such that R is computable in time $f(k) \cdot p(|x|)$.*

We write $Q \leq^{fpt} Q'$ if there is an FPT reduction from a parameterized problem Q to another parameterized problem Q'.

The breakthrough in measuring parameterized hardness depended on a measure of circuit (or formula) complexity called *weft*. Fix any constant $u \geq 2$; call a gate of a circuit "large" if it has fan-in more than u.

Definition 3. *Let C be a decision circuit. The* weft *of C is the maximum number of large gates on any path from the inputs to the output. A circuit is t-normalized iff the and- and or-gates alternate, with an and-gate at the output, and it has depth (and hence weft) at most t.*

Since we will deal only with circuits of bounded depth and polynomial size, circuits are equivalent to formulas (tree circuits) of the same depth and polynomial size. A formula in u-CNF has weft 1 and depth 2; any CNF formula is 2-normalized.

A central problem in the theory of parameterized complexity is the weighted satisfiability problem. An assignment to the variables of a circuit has weight k iff it sets esactly k variables to true. These following variations of weighted satisfiability formed the basis of original formulation of the W-hierarchy.

WEIGHTED SAT
   Instance: A Boolean formula φ.
   Parameter: An integer k.
   Question: Does φ have a weight-k satisfying assignment?

WEIGHTED c-SAT   (for $c \geq 2$)
   WEIGHTED SAT restricted to c-CNF formulas.

WEIGHTED WEFT-t-DEPTH-d SAT   (for $d \geq t \geq 2$)
   WEIGHTED SAT restricted to formulas of weft t and depth d.

WEIGHTED t-NORMALIZED SAT   (for $t \geq 2$)
   WEIGHTED SAT restricted to t-normalized formulas.

Firming the theory required establishing two key "collapses" in complexity.

Proposition 4.

1. WEIGHTED c-SAT $\leq^{fpt}$ WEIGHTED 2-SAT *for all constants $c \geq 2$.*
2. WEIGHTED WEFT-t-DEPTH-d SAT $\leq^{fpt}$ WEIGHTED t-NORMALIZED SAT, *for $d \geq t \geq 2$.*

Each of these intermediate results was obtained via a complicated combinatorial proof including an ancillary graph-theoretic problem. For the first case, the proof defined and used the novel RED/BLUE NONBLOCKER problem; for the second case, the classical DOMINATINGSET problem sufficed. The details of reductions from satisfiability to the graph problems were crucial to the proofs. Although the two properties seem related, their proofs were nevertheless quite distinct. Below, we obtain both results as direct corollaries of our proofs. Also, our proof for weft $t \geq 2$ is simply an extension of the proof for weft 1.

1.2   Machines and Algorithms

Recently,Chen, Flum, and Grohe [3,4] discovered a machine-based characterization of $W[t]$.

Definition 5. *A* W-RAM *is an alternating random-access machine with two sets of registers: standard registers $r_0, r_1, r_2, \ldots,$ and guess registers $g_0, g_1, g_2, \ldots.$ Any standard deterministic operation is allowed on the standard registers. Only the following operations may access the guess registers:*

Exists j: Store any natural number in the range $[0, r_0]$ *in register* g_j, *with existential branching.*

Forall j: Store any natural number in the range $[0, r_0]$ *in register* g_j, *with universal branching.*

JG= i j c: If $g_{r_i} = g_{r_j}$, *then jump to instruction location c.*

JG0 i j c: If $r_{\langle g_{r_i}, g_{r_j} \rangle} = 0$, *then jump to instruction location c.*

Here $\langle \cdot, \cdot \rangle$ *is any suitable pairing function.*

Limiting the access to guess registers simplifies the analysis of a computation.

Lemma 6. *If two executions of a W-RAM on the same input follow the same sequence of instructions, they have the same set of values in the standard registers.*

Definition 7. *A W-RAM program R is an* AW-program *if there are a computable function f and a polynomial p such that for every input* (x, k) *with* $|x| = n$ *the program R on every run,*

1. *performs at most* $f(k) \cdot p(n)$ *steps,*
2. *has at most* $f(k)$ *existential or universal steps,*
3. *accesses at most the first* $f(k) \cdot p(n)$ *standard registers, and*
4. *stores values at most* $f(k) \cdot p(n)$ *in any register at any time.*

We shall treat the following result of Chen, Flum and Grohe [4] as the definition of the classes $W[t]$.

Proposition 8. *Let Q be a parameterized problem and* $t \geq 1$. *Then Q is in* $W[t]$ *if and only if there are a positive integer u, a computable function h and an AW-program R for a W-RAM such that R decides Q and such that for every run of R on an instance* (x, k) *of Q as input,*

1. *all existential and universal steps are among the last* $h(k)$ *steps of the computation,*
2. *there are at most* $t - 1$ *alternations between existential and universal states, the first guess step is existential, and*
3. *every block without alternations, besides the first one, contains at most u guess steps.*[1]

Simple algorithms suffice to put some problems in the W-hierarchy.

Lemma 9.

– *The problems* INDEPENDENTSET, CLIQUE, SHORT NTM ACCEPTANCE *and* RED/BLUE NONBLOCKER *are in* $W[1]$.
– DOMINATINGSET *is in* $W[2]$.

---

[1] Omitting this condition gives apparently different classes denoted $A[t]$ [7]. With hindsight, one may consider that the difficulty of formulating this condition in terms of Turing machines or circuits lies at the heart of the difficulty in creating the original formulation of parameterized hardness.

Proof. For example, to determine whether a graph G has an independent set of size k, first construct the adjacency matrix A_G of G, using time $O(n^2)$. Then, in $O(k^2)$ further operations, guess k vertices and check that all of the corresponding entries in A_G are 0.

The other algorithms are similar. □

2   The Complexity of Weighted Satisfiability

2.1   AW-programs

Algorithms placing the variants of weighted satisfiability in the W-hierarchy are somewhat more complex than those given earlier, but follow the same basic plan.

Lemma 10. WEIGHTED 2-SAT *is in* $W[1]$.

Proof. The algorithm considers three categories of clauses. For each category, it creates a separate look-up table for the properties required by clauses of that category. It then guesses a single weight-k assignment and checks that the assignment satisfies the conditions of each table.

Anti-monotone clauses: $(\neg x_i \vee \neg x_j)$.

These clauses are equivalent to an instance of INDEPENDENTSET, with one node for each variable and an edge between two variables appearing in the same clause. The look-up table is simply the adjacency matrix.

Implications: $(\neg x_i \vee x_j)$.

We interpret these clauses as $x_i \Rightarrow x_j$. For each variable x_i, the algorithm computes and stores the set of variables $S_i = \{ x_j \mid (\neg x_i \vee x_j) \text{ appears in } \varphi \}$. Once an assignment A is guessed, for each variable x_i set to true by A, it looks up the corresponding S_i, and checks that each variable in S_i is also true. These checks require $O(k^2)$ steps for a weight-k assignment.

Monotone clauses: $(x_i \vee x_j)$.

If some variable x_i appears in more than k distinct monotone clauses, it must be set true in every weight-k satisfying assignment. The other clauses contain only variables that occur at most k times; if they are simultaneously satisfiable with weight k, there must be at most k^2 such clauses.

To check satisfaction by an assignment, the algorithm checks that each required variable is true and that the remaining true variables satisfy the k^2 other clauses.

After creating the data structure for each category of clauses, R nondeterministically guesses k variables to be true and accepts iff the guessed set of variables satisfies all three constraint structures. This check uses $O(k^2)$ steps, as required. □

The above monotone and anti-monotone clauses may also be considered as implications, of the respective forms $(true \Rightarrow (x_i \vee x_j))$ and $((x_i \wedge x_j) \Rightarrow false)$. We generalize the idea to an algorithm solving WEIGHTED c-SAT, for any constant c, and then further to higher weft.

Lemma 11. *For all $c \geq 2$,* WEIGHTED c-SAT $\in W[1]$.

Proof. For each $0 \leq r \leq c$, consider the clauses with r negated variables: $(\sum_{i=1}^{r} \neg x_i \vee \sum_{j=r+1}^{c} x_j)$ or equivalently, $\left((\prod_{i=1}^{r} x_i) \Rightarrow \sum_{j=r+1}^{c} x_j \right)$. Regroup all clauses with the same implicant together, producing an equivalent formula

$$\prod_{r=0}^{c} \prod_{x_1', \ldots, x_r'} \left((\prod_{i=1}^{r} x_i) \Rightarrow \psi_{x_1', \ldots, x_r'} \right) ,$$

where the second product is over all size-r subsets of the variables, and each $\psi_{x_1' \ldots x_r'}$ is a monotone $(c-r)$-CNF formula.

For each value of $r \leq c$, the algorithm constructs a table of size $\binom{n}{r}$, with one entry for each subset of size r. The entry corresponding to $x_1', \ldots, x_r'$ contains all information needed to check whether a guessed assignment satisfies $\psi_{x_1' \ldots x_r'}$. Since this formula is monotone, the data size and checking size are bounded by a functon of k and c, regardless of n. □

Lemma 12. WEIGHTED WEFT-t-DEPTH-d SAT $\in W[t]$, *for $d \geq t \geq 2$.*

Proof. Fix any constant d, and let φ be a weft-t, depth-d tree circuit with negations only at the inputs. The algorithm has three main phases: simplifying the form of the circuit, constructing tables to enable fast truth-checking, and finally evaluating the circuit at guessed values using alternation.

1. Simplify the form of the circuit, to have (starting from the output) at most two levels of small gates, followed by t levels of alternating large gates (no small gates), and then one level of small gates. The algorithm proceeds from the output of the circuit to the inputs, locally restructuring the circuit so as to combine each small gate with a neighbouring large gate.[2]
2. Construct satisfaction-checking tables (as in the algorithm for CNFSAT above) for the large gates closest to the input literals.
3. Guess k variables to be true, find inputs to $t-1$ levels of large gates using alternation, and finally check the satisfaction of the input gates using the tables constructed in phase 2.

Simplifying the Form of a Circuit. Let a "small sub-circuit" be a maximal connected sub-circuit comprising only small gates. Since each small sub-circuit has gates of fan-in two and has depth at most d, it receives at most 2^d "input" values (each either a literal or the output of a large gate). Therefore, its CNF and DNF equivalents each have size at most 2^{2^d}, which is a constant for our

---

[2] This process is essentially the one used as a (small) part of the original proof of the Normalization Theorem ([6], p. 293). The tables of phase 2, combined with the alternation of phase 3, replace the multiple changes of variables and bypass the DOMINATINGSET problem entirely. Our following proof of the $W[t]$-completeness of WEIGHTED WEFT-t SAT will complete the normalization.

purposes. We will select one of these equivalents depending on the surrounding circuit.

As a first step, convert the small sub-circuit (if any) at the output either to CNF or to DNF. Next consider a large gate G whose output is an input to this sub-circuit. We shall assume for simplicity of description that G is an and-gate. (If G is an or-gate, apply the De Morgan dual of the following.) Each input G_i of G comes from another small sub-circuit S_i. Convert each S_i to its CNF equivalent S_i', and merge the output gates of each S_i' into G, forming a larger and-gate G'.

Each or-gate S_{ij} of each S_i' has at most 2^d gates as inputs. Those that are or-gates may logically be merged with S_{ij}, but to do so would increase the weft if S_{ij} also has a large and-gate as input. To circumvent this problem, consider the partial circuit comprising gate S_{ij} and all of its large and-gate inputs. (The inputs to this partial circuit come from or-gates, small sub-circuits, and literals.) The partial circuit is in DNF form, with a constant number (2^d) of disjuncts. Therefore, its CNF equivalent circuit C_{ij} has size $n^{O(1)}$. Replace the partial circuit by C_{ij} and finally merge the output of C_{ij} with the original gate G.

At this point the gate G has only large gates and literals as inputs. Move to the next level of large gates, and modify them in the same manner. Upon reaching the input variables, the new circuit has the required form.

Creating the Tables for Satisfaction-Checking. Following the simplification, each depth-2 circuit at the inputs, comprising a single large gate whose inputs are small gates with literals as inputs, is in either 2^d-CNF (for a large and-gate) or 2^d-CNF (for a large or-gate) form. Each of these depth-2 circuits receives its own tables for checking satisfaction. In the CNF case, the tables are those from the previous algorithm. The DNF case requires a further technique.

For the DNF case, in addition to using variables as indices into the tables, we introduce a set of additional indices, $Y = \{\, y_{i,j} \mid 1 \le i \le j \le k \,\}$, where y_{ij} signifies that all variables x_ℓ, for $i \le \ell \le j$, are set false.[3] Let Z be any consistent set of at most 2^d indices from $X \cup Y$. The entry correponding to Z in the table for a large or-gate G is set to 1 if any single conjunct is made true by Z (which guarantees that any assignment consistent with Z satisfies G).

Evaluating the Circuit. To check satisfiability of the resulting weft-t formula by a weight-k assignment, existentially guess both the assignment and the values of the small monotone sub-circuit at the output. Proceed toward the inputs, guessing an input line for each large gate (existentially for an or-gate and universally for an and-gate), until an input depth-2 circuit is reached. Finally, check that the required conditions hold: each large gate computes its output correctly, the output sub-circuit produces *true*, and the truth-checking tables justify the guessed values. In the case of an or-gate, it is necessary to check all consistent

---

[3] These indices correspond to the "gap variables" of the original proof. They are the only remnant of the DOMINATINGSET problem.

combinations of indices; since there are $k^{O(2^d)}$ such combinations, the required bounds are met. □

2.2   Hardness for $W[1]$

For the hardness results, we use the SHORT NTM ACCEPTANCE problem. Two lemmas provide the connections to satisfiability and to AW-programs.

Lemma 13 (Chen, Flum). SHORT NTM ACCEPTANCE *is hard for $W[1]$.*

Lemma 14 (Cai, Chen, Downey, Fellows [2]).
SHORT NTM ACCEPTANCE $\leq^{fpt}$ WEIGHTED 2-AMSAT.

Together with the above algorithm for WEIGHTED c-SAT, these yield a new proof of the "Cook's Theorem" for $W[1]$.

Theorem 15. *The problems* SHORT NTM ACCEPTANCE, WEIGHTED c-SAT *(for any fixed $c \geq 2$) and* WEIGHTED 2-AMSAT *are complete for $W[1]$.*

Well-known reductions then give

Corollary 16. INDEPENDENTSET, CLIQUE, *and* RED/BLUE NONBLOCKER *are $W[1]$-complete.*

We reprise the original proofs of the lemmas (cf. [6]) in order to later extend them higher in the weft hierarchy.

Proof. (of Lemma 13.) Let $Q \in W[1]$ be a paramaterized problem solved by AW-program R, using $f(k)p(|x|)$ steps and ending with $h(k)$ guess steps, on any input (x, k). (Hereinafter we shall omit the arguments of f, p, and h.) The desired reduction, on input (x, k), will produce a nondeterministic Turing machine M that accepts the empty input in k steps if and only if $(x, k) \in Q$. M will have alphabet $\Sigma = [f \cdot p]$; that is, one symbol for each possible value of a register of R. The states of M will encode $(k+1)$-tuples $\langle S, i_1 \ldots, i_k \rangle$, where S is an instruction of R and $i_1, \ldots, i_k$ are indices of registers.

The start state of M encodes the configuration of $R(x)$ reached just before the first guess step. M begins by guessing the h values for the existential registers of R, writing one guess to each of h cells. It then checks, in h^2 steps, that R would indeed accept with these guesses. Thus M accepts the empty input in $h^2(k)$ steps iff R accepts (x, k), as required. □

Proof. (of Lemma 14.) Let M have state-set Q and alphabet Σ (including blank). Without loss of generality we assume that M has a unique halting state q_{yes}, that M reaches this state only at the left end of the tape, and that once M enters state q_{yes} it remains in the same state and cell indefinitely.

Consider the tableau associated with a k-step computation of M; it contains k rows and k columns, for a total of k^2 entries. The contents of the tableau are described using variables $Y_{tph\tau\sigma}$, where $1 \leq t \leq k$, $1 \leq p \leq k$, $h \leq k$, $\tau \in$

Q, and $\sigma \in \Sigma$. The reduction produces a formula φ that is satisfied by a weight-k^2 assignment A of the variables if and only if there is a computation of M such that for all t, m, h, q and σ, variable $Y_{tmhq\sigma}$ has value *true* in A exactly if at time t, position p of the tape contains symbol σ and the head scans position h in state q.

To obtain anti-monotone 2-CNF formulas, we do not directly require consistency among the parts of the computation; instead, we forbid inconsistency. All clauses will have the form $\neg w \vee \neg x$. For example, the condition that at most one state occurs at time t is expressed by the conjunction

$$\prod_{1 \leq p, p' \leq k} \prod_{1 \leq h, h' \leq k} \prod_{\sigma, \sigma' \in \Sigma} \prod_{q \neq q' \in Q} (\neg Y_{tphq\sigma} \vee \neg Y_{tp'h'q'\sigma'}) \ .$$

Similar conjunctions enforce the other constraints. We leave them to the reader.

The condition that some combination of state, head position and cell contents must hold at each step is not directly expressed by the formula. This condition is enforced by the consideration only of assignments of weight k^2 in combination with the expressed prohibitions.

The entire formula has size $f'(k)p'(n)$ and can be computed in proportional time. It has a satisfying assignment of weight $h(k)^2$ if and only if R accepts input (x, k), as required.   □

Looking ahead, we note that one can substitute monotone CNF formulas (with unbounded conjuncts) for the anti-monotone 2-CNF formulas of the proof. For example, the condition that at least one state occurs at time t is expressed by the formula

$$\prod_{1 \leq p \leq k} \prod_{1 \leq h \leq k} \prod_{\sigma \in \Sigma} \sum_{q \in Q} Y_{tphq\sigma} \ .$$

With all subformulas modified in this fashion, the restriction to assignments of weight $h(k)^2$ ensures that multiple states do not occur.

2.3   Hardness for $W[t]$, $t \geq 2$

To translate the above results to $W[t]$ for $t \geq 2$, we combine the proofs of Lemmas 13 and 14. The key additional technique is to handle all of the guess steps after the first existential block deterministically, as part of the reduction itself. In other words, for every configuration that could occur at the end of the first existential block, the reduction computes the conditions under which the machine will eventually accept from that configuration and incorporates them into the reduction formula.

Theorem 17. *For $t \geq 2$, WEIGHTED t-NORMALIZED SAT is complete for $W[t]$.*

Corollary 18 (Downey, Fellows [5]). DOMINATINGSET *is $W[2]$-complete.*

Proof. Let AW-program R accept Q in time $f(k)p(n)$, with guess steps only the last $h(k)$ steps and occurring in t alternating blocks, with an existential block first and all blocks after the first limited to u guesses. We consider a computation of R to have three parts: (1) the deterministic part, up until the first guess step, (2) the existential part, comprising the steps from the first guess step up to the first universal step, and (3) the decision part, comprising the remainder of the computation. We handle the first two parts by combining the constructions of Lemmas 13 and 14, producing a set φ_E of clauses that mimic the computation of R. However, we omit the clauses that require the (virtual) Turing machine to end in an accepting state. We will replace the omitted clauses with a formula φ_A that requires part 3 of R's computation to be accepting, as given below. As noted above, φ_E can be taken in either monotone or anti-monotone form, as required.

For simplicity, we start with the case $t = 2$; that is, part 3 comprises a single universal block with u guesses. For each possible outcome $G = \langle g_1, g_2, \ldots, g_u \rangle$ of the universal guesses and each possible branch b determined by outcome of the tests involving the earlier existential guesses, the reduction will produce a formula $\alpha_{G,b}$ described below. With any assignment to φ_E that represents a valid computation of the existential part, the formula $\alpha_{G,b}$ will evaluate to true if the universal choices specified by G lead to acceptance along path b or if some path other than b is followed, and will evaluate to false if the universal choices given by G lead to rejection along path b. Thus the conjunction of all formulas $\alpha_{G,b}$ gives the required acceptance condition for the computation of R. Since there are $(fp)^u$ possible outcomes of the u universal guesses and 2^h possible branches in h steps, the overall formula meets the required bounds on size.

If branch b ends by accepting, using $\alpha_{Gb} = \mathit{true}$ suffices. The case that b rejects requires a non-trivial formula, based on sub-formulas for each test instruction T on branch b that involves an existential register. Each subformula will evaluate to true if test T is satisfied. For a given test instruction T, we shall actually define four different formulas. The formulas π_T and σ_T will be satisfied if T is taken, and $\pi_{\neg T}$ and $\sigma_{\neg T}$ will be satisfied if T fails. The π formulas will be anti-monotone products-of-sums and the σ formulas monotone sums-of-products.

Let $Y_{j\ell}$ (in full, $Y_{kj\ell 1q_{yes}}$) be the variable that is true iff register j contains value ℓ at the end of part 2 of the computation by R. Note that when φ_E is satisfied by an assignment of the correct weight, exactly one Y_{jl} is true for each value of j. Thus we define the subformulas σ_T and π_T to be correct given this condition, as follows.

T is JG$= i\ j\ c$:
    If i is universal and j existential, let

$$\sigma_T = Y_{jg_i} \qquad \text{and} \qquad \pi_T = \prod_{\ell \neq g_i} \neg Y_{i\ell} \ .$$

If both i and j are existential, let

$$\sigma_T = \sum_{0 \le \ell \le fp} Y_{i\ell} \wedge Y_{j\ell} \qquad \text{and} \qquad \pi_T = \prod_{0 \le \ell \neq \ell' \le fp} \neg Y_{i\ell} \vee \neg Y_{j\ell'} \ .$$

T is JG0 $i\ j\ c$:

If i is universal and j existential, let

$$\sigma_T = \sum_{0 \le \ell \le fp} Y_{j\ell} \wedge Y_{\langle g_i, \ell \rangle 0}$$

and

$$\pi_T = \prod_{0 \le \ell \le fp} \prod_{1 \le m \le fp} \neg Y_{j\ell} \vee \neg Y_{\langle g_i, \ell \rangle m} \ .$$

If both i and j are existential, let

$$\sigma_T = \sum_{0 \le \ell \neq \ell' \le fp} Y_{i\ell} \wedge Y_{j\ell'} \wedge Y_{\langle \ell, \ell' \rangle 0}$$

and

$$\pi_T = \prod_{0 \le \ell \neq \ell' \le fp} \prod_{1 \le m \le fp} \neg Y_{i\ell} \vee \neg Y_{j\ell'} \vee \neg Y_{\langle \ell, \ell' \rangle m} \ .$$

For any test T, the formulas $\sigma_{\neg T}$ and $\pi_{\neg T}$ are obtained as the normal-form equivalents of $\neg \pi_T$ and of $\neg \sigma_T$, respectively.

The formula $\alpha_{G,b}$ is then $\sum_{T \text{ in } b} \sigma_{\neg T}$. Overall, the monotone CNF formula

$$\varphi_A = \prod_G \prod_b \alpha_{G,b}$$

evaluates to true on a weight-h^2 assignment corresponding to a part-2 computation by R if and only if R accepts after the corresponding existential guesses. The size of φ_A is $g(k)e(n)$ for a computable function g and a polynomial e (which depend on the constants t and u, as well as functions h and f).

At this point, we have obtained a weft of 2, but the depth is 3 due to the small conjunctions in the formulas $\sigma_{\neg T}$. To convert to depth 2, we introduce two new sets of variables Z_{cd} and Z_{cde} for $0 \le c, d, e \le fp$. Let φ_Z be the conjunction of the CNF equivalents of each bi-implication

$$Z_{\langle i_1, j_1 \rangle \langle i_2, j_2 \rangle} \Leftrightarrow (Y_{i_1 j_1} \wedge Y_{i_2 j_2})$$

and

$$Z_{\langle i_1, j_1 \rangle \langle i_2, j_2 \rangle \langle i_3, j_3 \rangle} \Leftrightarrow (Y_{i_1 j_1} \wedge Y_{i_2 j_2} \wedge Y_{i_3 j_3}) \ .$$

In formula φ_A, replace each compound disjunct of each α_{Gb} by the equivalent Z variable.

The final formula $\varphi_E \wedge \varphi_A \wedge \varphi_Z$ has depth 2, and its size remains polynomial in n and k. Further, program R accepts x iff the formula has a satisfying assignment of weight $h^2 + \binom{h^2}{2} + \binom{h^2}{3}$.

For higher values of t, the same idea applies, except that the α formulas are a depth-$(t-1)$ function of the σ or π formulas, giving an overall depth (and weft) of t. The formulas can be taken as monotone (using the π subformulas) for even values of t, and anti-monotone (using the σ subformulas) for odd values of t. $\square$

3   Remarks

Considering only the characterization of $W[t]$ by AW-programs, one may reasonably ask whether these classes really represent the "right" definition of hardness classes. The limited use of the guess registers and the restriction to a constant number of guesses in each block after the first seem rather arbitrary. To date, however, the classes $A[t]$ and $L[t]$ obtained by not imposing these restrictions [3,7] do not seem to have natural compete problems of independent interest. In contrast, $W[2]$ (for example) has importance as the class of languages FPT-reducible to DOMINATINGSET.

Cai and J. Chen [1] proposed a model of "guess-then-check" computations: class $GC(s(n), \mathcal{C})$ contains the problems that can be solved by existentially guessing a witness of $s(n)$ bits and then checking the witness within complexity class $\mathcal{C}$. The AW-programs considered here use a similar but more general paradigm that one might call "prepare, guess, and check." They have a preparatory deterministic part, followed by existential guesses and a check of their correctness. For example, $W[1]$ corresponds to parameterized polynomial time followed by a parametric constant number of guesses (of $O(\log n)$ bits each) with a parametric constant number of checking operations (with operands of $O(\log n)$ bits). Thus "$PGC(FPT, NTIME(O(\log n)))$" (with a suitable precise definition) is a suitable characterization of the class. The classes $W[t]$ arise from replacing the checking class by a version of bounded-alternation $LOGTIME$.

One can also allow other classes as checking classes. Some classes give analogues of classes above $W[t]$, for example $PGC(FPT, ATIME(O(\log n)))$ or $PGC(FPT, SPACE(O(\log n)))$. The former class is the same as $AW[*]$ [4]. The latter class contains the former, since $ATIME(O(\log n)) \subseteq SPACE(O(\log n))$, but with $SPACE(O(\log n)) \subseteq ATIME(O(\log^2 n))$ as our best result in the reverse direction, one might suspect the two to be unequal. This explains the difficulty of creating "$PSPACE$-analogue" classes as opposed to "AP-analogue" classes [4]. We do not know the relationship of $PGC(FPT, SPACE(O(\log n)))$ to $AW[P]$ or to other studied classes.

References

1. L. Cai and J. Chen, "On the Amount of Nondeterminism and the Power of Verifying," *SIAM J. Computing* 26,3 (1997) 733–750.
2. L. Cai, J. Chen, R.G. Downey, M.R. Fellows, "On the Complexity of Short Computation and Factorization," *Archiv. Math. Logic*.
3. Y. Chen and J. Flum, "Machine Characterization of the Classes of the W-hierarchy," in *Computer Science Logic: CSL 2003*, Lecture Notes in Computer Science, Springer, 2003, pp. 114–127.

4. Y. Chen, J. Flum and M. Grohe, "Bounded Nondeterminism and Alternation in Parameterized Complexity Theory," in *18th Ann. IEEE Conf. Computational Complexity*, 2003, pp. 18–29.

5. R.G. Downey and M.R. Fellows, "Fixed-Parameter Tractability and Completeness," *Congressus Numerantium* 87 (1992) 161–187.

6. R.G. Downey and M.R. Fellows, *Parameterized Complexity*, Springer, New York, 1999.

7. J. Flum and M. Grohe, "Fixed-Parameter Tractability, Definability, and Model-Checking," *SIAM J. Computing* 31,1 (2001) 113–145.

The Minimum Weight Triangulation Problem with Few Inner Points

Michael Hoffmann and Yoshio Okamoto[*]

Institute of Theoretical Computer Science, ETH Zurich, CH-8092 Zurich, Switzerland
{hoffmann,okamotoy}@inf.ethz.ch

Abstract. We propose to look at the computational complexity of 2-dimensional geometric optimization problems on a finite point set with respect to the number of inner points (that is, points in the interior of the convex hull). As a case study, we consider the minimum weight triangulation problem. Finding a minimum weight triangulation for a set of n points in the plane is not known to be NP-hard nor solvable in polynomial time, but when the points are in convex position, the problem can be solved in $O(n^3)$ time by dynamic programming. We extend the dynamic programming approach to the general problem and describe an exact algorithm which runs in $O(6^k n^5 \log n)$ time where n is the total number of input points and k is the number of inner points. If k is taken as a parameter, this is a fixed-parameter algorithm. It also shows that the problem can be solved in polynomial time if $k = O(\log n)$. In fact, the algorithm works not only for convex polygons, but also for simple polygons with k interior points.

1 Introduction

A lot of NP-hard optimization problems on graphs can be solved in polynomial time when the input is restricted to partial k-trees, that is, graphs with treewidth at most k, where k is fixed. In this sense, the treewidth is regarded as a natural parameter to measure the complexity of graphs. This is based on the observation that "some NP-hard optimization problems on graphs are easy when the class is restricted to trees."

We try to address the following question: What is a natural parameter that could play a similar role for geometric problems as the treewidth does for graph problems? One basic observation is that "some NP-hard optimization problems on a point set in the Euclidean plane are easy when the points are in convex position." Therefore, the number of inner points can be regarded as a natural parameter for the complexity of geometric problems. Here, an inner point is a point in the interior of the convex hull of the given point set.

In this paper, we concentrate on one specific problem: the minimum weight triangulation problem. The minimum weight triangulation is a triangulation with minimum total length of edges. Triangulations have numerous applications in finite element methods, interpolation and graphics, to name just a few. In applications one is usually interested in finding a triangulation that is optimal in a certain sense. Among several criteria, a minimum weight triangulation is one of the most natural ones.

---

[*] Supported by the Berlin-Zürich Joint Graduate Program "Combinatorics, Geometry, and Computation" (CGC), financed by ETH Zürich and the German Science Foundation (DFG).

R. Downey, M. Fellows, and F. Dehne (Eds.): IWPEC 2004, LNCS 3162, pp. 200–212, 2004.

The minimum weight triangulation problem is notorious as one of the problems which are not known to be NP-hard nor solvable in polynomial time for a long time [8]. However, when the points are in convex position, the problem can be solved in polynomial time by dynamic programming. The main result in this paper is an exact algorithm to compute the minimum weight triangulation in $O(6^k n^5 \log n)$ time, where n is the total number of input points and k is the number of inner points. From the viewpoint of parameterized complexity [7,14] this is a fixed-parameter algorithm if k is taken as a parameter.[1] Furthermore, the algorithm implies that the problem can be solved in polynomial time if $k = O(\log n)$.

Actually, our algorithm also works for simple polygons with inner points. Or, rather we should say that the algorithm is designed for such objects, and as a special case, we can compute a minimum weight triangulation of a point set. This digression to simple polygons is essential because our strategy is based on recursion and in the recursion we cannot avoid dealing with simple polygons.

Related work Since the literature on the minimum weight triangulation problem is vast, we just mention some articles that are closely related to ours. As already mentioned, finding a minimum weight triangulation of a finite point set is not known to be NP-hard nor solvable in polynomial time [8]. For an n-vertex convex polygon, the problem can be solved in $O(n^3)$ using dynamic programming. For an n-vertex simple polygon, Gilbert [9] and Klincsek [12] independently gave a dynamic-programming algorithm running in $O(n^3)$ time. But with inner points the problem seems more difficult. Another polynomial-time solvable case was discussed by Anagnostou and Corneil [3]: they considered the case where a given point set lies on a constant number of nested convex hulls. As for exact algorithms for the general case, Kyoda, Imai, Takeuchi & Tajima [11] took an integer programming approach and devised a branch-and-cut algorithm. Aichholzer [1] introduced the concept of a "path of a triangulation," which can be used to solve any kinds of "decomposable" problems (in particular the minimum weight triangulation problem) by recursion. These algorithms were not analyzed in terms of worst-case time complexity. As for approximation of minimum weight triangulations, Levcopoulos & Krznaric [13] gave a constant-factor polynomial-time approximation algorithm, but with a huge constant. As for the parameterization with respect to the number of inner points, the two-dimensional Euclidean traveling salesman problem was recently shown to be fixed-parameter tractable [6].

2   Preliminaries and Description of the Result

We start our discussion with introduction of some notations and definitions used in the paper. Then we state our result in a precise manner. From now on, we assume that input points are in general position, that is, no three points are on a single line and no two points have the same x-coordinate.

---

[1] A *fixed-parameter algorithm* has running time $O(f(k)\mathrm{poly}(n))$, where n is the input size, k is a parameter and $f : \mathbb{N} \to \mathbb{N}$ is an arbitrary function. For example, an algorithm with running time $O(440^k n)$ is a fixed-parameter algorithm whereas one with $O(n^k)$ is not.

The line segment connecting two points $p, q \in \mathbb{R}^2$ is denoted by $\overline{pq}$. The length of a line segment $\overline{pq}$ is denoted by $\mathsf{length}(\overline{pq})$, which is measured by the Euclidean distance. A *polygonal chain* is a planar shape described as $\gamma = \bigcup_{i=0}^{\ell-1} \overline{p_i p_{i+1}}$ where $p_0, \ldots, p_\ell \in \mathbb{R}^2$ are distinct points except that p_0 and p_ℓ can be identical (in such a case, the chain is *closed*). For a closed polygonal chain we assume in the following that all indices are taken modulo ℓ.

The length of γ is the sum of the lengths of the line segments, that is, $\mathsf{length}(\gamma) = \sum_{i=0}^{\ell-1} \mathsf{length}(\overline{p_i p_{i+1}})$. We say γ is *selfintersecting* if there exists $i, j \in \{0, \ldots, \ell-1\}$, $i \neq j$, such that $(\overline{p_i p_{i+1}} \cap \overline{p_j p_{j+1}}) \setminus \{p_i, p_{i+1}, p_j, p_{j+1}\} \neq \emptyset$. Otherwise, we say γ is *non-selfintersecting*. The points $p_0, \ldots, p_\ell$ are the *vertices* of γ. When γ is not closed, p_0 and p_ℓ are called the *endpoints* of γ. In this case, we say γ *starts from* p_0 (or p_ℓ).

A *simple polygon* P is a simply connected compact region in the plane bounded by a closed non-selfintersecting polygonal chain. A *vertex* of P is a vertex of the polygonal chain which is the boundary of P. We denote the set of vertices of P by $\mathsf{Vert}(P)$. A *neighbor* of a vertex $p \in \mathsf{Vert}(P)$ is a vertex $q \in \mathsf{Vert}(P)$ such that the line segment $\overline{pq}$ lies on the boundary of P.

Following Aichholzer, Rote, Speckmann & Streinu [2], we call a pair $\Pi = (S, P)$ a *pointgon* when P is a simple polygon and S is a finite point set in the interior of P. We call S the set of *inner points* of Π. The *vertex set* of Π is $\mathsf{Vert}(P) \cup S$, and denoted by $\mathsf{Vert}(\Pi)$. Fig. 1 shows an example of pointgons.

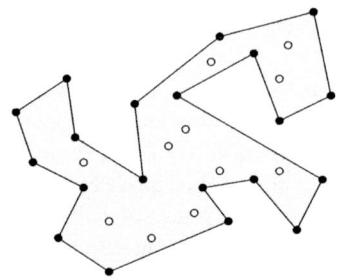

Fig. 1. A pointgon $\Pi = (S, P)$. In this paper, the points of S are drawn by empty circles and the points of $\mathsf{Vert}(P)$ are drawn by solid circles.

Let $\Pi = (S, P)$ be a pointgon. A *triangulation* $\mathcal{T}$ of a pointgon $\Pi = (S, P)$ is a subdivision of P into triangles whose edges are straight line segments connecting two points from $\mathsf{Vert}(\Pi)$ and which have no point from $\mathsf{Vert}(\Pi)$ in the interiors. The *weight* of $\mathcal{T}$ is the sum of the edge lengths used in $\mathcal{T}$. (Especially, all segments on the boundary of P are used in any triangulation and counted in the weight.) A *minimum weight triangulation* of a pointgon Π is a triangulation of Π which has minimum weight among all triangulations.

In this paper, we study the problem of computing a minimum weight triangulation of a given pointgon $\Pi = (S, P)$. The input size is proportional to $|\mathsf{Vert}(\Pi)|$. For a simple polygon, it is known that a minimum weight triangulation can be found in polynomial time [9,12]. However, in spite of the simplicity of the problem, the minimum weight triangulation problem for general pointgons is not known to be solvable in polynomial time nor to be NP-hard [8]. Our goal is to find an exact algorithm for a pointgon $\Pi = (S, P)$ where $|S|$ is small. The main theorem of this work is as follows.

Theorem 1. *Let $\Pi = (S, P)$ be a pointgon. Let $n := |\mathsf{Vert}(\Pi)|$ and $k := |S|$. Then we can find a minimum weight triangulation of Π in $O(6^k n^5 \log n)$ time. In particular, if $k = O(\log n)$ then a minimum weight triangulation can be found in polynomial time.*

This theorem shows that, in the terminology of parameterized complexity, the problem is fixed-parameter tractable, when the size of S is taken as a parameter.

In the next section, we prove this theorem by providing an algorithm.

3   A Fixed-Parameter Algorithm for Minimum Weight Triangulations

First, we describe a basic strategy for our algorithm. The details are then discussed in Sections 3.2 and 3.3.

3.1   Basic Strategy

In the sequel, for a given pointgon $\Pi = (S, P)$, we set $n := |\mathsf{Vert}(\Pi)|$ and $k := |S|$.

An *inner path* of a pointgon $\Pi = (S, P)$ is a polygonal chain $\gamma = \bigcup_{i=0}^{\ell-1} \overline{p_i p_{i+1}}$ such that $p_0, \ldots, p_\ell$ are all different, $p_i \in S$ for each $i \in \{1, \ldots, \ell-1\}$, $p_0, p_\ell \in \mathsf{Vert}(P)$, and $\gamma \setminus \{p_0, p_\ell\} \subseteq P$. An inner path $\bigcup_{i=0}^{\ell-1} \overline{p_i p_{i+1}}$ is called *x-monotone* if the x-coordinates of $p_0, \ldots, p_\ell$ are either increasing or decreasing.

The basic fact we are going to use is the following.

Observation 2. *Let $\Pi = (S, P)$ be a pointgon and p be a vertex of Π with the smallest x-coordinate. Denote by p', p'' the neighbors of p in P. Then, for every triangulation T of Π, either*

(1) there exists a non-selfintersecting x-monotone inner path starting from p and consisting of edges of T, or
(2) the three points p, p', p'' form a triangle of T.

The situation in Observation 2 is illustrated in Fig. 2.

We would like to invoke Observation 2 for our algorithm.

Let $\Pi = (S, P)$ be a pointgon, and $p \in \mathsf{Vert}(P)$ a vertex with the smallest x-coordinate. A non-selfintersecting x-monotone inner path divides a pointgon into two smaller pointgons. (See Fig. 2(a) and recall the general position assumption.) Hence, by looking at all non-selfintersecting x-monotone inner paths, we can recursively solve the minimum weight triangulation problem. To establish an appropriate recursive formula, denote by $\mathcal{D}(p)$ the set that consists of the line segment $\overline{p'p''}$ and of all non-selfintersecting x-monotone inner paths starting from p. Each non-selfintersecting inner path $\gamma \in \mathcal{D}(p)$ divides our pointgon Π into two smaller pointgons, say Π'_γ and Π''_γ. Then, we can see that

$$\mathsf{mwt}(\Pi) = \min_{\gamma \in \mathcal{D}(p)} (\mathsf{mwt}(\Pi'_\gamma) + \mathsf{mwt}(\Pi''_\gamma) - \mathsf{length}(\gamma)). \tag{1}$$

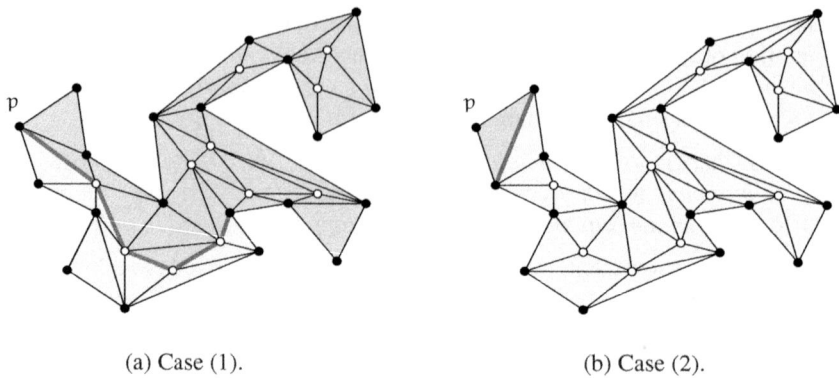

(a) Case (1).   (b) Case (2).

Fig. 2. Situations in Observation 2.

To see that Eq. (1) is really true, the following observation should be explicitly mentioned, although the proof is straightforward thus omitted.

Observation 3. *Let* $\Pi = (S, P)$ *be a pointgon and* $\mathcal{T}$ *be a minimum weight triangulation of* Π. *Choose an inner path* γ *which uses edges of* $\mathcal{T}$ *only, and let* Π' *and* Π'' *be two smaller pointgons obtained by subdividing* Π *with respect to* γ. *Then, the restriction of* $\mathcal{T}$ *to* Π' *is a minimum weight triangulation of* Π'. *The same holds for* Π'' *as well.*

Therefore, by solving Recursion (1) with an appropriate boundary (or initial) condition, we can obtain a minimum weight triangulation of Π. Note that even if Π is a convex pointgon, the pointgons Π'_γ and Π''_γ encountered in the recursion might not be convex. Thus, our digression to simple polygons is essential also for the minimum weight triangulation problem for a finite point set, i.e., a convex pointgon.

3.2   Outline of the Algorithm

Now, we describe how to solve Recursion (1) with the dynamic-programming technique.

First, let us label the elements of $\mathsf{Vert}(P)$ in a cyclic order, i.e., the order following the appearance along the boundary of P. According to this order, let us denote $\mathsf{Vert}(P) = \{p_0, p_1, \ldots, p_{n-k-1}\}$. Then, pick a vertex $p_i \in \mathsf{Vert}(P)$, and consider a non-selfintersecting x-monotone inner path γ starting from p_i. Let $p_j \in \mathsf{Vert}(P)$ be the other endpoint of γ. Note that $\mathsf{Vert}(\gamma) \setminus \{p_i, p_j\}$ consists of inner points of Π only. Therefore, such a path can be uniquely specified by a subset $T \subseteq S$. That is, we associate a triple (i, j, T) with an x-monotone inner path $\overline{p_i q_1} \cup \overline{q_1 q_2} \cup \cdots \cup \overline{q_{t-1} q_t} \cup \overline{q_t p_j}$ where $T = \{q_1, q_2, \ldots, q_t\}$. For the sake of brevity we write $\gamma(T)$ to denote the inner path associated with T when the endpoints p_i, p_j are clear from the context.

For two vertices $p_i, p_j \in \mathsf{Vert}(P)$ on the boundary of Π, and a set $T \subseteq S$ of inner points, let $\Pi(i, j, T)$ be the pointgon obtained from Π as follows: the boundary polygon is the union of the polygonal chains $\bigcup_{\ell=i}^{j-1} \overline{p_\ell p_{\ell+1}}$ and $\gamma(T)$. (Note that we only consider the case where $\gamma(T)$ is well-defined, that is, it does not intersect the boundary polygon.)

The inner points of $\Pi(i, j, T)$ consist of the inner points of Π contained in the boundary polygon specified above. Furthermore, denote by $\mathsf{mwt}(i, j, T)$ the weight of a minimum weight triangulation of the pointgon $\Pi(i, j, T)$. Then, Eq. (1) can be rewritten in the following way if we take p_0 for the role of p:

$$\mathsf{mwt}(\Pi) =$$

$$\min\Big\{ \min_{1 \leq i < n-k,\ T \subseteq S} \{\mathsf{mwt}(0, i, T) + \mathsf{mwt}(i, 0, T) - \mathsf{length}(\gamma(T))\},$$

$$\mathsf{mwt}(1, n-k-1, \emptyset, \mathsf{null}) + \mathsf{mwt}(n-k-1, 1, \emptyset, \mathsf{null}) - \mathsf{length}(\overline{p_1 p_{n-k-1}}) \Big\}. \quad (2)$$

The number of values considered in the right hand side of Eq. (2) is $O((n-k)2^k) = O(2^k n)$. Hence, for the computation of $\mathsf{mwt}(\Pi)$ it is sufficient to know $\mathsf{mwt}(i, j, T)$ for every triple (i, j, T) of two indices $i, j \in \{0, \dots, n-k-1\}$ and a subset $T \subseteq S$. Since the number of such triples is $O(2^k n^2)$, the efficient computation of each value results in fixed parameter tractability of the minimum weight triangulation problem.

Nevertheless, to compute these values, we have to generalize the class of pointgons under consideration. That is because pointgons we encounter in the recursion might not be of the form $\Pi(i, j, T)$. Therefore we introduce two additional types of pointgons.

The pointgon $\Pi(i, j, T)$ is bounded by two kinds of polygonal chains: a chain $\bigcup_{\ell=i}^{j-1} \overline{p_\ell p_{\ell+1}}$ from the boundary of the original pointgon and a non-selfintersecting x-monotone inner path $\gamma(T)$. Recall that T can be empty. We call such a pointgon a *type-1 pointgon* in the following. See Fig. 3(a) for illustration.

Another class of pointgons is defined for $i, j \in \{0, n-k-1\}$, two disjoint subsets $T_1, T_2 \subseteq S$, and a vertex $r \in \mathsf{Vert}(\Pi)$. Then, $\Pi(i, j, T_1, T_2, r)$ is a pointgon bounded by the x-monotone path connecting i and r through T_1, the x-monotone path connecting j and r through T_2, and $\bigcup_{\ell=i}^{j-1} \overline{p_\ell p_{\ell+1}}$. (Again we only consider those tuples which are well defined, that is, where the paths described above are indeed x-monotone and do not cross each other.) We call such a pointgon a *type-2 pointgon* of Π, and divide them into two subclasses according to whether r is a convex (type-2a) or reflex (type-2b) vertex of the pointgon. Fig. 3(b) & 3(c) illustrate the definition.

The last kind of pointgons uses at most one vertex of P. For a vertex $r \in \mathsf{Vert}(\Pi)$ and two subsets $T_1, T_2 \subseteq S$ with $T_1 \cap T_2 = \{s\}$, we define the pointgon $\Pi(T_1, T_2, r)$ as one which is bounded by two x-monotone paths connecting r and s through T_1 and through T_2, respectively. We call such a pointgon a *type-3 pointgon* of Π. See Fig. 3(d) for an example.

Let us count the number of these pointgons. The number of type-1 pointgons is $O(2^k n^2)$; the number of type-2 pointgons is $O(3^k n^3)$; the number of type-3 pointgons is $O(3^k n)$. Therefore, the total number of these pointgons is $O(3^k n^3)$. Our goal in the following is to compute the weights of minimum weight triangulations of these pointgons efficiently. Denote by $\mathsf{mwt}(i, j, T)$ the weight of a minimum weight triangulation of a type-1 pointgon $\Pi(i, j, T)$. Similarly, we define $\mathsf{mwt}(i, j, T_1, T_2, r)$ and $\mathsf{mwt}(T_1, T_2, r)$ for type-2 and type-3 pointgons, respectively.

Before describing the algorithm, let us discuss why we only encounter these three types of pointgons in the recursion. For this, we have to be careful which vertex to choose as p in the recursion step. Recall that in any step of Recursion (1) there are two

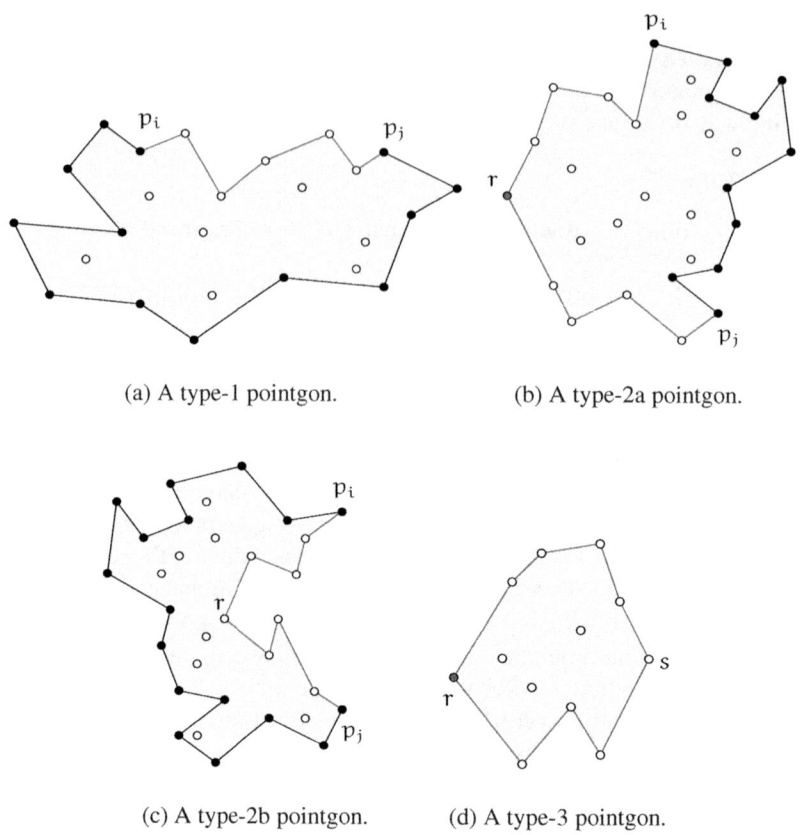

(a) A type-1 pointgon.   (b) A type-2a pointgon.

(c) A type-2b pointgon.   (d) A type-3 pointgon.

Fig. 3. The three types of subpointgons of Π. The vertex r can be either a solid or an empty circle.

cases: either p is cut off by joining its neighbors by an edge, or the pointgon is subdivided by an x-monotone inner path starting from p. Also recall that in Observation 2 we required p to be the leftmost point of the pointgon. If we apply the same argument as in Observation 2 to an arbitrary vertex of the pointgon, in the second case there appears an inner path starting from p that is *almost x-monotone*, i.e., x-monotone except for the first edge incident to p.

Initially we have a given pointgon $\Pi = (S, P)$ and choose the leftmost vertex as p. If p is cut off (Fig. 4(a)) the result is a type-1 pointgon where $T = \emptyset$. Any x-monotone inner path starting from p divides the pointgon into two type-1 pointgons (Fig. 4(b)).

When we apply Recursion (1) to a type-1 pointgon $\Pi(i, j, T)$, we choose as p the leftmost vertex of the inner path $\gamma(T)$ (which might consist just of a single edge joining p_i and p_j). If p is cut off, the result is either again a type-1 pointgon (Fig. 5(a)) or a type-2a pointgon (Fig. 5(b)). Otherwise, consider the vertex q on $\gamma(T)$ next to p. In every triangulation, the edge $\overline{pq}$ must belong to some triangle. To make such a triangle we need another vertex, say z. Let us choose z to be such that $\overline{pz}$ is the first edge of an

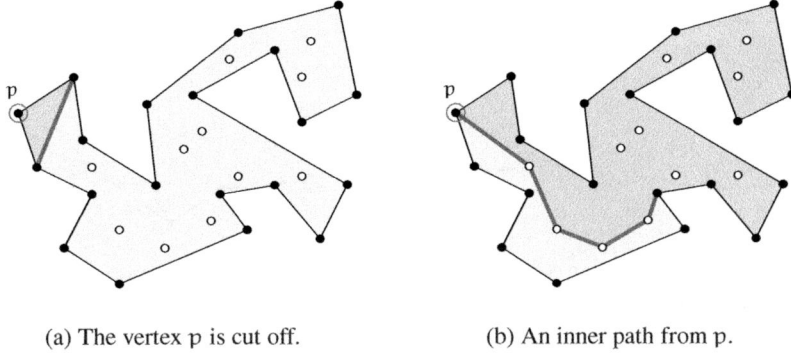

(a) The vertex p is cut off.   (b) An inner path from p.

Fig. 4. Subdivisions obtained from Π. From now on in the pictures, the vertex p is indicated by a larger circle.

almost x-monotone inner path γ' starting from p. If $z \in \mathsf{Vert}(P)$, then we get a type-1 pointgon, the triangle pqz and a type-2a pointgon when z is right of p (Fig. 5(c)), or a type-1 pointgon, the triangle pqz and a type-1 pointgon when z is left of p (Fig. 5(d)). If $z \in S$, then we have four subcases. When z is right of p and γ' ends at a vertex of $\gamma(T)$, we get a type-1 pointgon, the triangle pqz and a type-3 pointgon (Fig. 6(a)). When z is right of p and γ' ends at a vertex of P, we get a type-1 pointgon, the triangle pqz and a type-2a pointgon (Fig. 6(b)). When z is left of p and γ' ends at a vertex of $\gamma(T)$, we get a type-2b pointgon, the triangle pqz and a type-3 pointgon (Fig. 6(c)). When z is left of p and γ' ends at a vertex of P, we get a type-2b pointgon, the triangle pqz and a type-2a pointgon (Fig. 6(d)).

When we apply Recursion (1) to a type-2a pointgon $\Pi(i, j, T_1, T_2, r)$, we choose r as p. If p is cut off the result is either again a type-2a pointgon or a type-1 pointgon (Fig. 7(a)). Otherwise, consider an x-monotone inner path starting from p. If the path ends at a vertex of P, we get two type-2a pointgons (Fig. 7(b)). If, on the other hand, the inner path ends at a vertex in S, then it subdivides the pointgon into a type-2a and a type-3 pointgons (Fig. 7(c)).

When we apply Recursion (1) to a type-2b pointgon, we choose as p the leftmost vertex of the inner path. Since p is a reflex vertex, p cannot be cut off. So, every x-monotone inner path starting from p subdivides the pointgon into two type-1 pointgons (Fig. 7(d)).

When we apply Recursion (1) to a type-3 pointgon $\Pi(T_1, T_2, r)$, we choose r as p. Then, no matter how we divide the pointgon by the operations in the recursion, the result is again a type-3 pointgon (Fig. 8).

So much for preparation, and now we are ready to give the outline of our algorithm.

Step 1: Enumerate all possible type-1 pointgons $\Pi(i, j, T)$, type-2 pointgons $\Pi(i, j, T_1, T_2, r)$, and type-3 pointgons $\Pi(T_1, T_2, r)$.

Step 2: Compute the values $\mathsf{mwt}(i, j, T)$, $\mathsf{mwt}(i, j, T_1, T_2, r)$, and $\mathsf{mwt}(T_1, T_2, r)$ for some of them, which are sufficient for Step 3, by dynamic programming.

Step 3: Compute $\mathsf{mwt}(\Pi)$ according to Eq. (2).

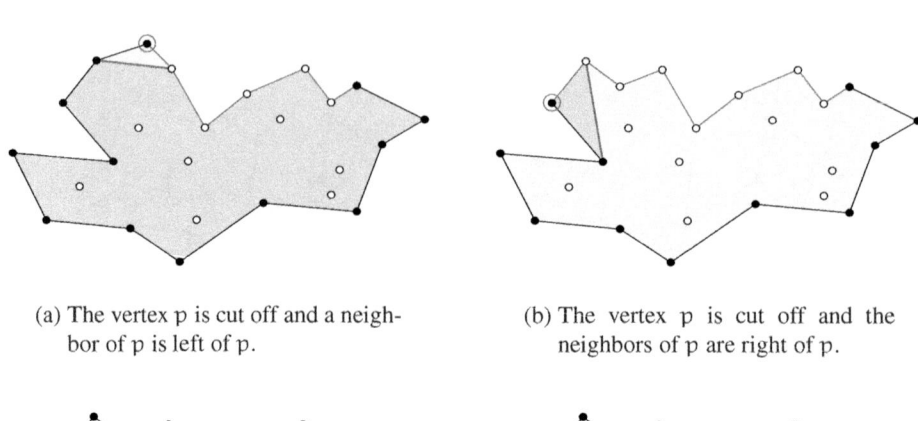

(a) The vertex p is cut off and a neighbor of p is left of p.

(b) The vertex p is cut off and the
neighbors of p are right of p.

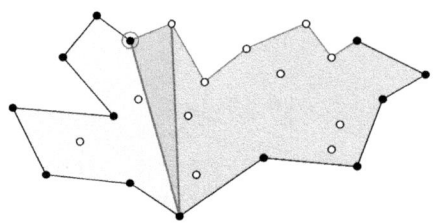

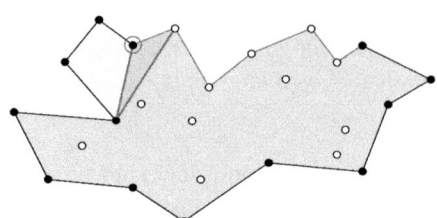

(c) The vertex z is a solid circle and z
is right of p.

(d) The vertex z is a solid circle and z
is left of p.

Fig. 5. Subdivisions obtained from a type-1 pointgon.

We already argued that Step 3 can be done in $O(2^k n)$ time. In the next section we will show that Steps 1 & 2 can be done in $O(6^k n^5 \log n)$ time, which dominates the overall running time.

3.3 Dynamic Programming

Now, we are going to compute the values of $\mathsf{mwt}(i, j, T)$, $\mathsf{mwt}(i, j, T_1, T_2, r)$, and $\mathsf{mwt}(T_1, T_2, r)$ for all possible choices of i, j, T_1, T_2, r.

First we enumerate all possibilities of i, j, T_1, T_2, r. Each of them can be enumerated in $O(1)$ time, and each of them can be identified as a well-defined pointgon or not (i.e., the inner paths do not intersect each other nor the boundary) in $O(n \log n)$ time. (Apply the standard line segment intersection algorithm [15].) Therefore, it takes $O(3^k n^3 \cdot 1 \cdot n \log n) = O(3^k n^4 \log n)$ time. This completes Step 1 of our algorithm.

Then, we perform the dynamic programming. Determine the vertex p and consider all possible subdivisions with respect to p as described in the previous section. Each subdivision replaces Π by two smaller pointgons. Then, as we saw, these two pointgons can be found among those enumerated in Step 1.

We can associate a parent-child relation between two pointgons Π_1, Π_2 in our enumeration: Π_1 is a parent of Π_2 if Π_2 is obtained as a smaller pointgon when we partition Π_1 by a path starting from p (which is fixed as in the previous section) or through the

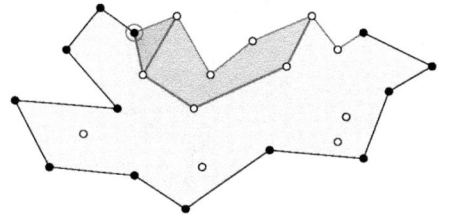 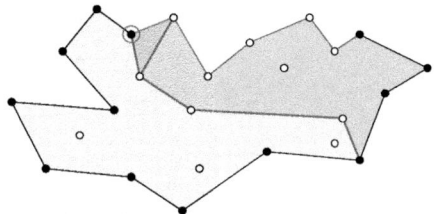

(a) The vertex z is right of p and the path γ' ends at an empty circle.

(b) The vertex z is right of p and the path γ' ends at a solid circle.

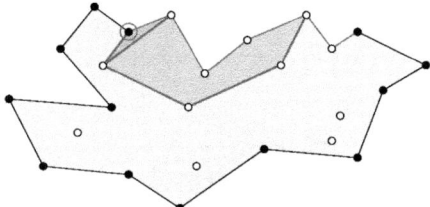 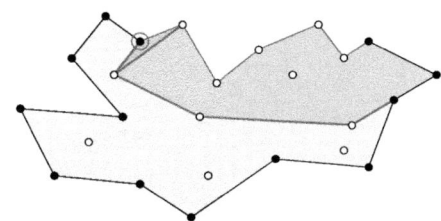

(c) The vertex z is left of p and the path γ' ends at an empty circle.

(d) The vertex z is left of p and the path γ' ends at a solid circle.

Fig. 6. Subdivisions obtained from a type-1 pointgon (continued). The vertex z is an empty circle.

edge cutting off p. It can also be thought as defining a directed graph on the enumerated pointgons: namely, draw a directed edge from Π_1 to Π_2 if the same condition as above is satisfied.

Observe that if Π_1 is a parent of Π_2, then the number of inner points in Π_2 is less than that in Π_1 or $|T_1| + |T_2|$ is smaller in Π_2 than in Π_1. Therefore, the parent-child relation is well-defined (i.e., there is no directed cycle in the directed-graph formulation).

Now, to do the bottom-up computation, we first look at the lowest descendants (or the sinks in the directed-graph formulation). They are triangles. So, the weights can be easily computed in constant time. Then, we proceed to their parents. For each parent, we look up the values of its children. In this way, we go up to the highest ancestor, which is Π. Thus, we can compute $\mathsf{mwt}(\Pi)$.

What is the time complexity of the computation? First, let us estimate the time for the construction of the parent-child relation. The number of enumerated pointgons is $O(3^k n^3)$. For each of them, the number of possible choices of non-selfintersecting x-monotone paths is $O(2^k n)$. For each of the paths, we can decide whether it really defines a non-selfintersecting path in $O(n \log n)$ time. Therefore, the overall running time for the construction is $O(3^k n^3 \cdot 2^k n \cdot n \log n) = O(6^k n^5 \log n)$.

In the bottom-up computation, for each pointgon we look up at most $O(2^k n)$ entries and compute the value according to Eq. (1). Therefore, this can be done in $O(3^k n^3 \cdot 2^k n) = O(6^k n^4)$.

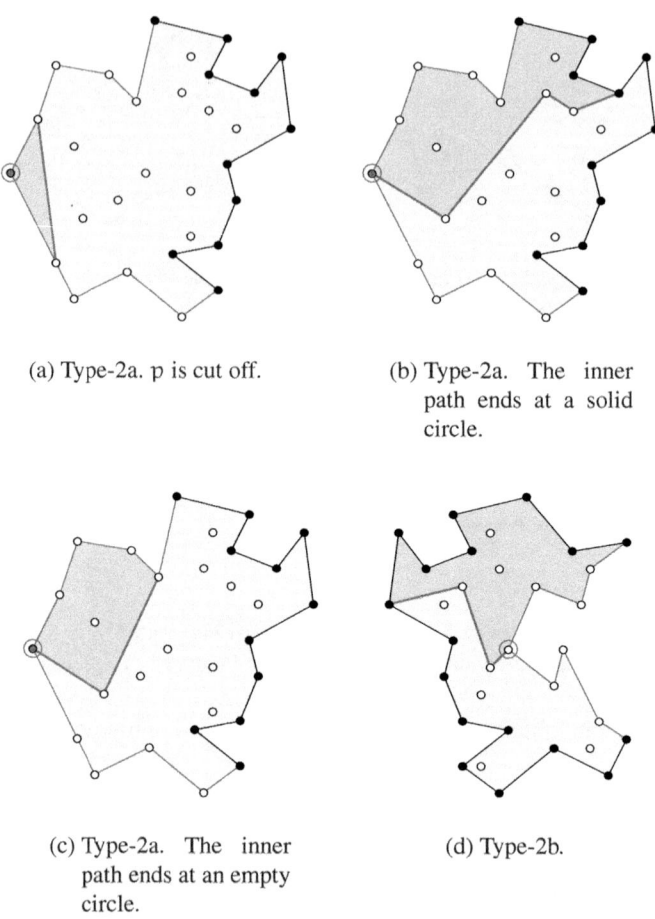

(a) Type-2a. p is cut off.

(b) Type-2a. The inner path ends at a solid circle.

(c) Type-2a. The inner path ends at an empty circle.

(d) Type-2b.

Fig. 7. Subdivisions obtained from a type-2 pointgon.

Hence, the overall running time of the algorithm is $O(3^k n^4 \log n + 6^k n^5 \log n + 6^k n^4) = O(6^k n^5 \log n)$. This completes the proof of Theorem 1.

4   Conclusion

In this paper, we studied the minimum weight triangulation problem from the viewpoint of fixed-parameter tractability. We established an algorithm to solve this problem for a simple polygon with some inner points which runs in $O(6^k n^6 \log n)$ time when n is the total number of input points and k is the number of inner points. Therefore, the problem is fixed-parameter tractable with respect to the number of inner points. We believe the number of inner points in geometric optimization problems plays a role similar to the treewidth in graph optimization problems.

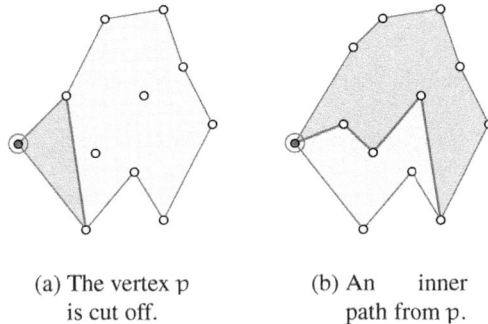

| (a) The vertex p is cut off. | (b) An inner path from p. |

Fig. 8. Subdivisions obtained from a type-3 pointgon.

Since our algorithm is based on a simple idea, it can be extended in several ways. For example, we can also compute a maximum weight triangulation in the same time complexity. (It seems quite recent that attention has been paid to a maximum weight triangulation [10,16].) To do that, we just replace "min" in Eqs. (1) and (2) by "max." Another direction of extension is to incorporate some heuristics. For example, there are some known pairs of vertices which appear as edges in all minimum weight triangulations, e.g. the β-skeleton for some β and the LMT-skeleton; see [4,5,17] and the references therein. Because of the flexibility of our algorithm, we can insert these pairs at the beginning of the execution as edges, and proceed in the same way except that we can use the information from the prescribed edges.

The framework proposed in this paper looks promising when we deal with the complexity of geometric problems concerning a finite point set on the plane. Study of another problem within the same framework is interesting. Recently, the traveling salesman problem was considered and it was shown that the problem can be solved in polynomial time when $k = O(\log n)$ [6].

The obvious open problem is to improve the time complexity of our algorithm. For example, is it possible to provide a polynomial-time algorithm for the minimum weight triangulation problem when $k = O(\log^2 n)$?

References

1. O. Aichholzer: The path of a triangulation. Proc. 15th SoCG (1999) 14–23.
2. O. Aichholzer, G. Rote, B. Speckmann and I. Streinu: The zigzag path of a pseudo-triangulation. Proc. 8th WADS, Lect. Notes Comput. Sci. **2748** (2003) 377–388.
3. E. Anagnostou and D. Corneil: Polynomial-time instances of the minimum weight triangulation problem. Comput. Geom. **3** (1993) 247–259.
4. P. Bose, L. Devroye and W. Evans: Diamonds are not a minimum weight triangulation's best friend. Internat. J. Comput. Geom. Appl. **12** (2002) 445–453.
5. S.-W. Cheng and Y.-F. Xu: On β-skeleton as a subgraph of the minimum weight triangulation. Theor. Comput. Sci. **262** (2001) 459–471.
6. V.G. Deĭneko, M. Hoffmann, Y. Okamoto and G.J. Woeginger: The traveling salesman problem with few inner points. Proc. 10th COCOON, Lect. Notes Comput. Sci. **3106** (2004). To appear.

7. R.G. Downey and M.R. Fellows: Parameterized Complexity. Springer, Berlin, 1999.
8. M. Garey and D. Johnson: Computers and Intractability. W.H. Freeman & Company, 1979.
9. P.D. Gilbert: New results in planar triangulations. Master Thesis, University of Illinois, Urbana, 1979.
10. S. Hu: A constant-factor approximation for maximum weight triangulation. Proc. 15th CCCG (2003) 150–154.
11. Y. Kyoda, K. Imai, F. Takeuchi, and A. Tajima: A branch-and-cut approach for minimum weight triangulation. Proc. 8th ISAAC, Lect. Notes Comput. Sci. **1350** (1997) 384–393.
12. G. Klincsek: Minimal triangulations of polygonal domains. Annals Discrete Math. **9** (1980) 121–123.
13. C. Levcopoulos and D. Krznaric: Quasi-greedy triangulations approximating the minimum weight triangulation. J. Algor. **27** (1998) 303–338.
14. R. Niedermeier: Invitation to fixed-parameter algorithms. Habilitation Thesis, Universität Tübingen, 2002.
15. M.I. Shamos and D. Hoey: Geometric intersection problems. Proc. 17th FOCS (1976) 208–215.
16. C.A. Wang, F.Y. Chin, and B.T. Yang: Maximum weight triangulation and graph drawing. Inform. Proccess. Lett. **70** (1999) 17–22.
17. C.A. Wang and B. Yang: A lower bound for β-skeleton belonging to minimum weight triangulations. Comput. Geom. **19** (2001) 35–46.

A Direct Algorithm for the Parameterized Face Cover Problem[*]

Faisal N. Abu-Khzam[1][**] and Michael A. Langston[2]

[1] Division of Computer Science and Mathematics, Lebanese American University,
Beirut, Lebanon
faisal.abukhzam@lau.edu.lb
[2] Department of Computer Science, University of Tennessee,
Knoxville, TN 37996, USA

Abstract. With respect to a given plane graph, G, a *face cover* is defined as a set of faces whose boundaries collectively contain every vertex in G. It is known that, when k is fixed, finding a cover of size k (if indeed any exist) can be accomplished in polynomial time. Recent improvements to face cover algorithms are based on the theory of fixed-parameter tractability and reductions to planar dominating set. A major goal has been to reduce the time required for branching, which is the most computationally-intensive aspect of fixed-parameter tractable methods. The fastest previously-known method for solving planar dominating set requires branching time $O(8^k n)$. The main contribution of this paper is a direct and relatively simple $O(5^k n)$ face cover branching algorithm. A direct $O(n^2)$ face cover kernelization algorithm is also presented.

1 Introduction

A set of faces whose boundaries contain all vertices in a plane graph G is said to be a *face cover* for G. For arbitrary $k > 0$, determining whether G has a face cover of size k or less is $\mathcal{NP}$-complete [4]. Algorithms are known that solve the face cover problem in $O(c^k n)$ time [2, 4]. Thus, the problem is fixed-parameter tractable (FPT) [5].

Algorithms for FPT problems very often use the following two steps: (i) kernelization, which reduces the problem size to a function of the parameter only and (ii) branching, which exhaustively searches for a solution in a tree whose height is bounded above by the parameter (or perhaps, say, a linear function of the parameter). Our work further illustrates this methodology, and elucidates how the planarity of a graph can be exploited in this context.

---

[*] This research has been supported in part by the following U.S. funding mechanisms: by the National Science Foundation under grants CCR–0075792 and CCR–0311500, by the Office of Naval Research under grant N00014–01–1–0608, and by the Department of Energy under contract DE–AC05–00OR22725.

[**] Communicating author.

R. Downey, M. Fellows, and F. Dehne (Eds.): IWPEC 2004, LNCS 3162, pp. 213–222, 2004.
© Springer-Verlag Berlin Heidelberg 2004

Let us use $FC(k)$ to denote the face cover problem with parameter k. Recent $FC(k)$ algorithms use a reduction to the parameterized planar dominating set problem ($PDS(k)$). The two most recent $PDS(k)$ algorithms are based on different approaches. One uses dynamic programming on tree decompositions, and can be used to solve $FC(k)$ in $O(c^{\sqrt{k}})$, where c is a very large constant (see [2], [7], and [8])[3]. Another addresses branching only (without kernelization), and for that phase runs in $O(8^k n)$ time [3]. Although face cover was not mentioned in [3], it is believed that an $FC(k)$ branching algorithm would follow from this work, and should have a run time of $O(8^k n)$ [6]. Although the first algorithm has a better asymptotic time bound, an algorithm with a run time of $O(8^k n)$ would be preferred in practice. To see this, note that face cover has a highly practical linear-time algorithm when $k = 1$, since this is just the recognition problem for outerplane graphs. Even for $k = 2$, however, an $O(3^{36\sqrt{(34k)}}n + n^2)$ algorithm does not seem very encouraging.

The main purpose of an effective branching algorithm is to achieve a low branching factor in the search tree used to explore the kernel exhaustively. The method presented in [3] tries to achieve this objective by exploiting the existence of low degree vertices in a planar graph. For a planar graph G, $\delta(G) \leq 5$, which implies that we can always find a vertex that belongs to the boundaries of at most five faces. So, if we use the search tree technique, branching at such vertex introduces at most five "smaller" instances of the problem. This suggests that any branching algorithm for face cover should attempt to attain a (possibly best possible) branching factor of 5. The challenge we face is illustrated by the following question: after performing the first (or first few) branch operation(s), is there any guarantee that another vertex of degree ≤ 5 is present in the resulting graph? The answer is typically "no" (as in the case of the PDS algorithm of [3]). We shall show, however, that certain reduction rules can be used to modify the graph and always guarantee (in the worst case) the existence of a vertex belonging to no more than five faces that qualify for membership in the cover. With this we devise a direct face cover branching algorithm that runs in $O(5^k n)$ time and guarantees a branching factor of 5. We also present a kernelization algorithm that runs in $O(n^2)$ time.

Our algorithm is fairly simple to implement, but its analysis is rather intricate. Most of this paper is involved with the proofs of correctness and time complexity analysis. The notation $G = (V, F)$ is used when referring to a plane graph. For a face $f \in F$, the set of vertices appearing in the ordered tuple associated with f is called the boundary of f and is denoted by $V(f)$. For a vertex $v \in V$, the set of faces whose boundaries contain v is denoted by $F(v)$.

Face cover is a special case of the hitting set problem (HS). To see this, note that a face cover of $G = (V, F)$ is a hitting set of the sets $\{F(u) : u \in V\}$. Unfortunately, $HS(k)$ is not FPT unless we fix the number of elements allowed in each of the input sets. This restricted form of $HS(k)$ is denoted by $d\text{-}HS(k)$, where d is the (fixed) upper bound on the size of input sets (see [10]. Thus, a $d\text{-}HS$ algorithm can be used to solve the face cover problem when the degree of

---

[3] c was 3^{36} in [2].

any vertex of the input plane graph is bounded by d. We shall, however, deal with the general face cover problem. Moreover, preprocessing techniques used for the general HS algorithm in [12] apply well to $FC(k)$ instances.

2 Preliminaries

We shall assume that our input is like the input of a HS algorithm. In fact, this is why we chose to characterize plane graphs by vertices and faces. The data structure used for a plane graph consists, essentially, of two lists corresponding to vertices and faces of the graph. The list associated with a vertex v starts with the number of faces in $F(v)$ then a list of the indices of these faces. Similarly, the list associated with face f contains the number of vertices in $V(f)$ followed by the indices of elements of $V(f)$. Figure 1 shows a plane graph and the corresponding input to our $FC(k)$ algorithm.

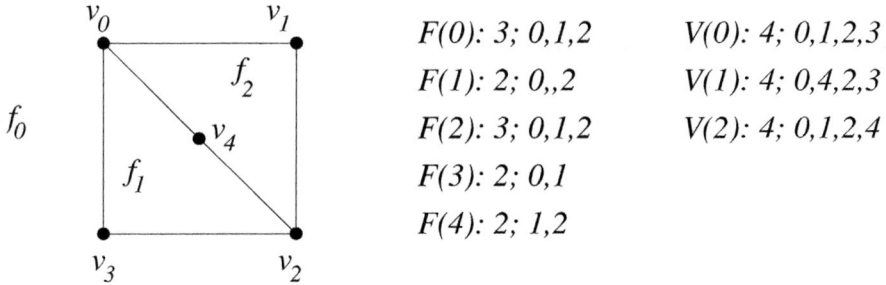

$F(0)$: 3; 0,1,2 $V(0)$: 4; 0,1,2,3

$F(1)$: 2; 0,,2 $V(1)$: 4; 0,4,2,3

$F(2)$: 3; 0,1,2 $V(2)$: 4; 0,1,2,4

$F(3)$: 2; 0,1

$F(4)$: 2; 1,2

Fig. 1. A plane graph and the corresponding input representation

The number of vertices of the given plane graph is denoted by n. When dealing with graphs, the size of input is often quadratic in n. This is due to the popular adjacency matrix representation. In our case, the graph is plane and, thanks to Euler's formula, has a linear number of edges and faces. We show that, according to our representation, the size of input is linear in n.

Lemma 1. *Let $G = (V, F)$ be a given plane graph, and let $V(f)$ and $F(v)$ be as defined above. Then the sets $\{F(v) : v \in V\}$ and $\{V(f) : f \in F\}$ have linear size.*

Proof. We fisrt note that $|\{F(v) : v \in V\}| = |\{V(f) : f \in F\}|$. To see this, consider the bipartite graph $H = (A, B)$ whose vertex sets A and B are the vertices and faces of G respectively. An edge of H connects a vertex v of A to a vertex f of B if and only if $f \in F(v)$ (if and only if $v \in V(f)$). Thus we have $|E(H)| = |\{F(v) : v \in V\}| = |\{V(f) : f \in F\}|$. Moreover, for each vertex v of G, $|F(v)|$ is bounded above by the degree of v in G. It follows that

$|\{F(v) : v \in V\}| \leq \Sigma_{v \in V} d_G(v) = 2|E(G)| \leq 2(3n-6)^4$. This proves that $|E(H)|$ is of linear size, which completes the proof.

We use the bounded search tree technique in our $FC(k)$ algorithm. During the search process, the vertex set is partially covered by the already selected faces. We shall, then, reduce the graph at each step by eliminating covered vertices. While this action is easy (and safe) when dealing with a general HS instance, it must be performed carefully in our case. Especially because we need to be assured that every node in the search tree has at most 5 children. Moreover, deleting a covered vertex from a plane graph might change the size of an optimal face cover and produce incorrect results. As an example, consider the graph shown in Figure 2. If the algorithm selects the outer face and deletes all its vertices, the resulting graph has a face cover of size one, while the original graph has a minimum face cover of size three. Our algorithm deals carefully with covered vertices. Before deleting a covered vertex, v, we modify all faces containing it by, simply, deleting v from their (ordered) lists. This face compaction will be called "the surgical operation" later in this paper. It is depicted in Figure 3.

2.1 Annotated Plane Graphs Representation

For a given plane graph $G = (V, F)$, vertices and faces are of two types: active and marked. Active vertices are those to be covered (i.e., not covered yet) and active faces are those that can be used to cover active vertices. A plane graph with active and marked vertices and/or faces will be called *annotated* in this paper.

A general version of the face cover problem was presented in [4], where not all vertices of the graph are to be covered. Our algorithm will be ready to deal with this version as well. In fact, if a vertex is not to be covered, then we may assume that it has been covered earlier during the process of finding the face cover and it will be marked.

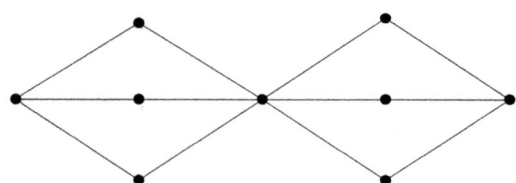

Fig. 2. Deleting the vertices that are covered by selecting the outer face produces a wrong answer.

Marking a face f is done by deleting $V(f)$ after replacing the index of f by -1 in the list(s) $F(v)$ of each vertex $v \in V(f)$. This process takes $O(|\{F(v) : v \in V(f)\}|)$. Thus it is $O(n)$. We will refer to this procedure by $MARK\text{-}FACE(f)$.

---

[4] It is known from [1] that, if G is a yes instance of $FC(k)$ then $|E(G)| \leq 2n - 3k + 6$.

Similarly, marking a vertex v is done by deleting $F(v)$ after replacing the index of v by -1 in the list(s) $V(f)$ of each $f \in F(v)$. This procedure, denoted by $MARK\text{-}VERTEX(v)$, takes time $O(|\{V(f) : f \in F(v)\}|)$ which is $O(n)$. To show that such simple operation is sound, we prove it to be equivalent to a surgical operation on the graph.

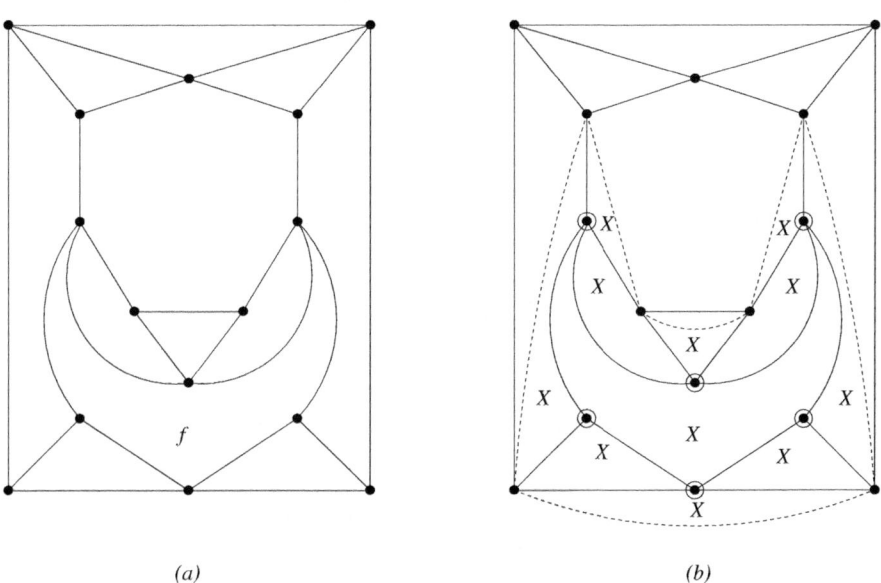

(a)   (b)

Fig. 3. (a) Before selecting face f. (b) After selecting f. The circled vertices are marked. For simplicity, edges between marked vertices are not contracted in this figure.

2.2   A Surgical Operation

Two neighbors, u and w, of a vertex, v, are consecutive if some face, f, contains the three vertices such that one if the two ordered tuples (u, v, w) and (w, v, u) is a sub-tuple of the ordered tuple of f.

If a vertex, v, is of degree at least 2, and is marked, then active faces that are adjacent to it will not be needed to cover it. Deleting v could lead to wrong answers as shown in Figure 2. So, prior to removing v, we make sure that the marking operation avoids such cases. The marking operation simply consists of (1) contracting edges between v and all marked neighbors of v, then (2) adding edges between consecutive active neighbors of v (even if they are adjacent) and marking all faces that contain v in the resulting graph. This action is safe in the following sense: every face that is adjacent to v and is used to cover a neighbor of v, must contain two consecutive neighbors of v. Thus, adding an edge between consecutive neighbors preserves active faces that might later be needed to cover neighbors of v (and possibly other vertices). We refer to this operation on the graph by the surgical operation. In our implementation of the algorithm, we did not need to perform the surgical operation. It is used only to show that the algorithm has the claimed performance and the operations of marking vertices and faces are sound.

Notice also the case where only one vertex of a face is left active at a certain point. In this case, the face is still active unless the vertex belongs to some other face. We shall see that preprocessing rules detects this case and deals with it, leaving no such faces in the graph. Figure 3, which shows (a snapshot of) the effect of the surgical operation, depicts some pendant faces.

3 Preprocessing

For simplicity, we assume the given graph is connected. Due to the chosen input representation, however, our algorithm works in both cases. Our preprocessing consists of two main rules:

> The Dominated Face Rule: If two faces, f and f', of G are such that $V(f) \subset V(f')$, then f is said to be dominated by f'. In such case, f can be marked since every face cover of G that contains f can be replaced by a (possibly better) cover that contains f'.
>
> The Dominated Vertex Rule: If two vertices, u and v, are such that $F(u) \subset F(v)$, then v is said to be dominated by u, and v is marked since any face that covers u will cover v.

Checking if a vertex (face) is dominated is done by a search through all the sets of $\{F(v) : v \in V\}$ ($\{V(f) : f \in F\}$). It thus takes linear time. It follows that checking if the graph has dominated vertices or faces can be accomplished in $O(n^2)$. Each dominated vertex or face is then marked in $O(n)$ time. Therefore, the total run time of preprocessing is $O(n^2)$. We shall refer to this preprocessing algorithm by procedure $PREPROCESS$ later in this paper.

The above two rules generalize many simple preprocessing techniques. For example, let P be a path in G whose length exceeds two. If all interior vertices of P are of degree two, then we can reduce the problem size by contracting all interior edges of P (edges between interior vertices), thus keeping one (necessary) interior vertex of degree two. This turns out to be a special case of the dominated vertex rule since two interior degree-two vertices have the same set of active faces. Another example is the pendant vertex rule, which simply adds the unique face containing a pendant vertex to the cover. Such rule is automatically accounted for, since it is also a special case of the dominated vertex rule (the unique neighbor of a pendant vertex is dominated).

4 Kernelization

This section is devoted to the proof of the following theorem.

Theorem 1. Let $G = (V, F)$ be an instance of $FC(k)$. There is an algorithm that runs in $O(n^2)$ and produces another, instance $G' = (V', F')$, of $FC(k')$ such that:

(i) G is a yes instance of $FC(k)$ $\iff$ G' is a yes instance of $FC(k')$,

(ii) $k' \leq k$, and

(iii) $|V(G')| \leq 2k'^3$.

Lemma 2. Three or more vertices of a plane graph, G, may not be common to more than two faces of G.

Proof. Assume vertices u_1, u_2, and u_3 belong to three faces, f_1, f_2, and f_3. Construct another plane graph, G', by drawing three vertices, v_1, v_2, and v_3 such that vertex v_i is placed inside face f_i and joined to all vertices that lie on the boundary of f_i. The subgraph of G' induced by vertices $\{u_i\}_{i=1}^3$ and $\{v_i\}_{i=1}^3$ is isomorphic to $K_{3,3}$. This is a contradiction.

Lemma 3. Let G be a yes instance of $FC(k)$. If two faces of G have more than $2k$ common vertices, then any face cover of size k contains at least one of them.

Proof. Let C be a face cover of size k. Assume faces f_1 and f_2 contain $2k + 1$ common vertices and are not elements of C. Then, by the pigeon hole principle, some element of C must cover at least three of the $2k + 1$ common vertices of f_1 and f_2. This is impossible by lemma 2.

Corollary 1. Let G be a yes instance of $FC(k)$. If no pair of faces of G have $2k + 1$ common vertices, then every face whose length exceeds $2k^2$ is in any optimal face cover of G.

Corollary 2. Let G be a yes instance of $FC(k)$. If $f \in F$ has more than $2k$ vertices with more than k faces, then f must be in any optimal face cover of G.

We now describe the kernelization process and use it in a proof of Theorem 1. This process is referred to, later, by procedure $KERNELIZE$.

The first step in $KERNELIZE$ is a simple search for all faces of length $> 2k'^2$, where k' is originally equal to k. This is done in $O(n)$ time since the length of a face is captured when reading input. All faces that are found whose length exceeds $2k'^2$ are kept in a list, L.

The second step is the following: For each face $f \in L$, the number of common vertices with all other faces of L is computed. If a face f' has more than $2k'$ common vertices with f then their common vertices are all added to a list M of vertices that are to be marked[5], and a virtual new vertex, v, is added to the list of vertices such that $F(v) = \{f, f'\}$. This operation is equivalent to adding a (virtual) degree-two vertex that dominates all common vertices of f and f'. If no such face f' exists, then, because of corollary 2, f is added to a list C of

---

[5] Note that, by the surgical operation, active faces do not have marked vertices on their boundaries.

faces that are in the cover and are to be marked. Moreover, if the number of faces that share more than $2k'$ common vertices with f exceeds k', then f is also selected in the cover and, thus added to list C. Otherwise, more than k' faces will have to be in the cover. Note that, because of Lemma 3, we stop the search if more than k' disjoint pairs of faces have (at least) $2k' + 1$ common vertices.

The last step deals with a cleanup of the lists and an update of the parameter: vertices in M are marked, faces in C are marked together with their boundary vertices, and the parameter k' is replaced by $k - |C|$ (k is unchanged while C, being originally empty, is filled with cover faces). Then list L is emptied and the process is repeated from step one until no more faces of length $> 2k'^2$ are found (or no solution can be found).

Proof of Theorem 1:

Algorithm $KERNELIZE$ just described satisfies the three conditions of the theorem. In particular, condition (iii) follows from corollary 1. As for the time complexity, we note the following:

(i) Step one takes $O(n)$ since it's a simple filling of list L by a one pass through the list of active faces.

(ii) Step (ii) takes $O(kn)$: for each face $f \in L$, we either find another face f' that shares $2k' + 1$ vertices with f or we add f to C. By Lemma 3, no more than k pairs of faces can share $2k + 1$ vertices, and no more than k faces can be added to C. Hence, the number of iterations in step two is at most k throughout the whole process.

(iii) The last (cleanup) step takes time $O(n^2)$ since it deals with marking vertices and faces that are in M and C respectively. The proof of Theorem 1 is now complete.

5   A Direct Face Cover Algorithm

Our direct algorithm is represented by the procedure $FACECOVER$ shown below. Subroutines $PREPROCESS$, $KERNELIZE$ and $MARK\text{-}FACE$ correspond (obviously) to the processes previously described in detail.

We shall prove that, at every call to $FACECOVER$, the selected vertex, v, has no more than five active faces in its $F(v)$ list. We know that the first call is guaranteed to select such vertex. As a remark, we note that, at least three such vertices are present in the graph. This is guaranteed by virtue of Euler's formula and the following lemma, which first appeared as a corollary in [11].

Lemma 4. If G is a planar graph, then G has at least three vertices of degree ≤ 5.

Proof. Let $m = |M = \{v \in V(G) : d(v) > 5\}|$ and $l = |L = \{v \in V(G) :$ $d(v) \leq 5\}|$. Then: $3n - 6 \geq e(G) \geq \dfrac{\sum_{v \in L} d(v) + \sum_{v \in M} d(v)}{2} \geq \dfrac{\sum_{v \in L} d(v) + 6m}{2} \geq$ $\dfrac{\sum_{v \in L} d(v) + 6(n-1)}{2} \geq \dfrac{\sum_{v \in L} d(v)}{2} + 3n - 3l$. Therefore $l \geq 2 + \dfrac{\sum_{v \in L} d(v)}{6} > 2$.

Procedure $FACECOVER$

Input: A plane graph $G = (V, F)$ given by $\{F(u) : u \in V\}$ and $\{V(f) : f \in F\}$, and an integer $k \geq 1$.

Output: A face cover, C, of size $\leq k$ of G if one exists. NULL otherwise.

Begin procedure
$k' \leftarrow k$
$(G, C, k') \leftarrow PREPROCESS(G, k')$
$(G, C, k') \leftarrow KERNELIZE(G, k')$
Select an active vertex v such that $|F(v)| = min\{F(u)|u$ is active in $V\}$
For every $f \in F(v)$ do
    $C_1 \leftarrow C \cup \{f\}$;
    $G_1 \leftarrow MARK\text{-}FACE(f)$;
    $(G_1, C_1, k') \leftarrow KERNELIZE(G_1, k' - 1)$;
    $C_2 \leftarrow FACECOVER(G_1, k')$;
    if$(C_2 \neq NULL)$
        return $C_1 \cup C_2$
return NULL
End procedure

If a vertex, v, is the only active vertex of face f, then f will only be selected (and marked) if v doesn't belong to any other face. Otherwise, it would be dominated (thus marked). We can, therefore, assume that every active face has at least two active vertices.

Faces of length two may exist due to the surgical operation which could introduce multiple edges between two vertices. This case is easily handled by $KERNELIZE$ since two vertices cannot belong to more than one face of length two (by the dominated face rule).

Lemma 5. Let v be an active vertex of an annotated plane graph, G. Then no marked neighbor of v belongs to an active face of v.

Proof. The lemma follows immediately from the surgical operation.

Theorem 2. $FACECOVER$ runs in $O(5^k + n^2)$ time.

Proof. At each call to $FACECOVER$, the (plane) subgraph induced by active vertices of G_1 must have a vertex, v, of degree ≤ 5. By Lemma 5, the active faces in $F(v)$ are faces that are common to v and its active neighbors. Thus the number of such active faces would exceed five only if v has multiple edges with at least one of its neighbors. Which means that v belongs to faces of length two. However, a face of length two will only be active if it's the unique face that is common to its two vertices. This proves that each node in the search tree has at most five children. Having no more than k levels, the search tree has at most $O(5^k)$ nodes. Thus pure branching would take $O(5^k k^3)$ (after the ($O(n^2)$) kernelization). since interleaving is used as in [9], the run time of branching reduces to $O(5^k + k^3)$ and the overall run time is $(5^k + k^3 + n^2)$. This completes the proof.

6   Remarks

As a preliminary test of efficiency, we have implemented our algorithm and tested it on random plane graphs of size up to 200. Answers were obtained in at most a few seconds. We plan larger and more systematic testing in the near future. We know of no other $FC(k)$ implementations, but would be most interested in conducting extensive experiments and comparisons.

References

[1] F. N. Abu-Khzam. *Topics in Graph Algorithms: Structural Results and Algorithmic Techniqu es, with Applications*. PhD thesis, Department of Computer Science, University of Tennessee, 2003.

[2] J. Alber, H. L. Bodlaender, H. Fernau, T.Kloks, and R. Niedermeier. Fixed parameter algorithms for dominating set and related problems on planar graphs. *Algorithmica*, 33(4):461–493, 2002.

[3] J. Alber, H. Fan, M. R. Fellows, H. Fernau, R. Niedermeier, F. Rosamond, and U. Stege. Refined search tree technique for DOMINATING SET on planar graphs. *Lecture Notes in Computer Science*, 2136:111–122, 2001.

[4] D. Bienstock and C. L. Monma. On the complexity of covering vertices by faces in a planar graph. *SIAM J. Sci. Comput.*, 17:53–76, 1988.

[5] R. G. Downey and M. R. Fellows. *Parameterized Complexity*. Springer-Verlag, 1999.

[6] M. Fellows. Private communication, 2003.

[7] F. V. Fomin and D. M. Thilikos. A simple and fast approach for solving problems on planar graphs. In *Proceedings of the 21st International Symposium on Theoretical Aspects of Computer Science (STACS 2004)*, volume 2996 of *Lecture Notes in Computer Science*, pages 56–67. Springer-Verlag, 2004.

[8] I. A. Kanj and L. Perković. Improved parameterized algorithms for planar dominating set. In *Proceedings of the 27th International Symposium on Mathematical Foundations of Computer Science (MFCS 2002)*, volume 2420 of *Lecture Notes in Computer Science*, pages 399–410. Springer-Verlag, 2002.

[9] R. Niedermeier and P. Rossmanith. A general method to speed up fixed-parameter tractacle algorithms. *Information Processing Letters*, 73:125–129, 2000.

[10] R. Niedermeier and P. Rossmanith. An efficient fixed parameter algorithm for 3-hitting set. *Journal of Discrete Algorithms*, 2001.

[11] C. Nishizeki. Planar graphs: Theory and algorithms. *Annals of Discrete Mathematics*, 32, 1988.

[12] K. Weihe. Covering trains by stations or the power of data reduction. In *Proceedings, International Conference on Algorithms and Experiments*, pages 1–8, 1998.

On Finding Short Resolution Refutations and Small Unsatisfiable Subsets

Michael R. Fellows[1,*], Stefan Szeider[2], and Graham Wrightson[1]

[1] School of Electrical Engineering and Computer Science, University of Newcastle,
Callaghan 2308 NSW, Australia
[2] Department of Computer Science, University of Durham,
Durham DH1 3LE, England

Abstract. We consider the parameterized problems of whether a given set of clauses can be refuted within k resolution steps, and whether a given set of clauses contains an unsatisfiable subset of size at most k. We show that both problems are complete for the class W[1], the first level of the W-hierarchy of fixed-parameter intractable problems. Our results remain true if restricted to 3-SAT formulas and/or to various restricted versions of resolution including tree-like resolution, input resolution, and read-once resolution.

Applying a metatheorem of Frick and Grohe, we show that restricted to classes of locally bounded treewidth the considered problems are fixed-parameter tractable. Hence, the problems are fixed-parameter tractable for planar CNF formulas and CNF formulas of bounded genus, k-SAT formulas with bounded number of occurrences per variable, and CNF formulas of bounded treewidth.

1 Introduction

Resolution is a fundamental method for establishing the unsatisfiability of a given formula in Conjunctive Normal Form (CNF) using one single rule of inference, the *resolution rule*. This rule allows to infer the clause $C \cup D$ from clauses $C \cup \{x\}$ and $D \cup \{\neg x\}$. A CNF formula is unsatisfiable if and only if the empty clause can be derived from it by repeated application of the resolution rule. Resolution is easy to implement and provides the basis for many Automated Reasoning systems.

It is well known that certain unsatisfiable CNF formulas require an exponential number of resolution steps in order to be refuted [11]. Iwama [12] shows that, given a CNF formula F together with an integer k, deciding whether F has a resolution refutation with at most k steps is NP-complete. This result is strengthened by Alekhnovich et al. [2] by showing that the minimum number of resolution steps cannot be approximated within a constant factor, unless P = NP (this result also holds for stronger proof systems like Frege systems). A closely related question is the "automatizability" of resolution: is there an

---

* Research has been partially supported by the Australian Research Council.

algorithm that finds a shortest resolution refutation R in polynomial time w.r.t. the number of steps in R? Alekhnovich and Razborov [3] show that resolution is not automatizable, assuming a parameterized intractability hypothesis regarding W[P]. For a survey of further results on the complexity of resolution, see, e.g., Beame and Pitassi [4] or Clote and Kranakis [6].

Parameterizing by the number of steps of a resolution refutation is of relevance if one has to deal with large CNF formulas which contain local inconsistencies. Evidently, one can use exhaustive search for finding a k-step resolution refutation of a CNF formula with n variables, yielding a time complexity of $n^{\mathcal{O}(k)}$. However, even if k is a small integer, say $k = 10$, exhaustive search becomes impractical for large n. The question rises whether one can find resolution refutations with a fixed number of steps significantly more efficient than by exhaustive search. The framework of parameterized complexity [8] offers a means for addressing this question. Here, problems are considered in two dimensions: one dimension is the usual size n of the instance, the second dimension is the parameter (usually a positive integer k). A parameterized problem is called *fixed-parameter tractable* (or *fpt*, for short) if it can be solved in time $f(k) \cdot n^{\mathcal{O}(1)}$ for some computable function f of the parameter. The parameterized complexity classes W[1] $\subseteq$ W[2] $\subseteq \cdots \subseteq$ W[P] contain problems which are believed to be not fpt (see [8]); since all inclusions are believed to be proper, the hierarchy provides a means for determining the degree of parameterized intractability. A parameterized problem P *fpt reduces* to a parameterized problem Q if we can transform an instance (x, k) of P into an instance $(x', g(k))$ of Q in time $f(k) \cdot |x|^{\mathcal{O}(1)}$ (f, g are arbitrary computable functions), such that (x, k) is a yes-instance of P if and only if $(x', g(k))$ is a yes-instance of Q.

As a main result of this paper, we show that SHORT RESOLUTION REFUTATION, that is, refutability within k resolution steps, is complete for the class W[1]. We also show that this result holds true for several resolution refinements including tree-like resolution, regular resolution, and input-resolution. We establish the hardness part of the result by an fpt-reduction of the parameterized clique problem. As it appears to be difficult to establish W[1]-membership by reducing the problem to the canonical W[1]-complete problem on circuit satisfiability, we use results from descriptive parameterized complexity theory.

We show that refutability within k resolution steps can be expressed as a statement in positive (i.e., negation-free and $\forall$-free) first-order logic. This yields W[1]-membership as it was shown by Papadimitriou and Yannakakis [16] in the context of query evaluation over databases, that the evaluation of statements in positive first-order logic over finite structures is W[1]-complete.

Along these lines, we also show W[1]-completeness of SMALL UNSATISFIABLE SUBSET, that is, the problem of whether at most k clauses of a given CNF formula form an unsatisfiable formula. Furthermore, we pinpoint that all our W[1]-completeness results remain valid if the inputs are confined to 3-CNF formulas.

The notion of *bounded local treewidth* for classes of graphs (see Frick and Grohe [10]) generalizes several graph classes, like planar graphs, graphs of bounded treewidth, or graphs of bounded degree. By means of *incidence graphs* (see Section 2.1) we can apply this notion to classes of CNF formulas. Special cases are *planar CNF formulas* (CNF formulas with planar incidence graphs) and of (k, s)-*CNF formulas* (CNF formulas with k literals per clause and at most s occurrences per variable). Frick and Grohe [10] show that the evaluation of first-order statements over classes of graphs with locally bounded treewidth is fixed-parameter tractable (the result holds also for finite structures whose Gaifman graphs have locally bounded treewidth). Applying this powerful result, we obtain fixed-parameter tractability of SHORT RESOLUTION REFUTATION and SMALL UNSATISFIABLE SUBSET restricted to classes of CNF formulas with locally bounded treewidth. Thus the problems are tractable for planar CNF formulas and for (k, s)-CNF formulas.

Note that satisfiability is NP-complete for planar CNF formulas (Lichtenstein [15]) and $(3, 4)$-CNF formulas (Tovey [18]), and even for the intersection of these two classes (Kratatochvíl [13]). However, satisfiability of CNF formulas of (globally) bounded treewidth is fixed-parameter tractable (Courcelle et al. [7], see also Szeider [17]).

2   Preliminaries and Notation

2.1   CNF Formulas

A *literal* is a propositional variable x or a negated variable $\neg x$; we also write $x^1 = x$ and $x^0 = \neg x$. A *clause* is a finite set of literals not containing a complementary pair $x, \neg x$. A *formula in conjunctive normal form* (or *CNF formula*, for short) F is a finite set of clauses. F is a k-*CNF formula* if the size of its clauses is at most k; F is a (k, s)-*CNF formula* if, additionally, every variable occurs in at most s clauses. The *length* of a CNF formula F is defined as $\sum_{C \in F} |C|$. For a CNF formula F, $var(F)$ denotes the set of variables x such that some clause of F contains x or $\neg x$. A literal x^ε is a *pure literal* of F if some clauses of F contain x^ε but no clause contains $x^{1-\varepsilon}$. F is *satisfiable* if there exists an assignment $\tau : var(F) \to \{0, 1\}$ such that every clause of F contains some variable x with $\tau(x) = 1$ or some negated variable $\neg x$ with $\tau(x) = 0$; otherwise, F is called *unsatisfiable*. F is called *minimal unsatisfiable* if F is unsatisfiable and every proper subset of F is satisfiable. Note that minimal unsatisfiable CNF formulas have no pure literals. A proof of the following lemma can be found in Aharoni and Linial [1], attributed there to Tarsi.

Lemma 1 *A minimal unsatisfiable CNF formula has more clauses than variables.*

The *incidence graph* $I(F)$ of a CNF formula F is a bipartite graph; variables and clauses form the vertices of $I(F)$, a clause C and variable x are joined by an edge if and only if $x \in C$ or $\neg x \in C$ (see Fig. 1 for an example). A *planar CNF formula* is a CNF formula with a planar incidence graph.

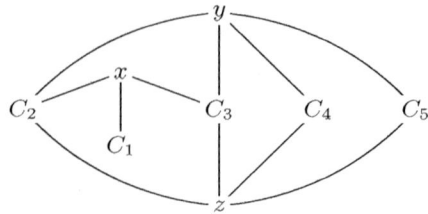

Fig. 1. The incidence graph $I(G)$ of the planar $(3,4)$-CNF formula $F = \{C_1, \ldots, C_5\}$ with $C_1 = \{x\}$, $C_2 = \{\neg x, y, z\}$, $C_3 = \{\neg x, y, \neg z\}$, $C_4 = \{\neg y, z\}$, $C_5 = \{\neg y, \neg z\}$.

2.2   Resolution

Let C_1, C_2 be clauses with $x \in C_1$, $\neg x \in C_2$, and $var(C_1) \cap var(C_2) = \{x\}$. The clause $C = (C_1 \cup C_2) \setminus \{x, \neg x\}$ is called the *resolvent* of C_1 and C_2. We also say that C is obtained by *resolving on* x, and we call C_1, C_2 *parent clauses* of C.

Recall that a vertex of a directed graph is called a *sink* if it has no successors, and it is called a *source* if it has no predecessors. A *resolution refutation* R is a directed acyclic graph whose vertices are labeled with clauses, such that

1. every non-source of R has exactly two predecessors and is labeled with the resolvent of the clauses labeling its predecessors;
2. R contains exactly one sink; the sink is labeled with the empty clause.

We call a non-source vertex of R a *step*. A clause labeling a source of R is called an *axiom* of R. R is a resolution refutation *of* a CNF formula F if all axioms of R are contained in F. It is well known that a CNF formula is unsatisfiable if and only if it has a resolution refutation (resolution is "refutationally complete").

In the sequel we will measure the size of resolution refutations in terms of the *number of steps*[3].

We refer to any decidable property of a resolution refutation as a *resolution refinement*. In particular, we will consider the following refinements:

- *Tree-like resolution*: The directed acyclic graph is a tree.
- *Regular resolution*: On any path from a source vertex to the sink, any variable is resolved at most once.
- *P-resolution*: at each resolution step, at least one of the parent clauses is a positive clause (i.e., a clause without negated variables);
- *Input resolution*: every vertex is either a source or has a predecessor which is a source.
- *Literal-once resolution*: distinct resolution steps resolve on distinct variables.
- *Read-once resolution*: distinct sources are labeled by distinct clauses.

---

[3] Another possible measure is the length of a refutation, defined as the total number of vertices (i.e., steps + source vertices). It is easy to verify that a resolution refutation with k steps has at most $k+1$ sources, and so its length is at most $2k+1$. Therefore, our results carry over if we bound the length instead of the number of steps.

Note that the first three refinements are refutationally complete, but the last three refinements are not. Note also that every literal-once resolution refutation is tree-like, read-once, and regular. Every input resolution refutation is tree-like.

2.3 Locally Bounded Treewidth

Treewidth, a popular parameter for graphs, was introduced by Robertson and Seymour in their series of papers on graph minors; see, e.g., Bodlaender's survey article [5] for definitions and references.

Let v be a vertex of a simple graph G and let r be some positive integer. $N_G^r(v)$ denotes the *r-neighborhood* of v, i.e., the set of vertices of G which can be reached from v by a path of length at most r. A class of graphs is said to have *locally bounded treewidth* if there exists a function f such that for all $r \geq 1$ and all vertices v of a graph G of that class, the treewidth of the subgraph included by $N_G^r(v)$ is at most $f(k)$. (Intuitively, the treewidth of the subgraph induced by an r-neighborhood of a vertex is a function of r and so less than the total number of vertices of G.) We give some examples of classes of graphs with locally bounded treewidth (see Frick and Grohe [10] for references).

- By trivial reasons, the class of graphs of treewidth $\leq t$ has locally bounded treewidth $(f(r) = t)$.
- The class of planar graphs has locally bounded treewidth $(f(r) = 3r)$; more generally, the class of graphs with genus $\leq g$ has locally bounded treewidth $(f(r) = \mathcal{O}(gr))$.
- The class of graphs with maximum degree $\leq d$ has locally bounded treewidth $(f(r) = d(d-1)^{r-1})$.

3   Statement of Main Results

Consider the following two parameterized problems.

SHORT RESOLUTION REFUTATION
Input: A CNF formula F.
Parameter: A positive integer k.
Question: Can F be refuted by at most k resolution steps? (i.e., can the empty clause be inferred from F by k applications of the resolution rule?).

SMALL UNSATISFIABLE SUBSET
Input: A CNF formula F.
Parameter: A positive integer k.
Question: Does F contain an unsatisfiable subset F' with at most k clauses?

Our main results are as follows.

Theorem 1 SHORT RESOLUTION REFUTATION *is* W[1]-*complete.*

The problem remains W[1]-*complete for the following resolution refinements: tree-like resolution, regular resolution, P-resolution, input resolution, read-once resolution, and literal-once resolution.*

Theorem 2 SMALL UNSATISFIABLE SUBSET *is* W[1]-*complete.*

Both theorems remain valid if inputs are confined to 3-CNF formulas.

We show fixed-parameter tractability for classes of CNF formulas whose incidence graphs have locally bounded treewidth:

Theorem 3 *For CNF formulas of locally bounded treewidth, the problems* SHORT RESOLUTION REFUTATION *and* SMALL UNSATISFIABLE SUBSET *are fixed-parameter tractable.*

Tractable cases include: planar CNF formulas, CNF formulas of bounded genus, and (k, s)-CNF formulas (k-CNF formulas with at most s occurrences per variable).

4   Proof of W[1]-hardness

We are going to reduce the following well-known W[1]-complete problem.

CLIQUE
Input: A graph G.
Parameter: A positive integer k.
Question: Is there a set $V' \subseteq V(G)$ of k vertices that induces a complete subgraph of G (i.e., a clique of size k)?

Given a simple graph $G = (V, E)$, $|V| = n$, and a positive integer k. We take distinct variables: x_i for $1 \leq i \leq k$, $y_{i,j}$ for $1 \leq i < j \leq k$, and $z_{v,i}$ for $v \in V$ and $1 \leq i \leq k$. We construct a CNF formula

$$F_G = \{C_{\text{start}}\} \cup F_{\text{edges}} \cup F_{\text{vertices}} \cup F_{\text{clean-up}}$$

where

$$
\begin{aligned}
C_{\text{start}} &= \{x_1, \dots, x_k\} \cup \{\, y_{i,j} : 1 \leq i < j \leq k \,\}, \\
F_{\text{edges}} &= \{\, \{\neg y_{i,j}, z_{u,i}, z_{v,j}\} : 1 \leq i < j \leq k, \ uv \in E \,\}, \\
F_{\text{vertices}} &= \{\, \{\neg x_i, z_{v,i}\} : 1 \leq i \leq k, \ v \in V \,\}, \\
F_{\text{clean-up}} &= \{\, \{\neg z_{v,i}\} : 1 \leq i \leq k, \ v \in V \,\}.
\end{aligned}
$$

We put

$$k' = \binom{k}{2} + 2k.$$

Lemma 2 *The following statements are equivalent.*

1. F_G *has an unsatisfiable subset* F' *with at most* $k' + 1$ *clauses;*
2. G *contains a clique on* k *vertices;*
3. F_G *has a resolution refutation with at most* k' *steps which complies with the resolution refinements mentioned in Theorem 1;*
4. F_G *has a resolution refutation with at most* k' *steps.*

Proof. $1{\Rightarrow}2$. We assume that F_G is unsatisfiable and choose a minimal unsatisfiable subset $F' \subseteq F_G$. First we show that

$$C_{\text{start}} \in F'. \tag{1}$$

Assume the contrary. Since F' has no pure literals, and since the variables x_i and $y_{i,j}$ occur positively only in C_{start}, we conclude that $F_{\text{vertices}} \cap F' = F_{\text{edges}} \cap F' = \emptyset$. Hence, in turn, $F_{\text{clean-up}} \cap F' = \emptyset$, thus $F' = \emptyset$. However, the empty formula is satisfiable, a contradiction. Thus C_{start} is indeed in F'. Since every clause in $F_{\text{edges}} \cup F_{\text{vertices}}$ contains the complement of exactly one variable of C_{start}, it follows that

$$|F_{\text{edges}} \cap F'| \geq \binom{k}{2}, \tag{2}$$

$$|F_{\text{vertices}} \cap F'| \geq k. \tag{3}$$

It also follows that for every $i \in \{1, \ldots, k\}$ there is some $v \in V$ such that $z_{v,i} \in var(F_{\text{vertices}} \cap F')$. The latter implies

$$|F_{\text{clean-up}} \cap F'| \geq k. \tag{4}$$

Since $|F'| \leq k' + 1$ by assumption, (1) and the estimations (2)–(4) yield $|F'| = k' + 1$. Hence the estimations (2)–(4) must be tight. Consequently, strengthening the above observation, we conclude that for every $i \in \{1, \ldots, k\}$, there is *exactly* one vertex $v \in V$ such that $z_{v,i} \in var(F_{\text{vertices}} \cap F')$. Let $\varphi : \{1, \ldots, k\} \to V$ be the map defined by

$$\varphi(i) = v \quad \text{if and only if} \quad z_{v,i} \in var(F_{\text{vertices}} \cap F').$$

In view of the tightness of the above estimations, we conclude that

$$var(F') = C_{\text{start}} \cup \{\, z_{\varphi(i),i} : 1 \leq i \leq k \,\}. \tag{5}$$

Consequently,

$$F_{\text{edges}} \cap F' = \{\, \{\neg y_{i,j}, z_{\varphi(i),i}, z_{\varphi(j),j}\} : 1 \leq i < j \leq k, \ \varphi(i)\varphi(j) \in E \,\}.$$

We conclude that the vertices $\varphi(1), \ldots, \varphi(k)$ are mutually distinct; thus $\varphi(1), \ldots, \varphi(k)$ induce a clique of size k in G.

$2{\Rightarrow}3$. Assume that G contains a clique on k vertices. Consequently, there is an injective map $\varphi : \{1, \ldots, k\} \to V$ such that $\varphi(i)\varphi(j) \in E$ for all $1 \leq i < j \leq k$. We devise an input resolution refutation R of F_G, proceeding in three phases:

1. For $1 \leq i < j \leq k$ we resolve C_{start} with the clauses $\{\neg y_{i,j}, z_{\varphi(i),i}, z_{\varphi(j),j}\} \in F_{\text{edges}}$. We end up with the clause $C' = \{x_i, z_{\varphi(i),i} : i = 1, \ldots, k\}$.
2. For $1 \leq i \leq k$ we resolve C' with the clauses $\{\neg x_i, z_{\varphi(i),i}\} \in F_{\text{vertices}}$. We end up with the clause $C'' = \{z_{\varphi(i),i} : i = 1, \ldots, k\}$.
3. For $1 \leq i \leq k$ we resolve C'' with the clauses $\{\neg z_{\varphi(i),i}\} \in F_{\text{clean-up}}$. We end up with the empty clause.

By construction, R complies with the resolution refinements as claimed. Moreover, R contains $\binom{k}{2} + k + k = k'$ resolution steps.

$3 \Rightarrow 4$. Trivial.

$4 \Rightarrow 1$. Assume that F_G has a resolution refutation R with at most k' steps. Let F' denote the set of axioms of R. Note that F' is necessarily unsatisfiable, and since R has at most $k' + 1$ sources, $|F'| \leq k' + 1$ follows. $\qquad \square$

The construction of F_G from G can be carried out in time $f(k)(|V| + |E|)^{\mathcal{O}(1)}$ for some function f. Thus Lemma 2 yields an fpt-reduction from CLIQUE to SHORT RESOLUTION REFUTATION with respect to the resolution refinements mentioned in Theorem 1, and an fpt-reduction from CLIQUE to SMALL UNSATISFIABLE SUBSET. Since CLIQUE is well-know to be W[1]-complete [8], we have established the hardness parts of Theorems 1 and 2.

4.1 3-CNF Formulas

Using a slight modification of the above construction, we can show that the above hardness results hold for 3-CNF formulas. By means of a variant of Tseitin Extension [19], we transform a CNF formula F with clauses of size ≥ 2 into a 3-SAT formula $t(F)$, applying the following operations.

- Replace a clause $\{w_1, \ldots, w_n\}$ of size $n > 3$ by the clauses $\{w_1, w_2, u_1\}$, $\{\neg u_{n-3}, w_{n-1}, w_n\}$, and $\{\neg u_i, w_{i+2}, u_{i+1}\}$ for $i = 1, \ldots, n-4$ where u_i are new variables.
- Replace a clause $\{w_1, w_2\}$ by the clauses $\{w_1, w_2, u\}$, $\{\neg u, w_1, w_2\}$, u is a new variable.
- Replace a clause $\{w\}$ by the four clauses $\{w, u_1, u_2\}$, $\{w, u_1, \neg u_2\}$, $\{w, \neg u_1, u_3\}$, $\{w, \neg u_1, \neg u_3\}$, u_i are new variables.

It is straightforward that F is satisfiable if and only if $t(F)$ is satisfiable. Moreover, if F is minimal unsatisfiable, then so is $t(F)$, and the difference between the number of clauses and the number of variables remains the same for F and $t(F)$.

In view of the first part of the proof of Lemma 2 it follows that a minimal unsatisfiable subset F'' of $t(F_G)$ contains all $\binom{k}{2} + k - 2$ clauses of $t(\{C_{\text{start}}\})$, $\binom{k}{2}$ clauses of $t(F_{\text{edges}})$, $2k$ clauses of $t(F_{\text{vertices}})$, and $4k$ clauses of $t(F_{\text{clean-up}})$. In summary, the number of clauses in F'' is exactly

$$k'' = 2\binom{k}{2} + 7k - 2.$$

The proof of Lemma 2 carries over to $t(F_G)$ using k'' instead of k'.

5 Membership in W[1] and FPT Results

Let S denote a finite relational structure and φ a first-order (FO) formula (we quietly assume that the vocabularies of φ and S are compatible). S is a *model* of φ (in symbols $S \models \varphi$) if φ is true in S in the usual sense (see, e.g., [9, 14] for further model theoretic definitions). Model-checking, the problem of deciding whether $S \models \varphi$, can be parameterized in different ways; in the sequel we will refer to the following setting.

FO MODEL CHECKING
Input: A finite structure S, a FO formula φ.
Parameter: The length of φ.
Question: Is S a model of φ?

Recall that a FO formula φ is *positive* if it does not contain negations or the universal quantifier $\forall$. We will use the following result of Papadimitriou and Yannakakis [16].

Theorem 4 FO MODEL CHECKING *for positive formulas is* W[1]*-complete.*

In [16] it is also shown that without the restriction to positive formulas, FO MODEL CHECKING is W[t]-hard for all t.

We associate to a relational structure S its *Gaifman graph* $G(S)$, whose vertices are the elements of the universe of S, and where two distinct vertices are joined by an edge if and only if they occur in the same tuple of some relation of S. By means of Gaifman graphs, one can speak of the treewidth of a relational structure and of classes of structures with locally bounded treewidth.

We shall use the following strong result of Frick and Grohe [10].

Theorem 5 FO MODEL CHECKING *for structures with locally bounded treewidth is fixed-parameter tractable.*

In the subsequent discussions, ρ denotes any of the resolution refinements mentioned in Theorem 1.

Let $y_1, y_2, \ldots$ be an infinite supply of variables. For $k \geq 1$ we define the following classes of CNF formulas.

- $\mathcal{F}^k$ denotes the set of CNF formulas F with $var(F) = \{y_1, \ldots, y_{k'}\}$ for some $k' \leq k$.
- $\mathcal{M}^k$ denotes the set of minimal unsatisfiable formulas in $\mathcal{F}^k$ with at most $k + 1$ clauses.
- $\mathcal{R}^k$ denotes the set of CNF formulas $F \in \mathcal{F}^k$ such that F is the set of axioms of some resolution refutation with at most k steps; $\mathcal{R}^k_\rho$ is $\mathcal{R}^k$ restricted to ρ-resolution.

Lemma 3 *Every formula $F \in \mathcal{R}^k$ has at most $k + 1$ clauses.*

Proof. We proceed by induction on k. If $k \leq 1$ then the lemma holds trivially, since either $F = \{\emptyset\}$ or $F = \{\{y_1\}, \{\neg y_1\}\}$. Assume that $k \geq 2$ and $F \in \mathcal{R}^k \setminus \mathcal{R}^{k-1}$. Consequently, there is a resolution refutation R with exactly k steps such that F is the set of axioms of R. We observe that R must contain a step v_0 where both predecessors v_1, v_2 of v_0 are sources. Let C_i denote the clause which labels v_i, $0 \leq i \leq 2$. We remove v_1 and v_2 from R and obtain a resolution refutation R' with $k - 1$ steps. The vertex v_0 is now a source of R'. Let a and a' denote the number of axioms of R and R', respectively. Observe that a' is minimal if (1) C_0 is an axiom of R and (2) C_1, C_2 are not axioms of R'. Thus $a' \geq a - 2 + 1$. Since the set of axioms of R' belongs to $\mathcal{R}^{k-1}$, we have $a' \leq k$ by induction hypothesis, hence $|F| = a \leq k + 1$ follows. $\square$

Since there are less than 3^k clauses over the variables $\{y_1, \ldots, y_k\}$ (a variable appears positively, appears negatively, or does not appear in a clause), we conclude the following.

Lemma 4 *The sets $\mathcal{M}^k$ and $\mathcal{R}_\rho^k$ are finite and computable.*

We represent a CNF formula F by a relational structure $S_F = (P, N, V)$ as follows. For every variable x of F and every clause C of F, the universe of S_F contains distinct elements a_x and a_C, respectively. The relations of S_F are

$$P = \{\, (a_x, a_C) : x \in var(F),\ C \in F,\ x \in C \,\} \text{ (positive occurrence)},$$
$$N = \{\, (a_x, a_C) : x \in var(F),\ C \in F,\ \neg x \in C \,\} \text{ (negative occurrence)},$$
$$V = \{\, a_x : x \in var(F) \,\} \text{ (being a variable)}.$$

For example, the formula of Fig. 1 is represented by the structure $S_F = (P, N, V)$ with $P = \{(a_x, a_{C_1}), (a_y, a_{C_2}), (a_y, a_{C_3}), (a_z, a_{C_2}), (z, a_{C_4})\}$, $N = \{(a_x, a_{C_2}), (a_x, a_{C_3}), (a_y, a_{C_4}), (a_y, a_{C_5}), (a_z, a_{C_3}), (a_z, a_{C_5})\}$ and $V = \{a_x, a_y, a_z\}$.

In order to express that two variables are distinct without using negation, we also consider the structure $S_F^+ = (P, N, V, D)$ with the additional relation

$$D = \{\, (a_x, a_{x'}) : x, x' \in var(F),\ x \neq x' \,\} \text{ (distinctness)}.$$

The next lemma is a direct consequence of the definitions (cf. Fig. 1).

Lemma 5 *The incidence graph $I(F)$ and the Gaifman graph $G(S_F)$ are isomorphic for every CNF formula F.*

Let $k \geq 1$ and take two sequences of distinct FO variables $\vec{v} = v_1, \ldots, v_k$ and $\vec{w} = w_1, \ldots, w_{k+1}$. For a CNF formula $F \in \mathcal{F}^k$ with $F = \{C_1, \ldots, C_{k''}\}$, $k'' \leq k + 1$, and $|var(F)| = k' \leq k$ we define the quantifier-free formula

$$\varphi[F] = \bigwedge_{1 \leq i < j \leq k'} \neg v_i = v_j \wedge \bigwedge_{j=1}^{k''} \left(\bigwedge_{y_i \in C_j} P(v_i, w_j) \wedge \bigwedge_{\neg y_i \in C_j} N(v_i, w_j) \right).$$

Furthermore, for $\mathcal{X}^k \in \{\mathcal{M}^k, \mathcal{R}_\rho^k\}$ we define

$$\varphi[\mathcal{X}^k] = \exists \vec{v}\, \exists \vec{w} \left(\bigwedge_{i=1}^{k} V(v_i) \wedge \bigvee_{F \in \mathcal{X}^k} \varphi[F] \right).$$

Similarly we define positive formulas $\varphi^+[F]$ using "$D(v_i, v_j)$" instead of "$\neg v_i = v_j$" and $\varphi^+[\mathcal{X}^k]$ using $\varphi^+[F]$ instead of $\varphi[F]$.

Lemma 6 *For every CNF formula F the following holds true.*

1. *F has a ρ-resolution refutation with at most k steps if and only if $S_F \models \varphi[\mathcal{R}_\rho^k]$ (i.e., $S_F^+ \models \varphi^+[\mathcal{R}_\rho^k]$).*
2. *F contains an unsatisfiable subset of size at most $k + 1$ if and only if $S_F \models \varphi[\mathcal{M}^k]$ (i.e., $S_F^+ \models \varphi^+[\mathcal{M}^k]$).*

Proof. Let R be a ρ-resolution refutation of F with at most k steps, and let F' denote the set of axioms of R. Since all variables occurring in axioms of R are resolved in some of the resolution steps, $|var(F')| \leq k$ follows. We put $k' = |var(F')|$ and pick arbitrarily a bijection $r : var(F') \rightarrow \{y_1, \dots, y_{k'}\}$. Renaming the variables in F' according to r yields a formula $r(F')$ which belongs to $\mathcal{R}_\rho^{k'} \subseteq \mathcal{R}_\rho^k$ (observe that $|F'| = |r(F')| \leq k + 1$ by Lemma 3). It follows now from the definition of $\varphi[\mathcal{R}_\rho^k]$ that $S_F \models \varphi[\mathcal{R}_\rho^k]$ (equivalently, that $S_F^+ \models \varphi^+[\mathcal{R}_\rho^k]$).

Now assume that F contains an unsatisfiable subset F' with at most $k + 1$ clauses; we may assume that F' is minimal unsatisfiable. By Lemma 1 it follows that $|var(F')| \leq k$. Consequently, as in the previous case, we obtain from F' by renaming a formula $r(F') \in \mathcal{M}^k$, establishing $S_F \models \varphi[\mathcal{M}^k]$ and $S_F^+ \models \varphi^+[\mathcal{M}^k]$.

The converse directions follow directly from the respective definitions of $\mathcal{R}_\rho^k$ and $\mathcal{M}^k$. □

To complete the proofs of Theorems 1, 2, and 3, it only remains to join together the above results: In view of Theorem 4, Lemma 6 implies directly the W[1]-membership part of Theorems 1 and 2. Whence Theorems 1 and 2 are shown true. Furthermore, Theorem 3 follows directly from Theorem 5 by Lemmas 5 and 6.

6 Concluding Remarks

Numerous parameterized problems have been identified as being W[1]-complete, for example, the Halting Problem for nondeterministic Turing machines, parameterized by the number of computation steps. Our Theorem 1 links parameterized complexity with the length of resolution refutations, another fundamental concept of Logic and Computer Science; thus our result provides additional evidence for the significance of the class W[1].

Our positive results, the fp-tractability of SHORT RESOLUTION REFUTATION and SMALL UNSATISFIABLE SUBSET for classes of CNF formulas of locally bounded tree-width, are obtained by application of Frick and Grohe's metatheorem which does not provide practicable algorithms. However, the results show that fp-tractability can be achieved in principle, and so that further efforts for finding more practicable algorithms based on the particular combinatorics of the problems are encouraged. We think that the classes of planar CNF formulas and (k, s)-CNF formulas are good candidates for such an approach.

References

[1] R. Aharoni and N. Linial. Minimal non-two-colorable hypergraphs and minimal unsatisfiable formulas. *J. Combin. Theory Ser. A*, 43:196–204, 1986.

[2] M. Alekhnovich, S. Buss, S. Moran, and T. Pitassi. Minimum propositional proof length is NP-hard to linearly approximate. *J. Symbolic Logic*, 66(1):171–191, 2001.

[3] M. Alekhnovich and A. A. Razborov. Resolution is not automatizable unless W[P] is tractable. In *42nd IEEE Symposium on Foundations of Computer Science (FOCS 2001)*, pages 210–219. IEEE Computer Soc., 2001.

[4] P. Beame and T. Pitassi. Propositional proof complexity: past, present, and future. In *Current trends in theoretical computer science*, pages 42–70. World Sci. Publishing, River Edge, NJ, 2001.

[5] H. L. Bodlaender. A partial k-arboretum of graphs with bounded treewidth. *Theoret. Comput. Sci.*, 209(1-2):1–45, 1998.

[6] P. Clote and E. Kranakis. *Boolean functions and computation models*. Springer Verlag, 2002.

[7] B. Courcelle, J. A. Makowsky, and U. Rotics. On the fixed parameter complexity of graph enumeration problems definable in monadic second-order logic. *Discr. Appl. Math.*, 108(1-2):23–52, 2001.

[8] R. G. Downey and M. R. Fellows. *Parameterized Complexity*. Springer Verlag, 1999.

[9] H.-D. Ebbinghaus and J. Flum. *Finite model theory*. Perspectives in Mathematical Logic. Springer Verlag, second edition, 1999.

[10] M. Frick and M. Grohe. Deciding first-order properties of locally tree-decomposable structures. *Journal of the ACM*, 48(6):1184–1206, 2001.

[11] A. Haken. The intractability of resolution. *Theoret. Comput. Sci.*, 39:297–308, 1985.

[12] K. Iwama. Complexity of finding short resolution proofs. In *Mathematical Foundations of Computer Science (MFCS 1997)*, volume 1295 of *Lecture Notes in Computer Science*, pages 309–318. Springer Verlag, 1997.

[13] J. Kratochvíl. A special planar satisfiability problem and a consequence of its NP-completeness. *Discr. Appl. Math.*, 52:233–252, 1994.

[14] L. Libkin. *Elements of Finite Model Theory*. Texts in Theoretical Computer Science. Springer Verlag, 2004.

[15] D. Lichtenstein. Planar formulae and their uses. *SIAM J. Comput.*, 11(2):329–343, 1982.

[16] C. H. Papadimitriou and M. Yannakakis. On the complexity of database queries. *J. of Computer and System Sciences*, 58(3):407–427, 1999.

[17] S. Szeider. On fixed-parameter tractable parameterizations of SAT. In E. Giunchiglia and A. Tacchella, editors, *Theory and Applications of Satisfiability, 6th International Conference, SAT 2003, Selected and Revised Papers*, volume 2919 of *Lecture Notes in Computer Science*, pages 188–202. Springer Verlag, 2004.

[18] C. A. Tovey. A simplified NP-complete satisfiability problem. *Discr. Appl. Math.*, 8(1):85–89, 1984.

[19] G. S. Tseitin. On the complexity of derivation in propositional calculus. *Zap. Nauchn. Sem. Leningrad Otd. Mat. Inst. Akad. Nauk SSSR*, 8:23–41, 1968. Russian. English translation in J. Siekmann and G. Wrightson (eds.) *Automation of Reasoning. Classical Papers on Computer Science 1967–1970*, Springer Verlag, 466–483, 1983.

Parameterized Algorithms for Feedback Vertex Set

Iyad Kanj[1], Michael Pelsmajer[2], and Marcus Schaefer[1]

[1] School of Computer Science, Telecommunications and Information Systems,
DePaul University, 243 S. Wabash Avenue, Chicago, IL 60604-2301.
{ikanj,mschaefer}@cs.depaul.edu*
[2] Department of Applied Mathematics, Illinois Institute of Technology,
Chicago, IL 60616.
pelsmajer@iit.edu

Abstract. We present an algorithm for the parameterized feedback vertex set problem that runs in time $O((2 \lg k + 2 \lg \lg k + 18)^k n^2)$. This improves the previous $O(max\{12^k, (4 \lg k)^k\} n^\omega)$ algorithm by Raman et al. by roughly a 2^k factor ($n^\omega \in O(n^{2.376})$ is the time needed to multiply two $n \times n$ matrices). Our results are obtained by developing new combinatorial tools and employing results from extremal graph theory. We also show that for several special classes of graphs the feedback vertex set problem can be solved in time $c^k n^{O(1)}$ for some constant c. This includes, for example, graphs of genus $O(\lg n)$.

1 Introduction

Given an undirected graph G, a *feedback vertex set* in G is a subset of vertices F in G such that $G - F$ is acyclic. The *size* of a feedback vertex set F is $|F|$. The FEEDBACK VERTEX SET problem (FVS) is: given a graph G and a positive integer k, decide if G has a feedback vertex set of size at most k. It is well-known that the FVS problem is NP-complete on both directed and undirected graphs [17]. The minimization version of this problem has been studied intensively from the approximability point of view [1, 2, 13, 16, 24] due to its important applications in fields like circuit testing, deadlock resolution, and analyzing manufacturing processes [13, 18, 21]. For example, in the field of circuit testing, a small set of registers (vertices) needs to be identified in the circuit (graph) whose removal makes the circuit testable (i.e., the circuit needs to be acyclic) [18].

The FVS problem has also received considerable attention from the parameterized complexity point of view [3, 4, 10, 11, 23]. A parameterized problem is said to be *fixed-parameter tractable* if the problem can be solved in time $f(k)n^{O(1)}$ for some function f which is independent of n [11]. The class FPT denotes the class of all fixed-parameter tractable problems [11]. It was shown that the FVS problem on undirected graphs is in FPT [4, 10, 11], whereas it remains an open question whether the FVS problem on directed graphs is in FPT [11].

---

* The first author was supported in part by DePaul University Competitive Research Grant.

R. Downey, M. Fellows, and F. Dehne (Eds.): IWPEC 2004, LNCS 3162, pp. 235–247, 2004.
© Springer-Verlag Berlin Heidelberg 2004

Once a problem has been shown to be in FPT, the search for better algorithms for the problem continues; that is, algorithms that remain practical for larger values of the parameter k. Successful examples of such developments include the VERTEX COVER and PLANAR DOMINATING SET problems (see [7, 15] and their references). The same holds true for the FVS problem on undirected graphs. Bodlaender [4], and Downey and Fellows [10], were the first to show that the problem is in FPT. In [11], Downey and Fellows presented an $O((2k+1)^k n^2)$ time algorithm for the problem. Becker et al. [3] gave a randomized algorithm running in time $O(4^k kn)$ that finds a minimum feedback vertex set of size k with probability at least $1 - (1 - 4^{-k})^{c4^k}$ for an arbitrary constant c. By observing that every undirected graph on n vertices with minimum degree at least 3 has a cycle of length bounded by $2 \lg n + 1$, Raman presented a very simple algorithm for the problem running in time $O((6k \lg k)^{k+1} n^\omega)$ [22], where $n^\omega \in O(n^{2.376})$ is the running time of the best algorithm for multiplying two $n \times n$ matrices [8]. More recently, using some nice combinatorial techniques, Raman et al. [23] presented an algorithm for the problem running in time $O(max\{12^k, (4 \lg k)^k\} n^\omega)$ improving significantly the $O((2k + 1)^k n^2)$ time algorithm given in [11] (when k is sufficiently large).

In this paper we continue the efforts towards reducing the running time of the algorithms for the FVS problem. We develop new combinatorial tools and employ known results from extremal graph theory to show that the size of a minimum feedback vertex set is $\Omega(n/\lg n)$ in a graph with no cycles of length bounded by $\lg n$. This allows us to obtain an $O((2 \lg k + 2 \lg \lg k + 18)^k n^2)$ time algorithm for the FVS problem, improving the previous $O(max\{12^k, (4 \lg k)^k\} n^\omega)$ time algorithm in [23] by roughly a 2^k factor. Obviously, the running time of this algorithm is still far from being practical, and the question of whether the FVS problem can be solved in time $c^k n^{O(1)}$ remains open.

We also consider the FVS problem on special classes of graphs. We show that the problem on graphs of genus $O(\lg n)$ and on K^r-minor free graphs is solvable in time $c^k n^{O(1)}$ for some constant c. We also show that the problem on bipartite graphs, graphs of genus $O(n^\epsilon)$ for any $\epsilon > 0$, and constant average degree graphs, can be solved in time $c^k n^{O(1)}$ for some constant c if and only if the FVS problem on general graphs can.

2   Preliminaries

Let $G = (V, E)$ be an undirected graph. For a set of vertices S in G we denote by $G - S$ the subgraph of G that results from removing the vertices in S, together with all edges incident to them. A *minimum feedback vertex set* is a feedback vertex set of minimum size. We denote by $\gamma(G)$ the size of a minimum feedback vertex set in G. For a vertex v, we denote by $deg(v)$ the degree of v in G.

Let v be a vertex in G such that $deg(v) \leq 2$. We define the following operation, which is standard in the literature. If $deg(v) = 1$ then remove v (together with its incident edge) from G; if $deg(v) = 2$ and the two neighbors x and y of v are not adjacent, then remove v and add an edge between x and y. Let us denote this operation by **Short-Cut**(v). We say that the operation **Short-Cut**() is *applicable* to a vertex v, if either $deg(v) = 1$, or $deg(v) = 2$ and the two neighbors of v are non-adjacent. A variation of the following proposition appears in [23] (see [2] for a proof).

Proposition 1. *(Lemma 1, [23]) Let G be an undirected graph and let v be a vertex in G to which the operation **Short-Cut**() is applicable. Let G' be the graph resulting from applying **Short-Cut**(v). Then $\gamma(G') = \gamma(G)$.*

We assume that we have a subroutine **Clean**(G) which applies the operation **Short-Cut**() repeatedly to G until the operation is no longer applicable. It is clear from Proposition 1 that if G' is the resulting graph from applying **Clean**(G), then $\gamma(G') = \gamma(G)$. The graph G is said to be *clean*, if **Clean**(G) is not applicable. Note that any degree-2 vertex in a clean graph must be on a cycle of length three.

An *almost shortest* cycle in G is a cycle whose length is at most the length of a shortest cycle in G plus one. It is well-known that an almost shortest cycle in an undirected graph with n vertices can be found in time $O(n^2)$ [19]. It is also well-known that any undirected graph with minimum degree at least 3 has a cycle of length at most $2 \lg n + 1$ [12].

3   The Algorithm

The basic idea behind most of the parameterized algorithms for the FVS problem presented so far has been to *branch* on short cycles and use the search tree method [11, 22, 23]. Suppose we are trying to determine if there exists a feedback vertex set in G of size bounded by k. Let C be a cycle in G of length l. Then every feedback vertex set of G must contain at least one vertex of C. For every vertex v on C, we can include v in the feedback vertex set, and then recurse to determine if $G - v$ has a feedback vertex set of size $k - 1$. Let us call such a process: *branching on* the cycle C. If we let $T(k)$ be the number of nodes in the search tree of such an algorithm that looks for a feedback vertex set of size bounded by k, then when the algorithm branches on a cycle of length l, $T(k)$ can be expressed using the recurrence relation $T(k) \leq l \cdot T(k-1) + 1$. The number of nodes in the search tree corresponding to the algorithm is $O((l_{max})^k)$, where l_{max} is the length of the longest cycle the algorithm branches on [11]. The running time of the algorithm is now proportional to the number of nodes in the search tree multiplied by the time we spend at every node of the search tree to find a cycle and process the graph. Thus, to reduce the running time of the algorithm, it is desirable to branch on short cycles. Most parameterized algorithms so far hinge on this approach [11, 22, 23].

In this section we develop another algorithm that uses this approach (based on the algorithm in [23]). We present the algorithm in Figure 1 below. We prove its correctness and analyze its running time in the next section.

FVS-solver

Input: an instance (G, k) of FVS
Output: a feedback vertex set F of G of size bounded by k in case it exists

0. $F = \emptyset$;
1. **if** G is acyclic **then return**(F);
2. **if** $k = 0$ and G contains a cycle **then return**('NO');
3. apply **Clean(G)**;
4. let C be an almost shortest cycle in G of length l;
5. **if** $l > 13$ and $k \leq 3\sqrt{n}$ **then return**('NO');
    else if $l > \lg n + 1$ and $(\lg n > \lg k + \lg \lg k + 13)$ **then return**('NO');
        else if $l > max\{2\lg k + 18, 2\lg n - 9\}$ **then return**('NO');
            else branch on C and update k, F, and G accordingly;

Fig. 1. The algorithm FVS-solver

4   Analysis of FVS-solver

The main idea behind the analysis of the algorithm presented in [23] is that if the graph does not contain a cycle of constant length, then the size of the feedback vertex set k must be large. In particular, the following structural result immediately follows from [23].

Lemma 1. *(Theorem 2, [23]) Let G be a graph on n vertices with minimum degree at least 3. If there is no cycle in G of length at most 12 then $\gamma(G) > 3\sqrt{n}$.*

The above result shows that when a clean graph has no cycle of length bounded by 12 (note that no degree-2 vertex exists in G at this point since G does not contain a cycle of length 3), $k > 3\sqrt{n}$, and hence, $\lg n < 2\lg k - 3$. Since every undirected graph with minimum degree at least 3 must have a cycle of length bounded by $2\lg n + 1$, G has a cycle of length bounded by $4\lg k$. An algorithm that branches on a shortest cycle will then either branch on a cycle of length at most 12, or of length at most $4\lg k$. According to the discussion in the previous section, this gives a search tree of size $O((max\{12, 4\lg k\})^k)$.

In this section we develop new combinatorial techniques and employ results from extremal graph theory to improve this analysis. The structural results obtained in this section will allow us to prove an upper bound of $O((2\lg k + 2\lg \lg k + 18)^k)$ on the size of the search tree of the algorithm **FVS-solver** presented in the previous section.

Let T be a tree. For a vertex v in T we denote by $d_T(v)$ the degree of v in T. A vertex $v \in T$ is said to be *good* if $d_T(v) \leq 2$. The statement of the following lemma can be easily proved.[3]

---

[3] An easy way to see why the statement is true is to note that the average degree of a tree is bounded by 2.

Lemma 2. *There are at least $\lfloor q/2 + 1 \rfloor$ good vertices in a tree on q vertices.*

For a good vertex u in T, $\{u\}$ is said to be a *nice* set if $d_T(u) \leq 1$, and for two good vertices u and v in T, the set $\{u, v\}$ is said to be a *nice* set if (u, v) is an edge in T.

Lemma 3. *Let T be a tree, and let n_g be the number of good vertices in T. Then there exists at least $\lfloor (n_g + 1)/2 \rfloor$ nice sets in T that are pairwise disjoint.*

Proof. Without loss of generality, we assume that T is rooted at a vertex r. We define the natural parent-child, and ancestor-descendent relationships, between the vertices in T. Note that each vertex in T is either the root vertex or has exactly one parent in T. For every vertex v in T, let T_v denote the subtree of T rooted at v and containing all its descendents. We will prove the following statement: if T_v contains q good vertices of T, then the number of pairwise disjoint nice sets in T_v is at least $\lfloor (q+1)/2 \rfloor$. It is clear that the previous statement will imply the statement of the lemma because $T = T_r$. (Observe that a good vertex in T_v might not be a good vertex in T, and this is why we require the q vertices to be good in T.) To prove the statement, let T_v be a rooted subtree of T, and proceed by induction on q. T_v must contain at least one leaf of T, so $q > 0$. If $q \leq 2$, then T_v must contain a leaf u in T, and $\{u\}$ is a nice set. Therefore the number of (pairwise disjoint) nice sets in this case is at least $1 \geq \lfloor (q+1)/2 \rfloor$. Suppose now that the number of pairwise disjoint nice sets in any rooted tree containing p good vertices from T, for $2 \leq p < q$, is at least $\lfloor (p+1)/2 \rfloor$. We distinguish two cases.

Case 1. v has at least two children. Let $v_1, \ldots, v_d$, be the children of v in T_v. Let q_i be the number of good vertices in T that are in T_{v_i}, $i = 1 \ldots d$, and note that $1 \leq q_i < q$, and that $q \leq q_1 + \ldots q_d + 1$ (v might be good, in which case it is the root of the tree and $d = 2$). By the inductive hypothesis, each T_{v_i} contains at least $\lfloor (q_i + 1)/2 \rfloor$ pairwise disjoint nice sets in T. Since every nice set in T_{v_i} is disjoint from every nice set in T_{v_j} for $1 \leq i \neq j \leq d$, it follows that T_v contains at least $\lfloor (q_1 + \ldots + q_d + d)/2 \rfloor = \lfloor (q_1 + \ldots q_d + 1 + d - 1)/2 \rfloor \geq \lfloor (q+1)/2 \rfloor$ pairwise disjoint nice sets in T (note that $d \geq 2$).

Case 2. v has exactly one child v'. In this case v must be a good vertex in T. If v' is good, let v'' be the child of v' in T_v (note that v'' must exist since $q > 2$). Now $T_{v''}$ contains $q - 2$ good vertices in T. By induction, the number of pairwise disjoint nice sets in $T_{v''}$ is at least $\lfloor (q - 1)/2 \rfloor$. Since $\{v, v'\}$ is a nice set which is disjoint from all the nice sets in $T_{v''}$, it follows that the number of pairwise disjoint nice sets in T_v is at least $\lfloor (q+1)/2 \rfloor$. If v' is bad, then v' must have at least two children in $T_{v'}$. The proof now is identical to that of **Case 1** by applying induction on the trees rooted at the children of v'.

This completes the induction and the proof. ☐

This lemma follows directly from Lemma 2 and Lemma 3.

Lemma 4. *Let T be a tree on q vertices. There are at least $q/4$ pairwise disjoint nice sets in T.*

Let (G, k) be an instance of FVS, where G is clean and has n vertices, and assume that G does not have a cycle of length bounded by 12. Since G is clean and has no cycles of length 3, G has minimum degree at least 3. Let F be a minimum feedback set of G, let $f = |F|$, and $\mathcal{F} = G - F$. Applying Lemma 4 to every tree in $\mathcal{F}$, we get that $\mathcal{F}$ (i.e., the trees in $\mathcal{F}$) contains at least $(n - f)/4$ pairwise disjoint nice sets. So if we let S be the set of pairwise disjoint nice sets in $\mathcal{F}$, then $|S| \geq (n-f)/4$. We construct a graph G_F as follows. The set of vertices of G_F is F. The edges of G_F are defined as follows. Let $\{a, b\}$ be a nice set in S. Since a and b are good in $\mathcal{F}$ (i.e., in the tree of $\mathcal{F}$ that they belong to) and both have degree greater or equal to 3 in G, a must have at least one neighbor in F, and b must have at least one neighbor F. Pick exactly one neighbor a_1 of a in F, and exactly one neighbor b_1 of b in F. Note that a_1 and b_1 must be distinct since G has no cycles of length 3. Add the edge (a_1, b_1) to G_F. Now let $\{a\}$ be a nice set in S, then a must have at least two neighbors in F. We pick exactly two neighbors a_1 and a_2 of a in F, and we add the edge (a_1, a_2) in G_F. This completes the construction of G_F. Since G has no cycles of length bounded by 6, for any two distinct nice sets in S, the edges associated with them in G_F are distinct. This means that G_F is a simple graph with at least $(n - f)/4$ edges. We have the following lemma.

Lemma 5. *If G_F has a cycle of length l then G has a cycle of length bounded by $3l$.*

Proof. Let $(v_1, \ldots, v_l, v_1)$ be a cycle in G_F. Since each (v_i, v_{i+1}) (the index arithmetic is taken modulo l) is an edge in G_F, (v_i, v_{i+1}) is associated with a nice set S_i in S. If $S_i = \{a\}$, let P_i be the path (v_i, a, v_{i+1}) in G; if $S_i = \{a, b\}$, let P_i be the path (v_i, a, b, v_{i+1}) in G. Since the nice sets in S are disjoint, the paths P_i, $i = 1, \ldots, l$, are internally vertex-disjoint in G, and they determine a cycle of length bounded by $3l$. □

The following result is known in extremal graph theory (see [5, 6, 14]).

Lemma 6. *Let G be a graph with n vertices where $n \geq 3$. If G does not contain a cycle of length at most $2l$, then the number of edges in G is bounded by $90ln^{1+1/l}$.*

Theorem 1. *Let G be a graph with $n \geq 3$ vertices and with no cycles of length bounded by $max\{12, \lg n\}$. Then $\gamma(G) \geq (n/(61 \lg n))^{1-6/(\lg n+6)}$.*

Proof. Suppose that G does not have a cycle of length bounded by $max\{12, \lg n\}$. Let F be a minimum feedback vertex set of G, and let $f = |F| = \gamma(G)$. Since G has no cycles of length bounded by 12, if we let $\mathcal{F}$ and G_F be as defined in the above discussion, then it follows from above that the number of edges in G_F is at least $(n - f)/4$. Since G does not have a cycle of length bounded by $\lg n$, by Lemma 5, G_F has no cycle of length bounded by $(\lg n)/3$. Applying

Lemma 6 with $l = (\lg n)/6$, we get that the number of edges in G_F is bounded by $15(\lg n)f^{1+6/\lg n}$. Thus

$$(n - f)/4 \leq 15(\lg n)f^{1+6/\lg n}$$
$$n \leq 60(\lg n)f^{1+6/\lg n} + f \leq 60(\lg n)f^{1+6/\lg n} + f^{1+6/\lg n}$$
$$n \leq 61(\lg n)f^{1+6/\lg n} \tag{1}$$

Manipulating (1) we get $f \geq (n/(61 \lg n))^{1-6/(\lg n+6)}$ completing the proof.
$\square$

Theorem 1 implies that the size of a minimum feedback vertex set in a graph with minimum degree at least 3 and no cycles of length bounded by $\lg n$ must be $\Omega(n/\lg n)$.

Corollary 1. *Let G be a graph with $n \geq 3$ vertices and with no cycles of length bounded by $max\{12, \lg n\}$. Then $\lg n \leq \lg \gamma(G) + \lg \lg \gamma(G) + 13$.*

Proof. Applying the $\lg()$ function on both sides of (1) in Theorem 1 we get

$$\lg n \leq \lg 61 + \lg \lg n + \lg f + 6 \lg f/\lg n \tag{2}$$

Since G has no cycles of length bounded by 12, from Lemma 1 we get $\lg n < 2 \lg f$, and hence $\lg \lg n < 1 + \lg \lg f$. Now noting that $\lg 61 \leq 6$ and $\lg f \leq \lg n$, it follows from (2) that $\lg n \leq \lg f + \lg \lg f + 13$. $\square$

Lemma 7. *Let G be a graph with n vertices and minimum degree at least 3. For any integer constant $d > 0$, there is a cycle in G of length at most $max\{2 \lg \gamma(G) + 2 \lg(d-1) + 3, 2 \lg n - 2 \lg(d+1) + 4\}$.*

Proof. Let $d > 0$ be given, and let Δ be the maximum degree of G. By Lemma 4 in [23], $\gamma(G) > (\delta-2)n/(2(\Delta-1))$, where δ is the minimum degree of the graph. Since $\delta \geq 3$, we get $n < 2(\Delta - 1)\gamma(G)$, and hence

$$\lg n < \lg \gamma(G) + \lg(\Delta - 1) + 1$$
$$2 \lg n + 1 < 2 \lg \gamma(G) + 2 \lg(\Delta - 1) + 3 \tag{3}$$

Let r be a vertex in G of degree Δ. Perform a breadth-first search starting at r until a shortest cycle is first encountered in the graph. Let l be the length of the shortest cycle first encountered in this process. Since r has degree Δ and every other vertex has degree at least 3, it is not difficult to show, using a counting argument, that the number of vertices in the graph is at least $\Delta(2^{(l-2)/2} - 1) + 1$ if l is even, and $\Delta(2^{(l-1)/2} - 1) + 1$ if l is odd.

Suppose that $l \geq 4$. We have

$$n \geq \Delta(2^{(l-2)/2} - 1) + 1$$
$$n > \Delta(2^{(l-2)/2} - 1)$$
$$\lg n > \lg \Delta + \lg (2^{(l-2)/2} - 1) \tag{4}$$

Also $l \geq 4$ implies that $2^{(l-2)/2} \geq 2$. Using the inequality $\lg (x - 1) \geq \lg x - 1$ for $x \geq 2$ in (4), and manipulating (4), we get

$$l \leq 2 \lg n - 2 \lg \Delta + 4 \tag{5}$$

If $l < 4$, then $l = 3$ and since $n > \Delta$, (5) is still true.

Now if $\Delta \leq d$, then from the fact that G has a cycle of length bounded by $2 \lg n + 1$, and from (3), we conclude that there is a cycle in G of length at most $2 \lg \gamma(G) + 2 \lg (\Delta - 1) + 3 \leq 2 \lg \gamma(G) + 2 \lg (d - 1) + 3$. Otherwise, $\Delta \geq d + 1$, and by (5), there is a cycle in G of length at most $2 \lg n - 2 \lg \Delta + 4 \leq 2 \lg n - 2 \lg (d + 1) + 4$. It follows that G has a cycle of length at most $max\{2 \lg \gamma(G) + 2 \lg (d - 1) + 3, 2 \lg n - 2 \lg (d + 1) + 4\}$. This completes the proof. □

Corollary 2. *Let G be a graph with minimum degree at least 3. There exists a cycle in G of length at most $max\{2 \lg \gamma(G) + 17, 2 \lg n - 10\}$*

Proof. Apply Lemma 7 with $d = 127$. □

Theorem 2. *Let G be a graph with n vertices. In time $O((2 \lg k + 2 \lg \lg k + 18)^k n^2)$ we can decide if G has a feedback vertex set of size bounded by k.*

Proof. It is not difficult to see that the algorithm **FVS-solver** solves the FVS problem. In particular, if **FVS-solver** returns a NO answer in step 5, then either (*i*) $l > 13$ and $k \leq 3\sqrt{n}$, or (*ii*) $l > \lg n + 1$ and $\lg n > \lg k + \lg \lg k + 13$, or (*iii*) $l > max\{2 \lg k + 18, 2 \lg n - 9\}$. Since l is the length of an almost shortest cycle, if (*i*) holds then G does not have a cycle of length bounded by 12, and hence no feedback vertex set of size bounded by k by Lemma 1. If (*ii*) holds, then G does not have a cycle of length bounded by $max\{12, \lg n\}$, and hence no feedback vertex set of size bounded by k by Corollary 1 (note that in this case $n \geq 3$, and l must also be greater than 13 since $\lg n > 13$). Finally if (*iii*) holds, then G has no cycle of length bounded by $\{2 \lg k + 17, 2 \lg n - 10\}$. From Corollary 2 (note that G has minimum degree at least three at this point since G is clean and has no cycles of length three) we conclude that k must be smaller than $\gamma(G)$, and hence G has no feedback vertex set of size bounded by k.

To analyze the running time of the algorithm, let l be the length of a cycle C that the algorithm branches on. By looking at step 5 in the algorithm, we see that if the algorithm branches on C then one of the following cases must hold: (a) $l \leq 13$, or (b) $l > 13$, $k > 3\sqrt{n}$, and $l \leq \lg n + 1$, or (c) $\lg n \leq \lg k + \lg \lg k + 13$ and $l \leq max\{2\lg k + 18, 2\lg n - 9\}$.

If (b) holds, then the conditions in (b) give that $l < 2\lg k$. If (c) holds, then combining the two inequalities in (c) we get $l \leq max\{2\lg k + 18, 2\lg k + 2\lg \lg k + 17\} \leq 2\lg k + 2\lg \lg k + 18$. It follows that in all cases (a), (b), and (c), the algorithm branches on a cycle of length at most $2\lg k + 2\lg \lg k + 18$. Thus, according to the discussion at the beginning of this section, the size of the search tree corresponding to the algorithm is $O((2\lg k + 2\lg \lg k + 18)^k)$. Now at each node of the search tree the algorithm might need to find an almost shortest cycle, call **Clean()**, check if the graph is acyclic, and process the graph. Finding an almost shortest cycle takes $O(n^2)$ time. When **Clean()** is applied, since every vertex that **Clean()** removes has degree bounded by 2, **Clean()** can be implemented to run in linear time in the number of vertices it removes, and hence, in $O(n)$ time. Checking if the graph is acyclic and processing the graph takes no more than $O(n^2)$ time. It follows that the running time of the algorithm is $O((2\lg k + 2\lg \lg k + 18)^k n^2)$. ☐

According to the above theorem, the algorithm **FVS-solver** improves the algorithm in [23] by roughly a 2^k factor.

5   FVS on Special Classes of Graphs

In this section we consider the FVS problem on special classes of graphs. We look at the following classes: bipartite graphs, bounded genus graphs, and K^r-minor free graphs for fixed r.

Bipartite Graphs

Let G be a graph with n vertices and m edges. Consider the operation of subdividing an edge e in G by introducing a degree-2 vertex v. Then this operation is precisely the inverse operation of **Short-Cut**(v). Therefore, if we let G' be the graph obtained from G by subdividing an edge $e \in G$, then it follows from Proposition 1 that $\gamma(G') = \gamma(G)$. Subdividing each edge in G yields a bipartite graph G', which according to the previous statement satisfies $\gamma(G') = \gamma(G)$. The graph G' has $n + m$ vertices and $2m$ edges. This shows that the FVS problem on general graphs can be solved in time $f(k)n^{O(1)}$ if and only if the FVS problem on bipartite graphs can be solved in $f(k)n^{O(1)}$ time. In particular, an algorithm of running time $c^k n^{O(1)}$ (for some constant c) for the FVS problem on bipartite graph implies an algorithm of running time $c^k n^{O(1)}$ for the FVS problem on general graphs.

Bounded Genus Graphs

The following lemma follows from a standard Euler-formula argument.

Lemma 8. *Let G be a graph with n vertices, minimum degree at least 3, and genus g. If $n \geq 8g$ then there is a cycle in G of length at most 12.*

Let G be a graph with n_0 vertices and genus g_0 satisfying $g_0 \leq c \lg n_0$ for some constant c. Note that if we branch on a cycle in G or process G as in the algorithm **FVS-solver**, the number of vertices and genus of G change, and the relationship $g \leq c \lg n$, where n and g are the number of vertices and genus, respectively, in the resulting graph may not hold. However, since the operations in **FVS-solver** do not increase the genus of the graph, the genus g of the resulting graph satisfies $g \leq g_0 \leq c \lg n_0$. Consider the algorithm **BGFVS-solver** given in Figure 2, which is a modification of the algorithm **FVS-solver** presented in Section 1, that will solve the FVS problem on graphs of genus bounded by $c \lg n$.

BGFVS-solver

Input: an instance (G, k) of FVS where G has n_0 vertices and genus g_0
Output: a feedback vertex set F of G of size bounded by k in case it exists
        and g_0 satisfies $g_0 \leq c \lg n_0$

0. $F = \emptyset$;
1.   **if** G is acyclic **then return**(F);
2.   **if** $k = 0$ and G contains a cycle **then return** ('NO');
3.   apply **Clean(G)**;
4.   **if** the number of vertices n of G is bounded by $8c \lg n_0$ **then**
        solve the problem by brute force and **STOP**;
5.   let C be an almost shortest cycle in G of length l;
6.   **if** $l > 13$ **then return**('Invalid Instance');
        else branch on C and update k, F, and G accordingly;

Fig. 2. The algorithm BGFVS-solver

It is not difficult to see the correctness of the algorithm **BGFVS-solver**. The only step that needs clarification is when the algorithm returns in step 6 that the instance is invalid. If this happens, then the number of vertices n in the resulting graph G must satisfy $n > 8c \lg n_0$, where n_0 is the number of vertices in the original graph, and G does not have a cycle of length bounded by 12 (note that the algorithm finds an almost shortest cycle). Since the genus of the resulting graph g is bounded by the genus g_0 of the original graph, if g_0 satisfies the condition $g_0 \leq c \lg n_0$ then we would have $8g \leq 8g_0 \leq 8c \lg n_0 \leq n$, which according to Lemma 8, would imply the existence of a cycle of length at most 12 in G (since G has minimum degree at least 3 at this point). Since no such cycle exists in G, the genus g_0 in the original graph does not satisfy the condition $g_0 \leq c \lg n_0$, and hence the input instance does not satisfy the genus bound requirement. Therefore the algorithm rejects the instance in this case. This shows actually that **BGFVS-solver** does not need to know in advance if the input instance satisfies the genus bound requirement or not. As long as the input instance satisfies the genus bound requirement, the algorithm solves the problem.

Now to analyze the running time of the algorithm, we note that step 4 can be carried out in time $O(n_0^{8c+1})$ by enumerating every subset of vertices in the graph and checking whether it is a feedback vertex set of size bounded by k. The algorithm never branches on a cycle of length greater than 13. It follows that the running time of the algorithm is $O(13^k n_0^{8c+\omega+1})$.

Theorem 3. *The* FVS *problem on graphs of genus* $O(\lg n)$ *can be solved in time* $13^k n^{O(1)}$.

Let G be a graph with n_0 vertices and genus g_0, and let $\epsilon > 0$ be given. We construct a graph G' as follows. Let W be a wheel on $n_0^{2/\epsilon}$ vertices. Add W to G by linking a vertex in W other than the center to any vertex in G. It is easy to verify that $\gamma(G') = \gamma(G) + 2$, and that a feedback vertex set of G can be constructed from a feedback vertex set of G' easily. Since a wheel is planar, it follows that the genus g' of G is equal to g. The number of vertices n' of G' is $n' = n_0 + n_0^{2/\epsilon}$. Therefore $g' = g \leq n_0^2 \leq n'^\epsilon$. This shows that we can reduce the FVS problem on general graphs to the FVS problem on graphs of genus $O(n^\epsilon)$ (for any $\epsilon > 0$) in polynomial time such that if the FVS problem on graphs of genus $O(n^\epsilon)$ can be solved in $f(k)n^{O(1)}$ time, then so can the FVS problem on general graphs. In particular, if the FVS problem on graphs of genus $O(n^\epsilon)$ can be solved in time $c^k n^{O(1)}$ then so can the FVS problem on general graphs.[4]

K^r-Minor Free Graphs

Let r be a constant, and let $\mathcal{G}_r$ be the class of graphs with minimum degree at least 3 and no K^r-minor. Combining theorems from [20, 25] and [26] as described in [9, p180], we have the following: There exists a constant c such that, for all r, each $G \in \mathcal{G}_r$ has a cycle of length less than $cr\sqrt{\lg r}$.

Let G be a K^r-minor free graph with minimum degree at least 3. According to the result described above, there is a cycle in G of length bounded by $c_1 = cr\sqrt{\lg r}$, which is a constant. Therefore, if we modify the algorithm **FVS-solver** so that it branches on an almost shortest cycle after applying **Clean(G)**, then the algorithm always branches on a cycle of length at most $c' = 1 + max\{3, c_1\}$.[5] This shows that the running time of the modified algorithm is $O(c'^k n^2)$.

Theorem 4. *For any constant* r, *the* FVS *problem on* K^r-*minor free graphs can be solved in time* $O(c'^k n^2)$, *where* c' *is a constant.*

---

[4] Using the same technique, one can prove that if the FVS problem on graphs with average degree bounded by a constant, can be solved in time $f(k)n^{O(1)}$, then so can the FVS problem on general graphs.

[5] Note that if the length of the almost shortest cycle that the algorithm computes is not bounded by c', the algorithm rejects the instance.

References

[1] V. Bafna, P. Berman, and T. Fujito. A 2-approximation algorithm for the Undirected Feedback Vertex Set problem. *SIAM Journal on Discrete Mathematics*, 12(3):289–297, 1999.

[2] Reuven Bar-Yehuda, Dan Geiger, Joseph Naor, and Ron M. Roth. Approximation algorithms for the feedback vertex set problem with applications to constraint satisfaction and bayesian inference. *SIAM Journal on Computing*, 27(4):942–959, 1998.

[3] A. Becker, R. Bar-Yehuda, and D. Geiger. Random algorithms for the Loop Cutset problem. *Journal of Artificial Intelligence Research*, 12:219–234, 2000.

[4] H. Bodlaender. On disjoint cycles. *International Journal of Foundations of Computer Science*, 5:59–68, 1994.

[5] B. Bollobás. *Extremal Graph Theory*. Academic Press, London, 1978.

[6] J. A. Bondy and M. Simonovits. Cycles of even length in a graph. *J. Combinatorial Theory (B)*, 16:97–105, 1974.

[7] J. Chen, I. Kanj, and W. Jia. Vertex cover: further observations and further improvements. *Journal of Algorithms*, 41:280–301, 2001.

[8] D. Coppersmith and S. Winograd. Matrix multiplication via arithmetic progression. *Journal of Symbolic Computation*, 9:251–280, 1990.

[9] R. Diestel. *Graph Theory (2nd Edition)*. Springer-Verlag, New York, 2000.

[10] R. Downey and M. Fellows. Fixed-parameter tractability and completeness. *Congressus Numerantium*, 87:161–187, 1992.

[11] R. Downey and M. Fellows. *Parameterized Complexity*. Springer, New York, 1999.

[12] P. Erdos and L. Posa. On the maximal number of disjoint circuits of a graph. *Pubbl. Math. Debrecen*, 9:3–12, 1962.

[13] G. Even, J. Naor, B. Schieber, and M. Sudan. Approximating minimum feedback sets and multicuts in directed graphs. *Algorithmica*, 20(2):151–174, 1998.

[14] R. J. Faudree and M. Simonovits. On a class of degenerate extremal graph problems. *Combinatorica*, 3 (1):83–93, 1983.

[15] F. Fomin and D. Thilikos. Dominating sets in planar graphs: branch-width and exponential speed-up. In *Proceedings of the 14th ACM-SIAM Symposium on Discrete Algorithms*, pages 168–177, 2003.

[16] T. Fujito. A note on approximation of the Vertex Cover and Feedback Vertex Set problems-unified approach. *Information Processing Letters*, 59(2):59–63, 1996.

[17] M. Garey and D. Johnson. *Computers and Intractability: A Guide to the Theory of NP-Completeness*. W. H. Freeman, New York, 1979.

[18] R. Gupta, R. Gupta, and M. Breuer. Ballast: A methodology for partial scan design. *IEEE Transactions on Computers*, 39(4):538–544, 1990.

[19] A. Itai and M. Rodeh. Finding a minimum circuit in a graph. *SIAM Journal on Computing*, 7(4):413–423, 1978.

[20] A. V. Kostochka. The minimum Hadwiger number for graphs with a given mean degree of vertices. *Metody Diskret. Analiz.*, 38:37–58, March 1982.

[21] A. Kunzmann and H. Wunderlich. An analytical approach to the partial scan problem. *Journal of Electronic Testing: Theory and Applications*, 1:163–174, 1990.

[22] V. Raman. Parameterized complexity. In *Proceedings of the 7th National Seminar on Theoretical Computer Science*, pages 1–18, 1997.

[23] V. Raman, S. Saurabh, and C. Subramanian. Faster fixed-parameter tractable algorithms for Undirected Feedback Vertex Set. In *Proceedings of the 13th Annual International Symposium on Algorithms and Computation*, volume 2518 of *Lecture Notes in Computer Science*, pages 241–248, 2002.

[24] P. Seymour. Packing directed circuits fractionally. *Combinatorica*, 15(2):281–288, 1995.

[25] A. Thomason. An extremal function for contractions of graphs. *Math. Proc. Cambridge Philos. Soc.*, 95(2):261–265, 2001.

[26] C. Thomassen. Girth in graphs. *J. Combin. Theory Ser. B*, 35:129–141, March 1983.

Automated Proofs of Upper Bounds on the Running Time of Splitting Algorithms

Sergey S. Fedin and Alexander S. Kulikov

St.Petersburg State University, Department of Mathematics and Mechanics,
St.Petersburg, Russia
http://logic.pdmi.ras.ru/∼{fedin, kulikov}

Abstract. The splitting method is one of the most powerful and well-studied approaches for solving various NP-hard problems. The main idea of this method is to split an input instance of a problem into several simpler instances (further simplified by certain *simplification rules*), such that when the solution for each of them is found, one can construct the solution for the initial instance in polynomial time. There exists a huge number of articles describing algorithms of this type and usually a considerable part of such an article is devoted to case analysis. In this paper we show how it is possible to write a program that given simplification rules would automatically generate a proof of an upper bound on the running time of a splitting algorithm that uses these rules. As an example we report the results of experiments with such a program for the SAT, MAXSAT, and $(n, 3)$-MAXSAT (the MAXSAT problem for the case where every variable in the formula appears at most three times) problems.

1   Introduction

There are several approaches for dealing with NP-hard problems. For example, for some problems the algorithms that find approximate solutions in polynomial time are known, but there are also algorithms that find exact solutions with some probability. However, in practice many problems have to be solved exactly, so in this paper we consider the most popular type of algorithms, so-called *splitting algorithms*. A typical example of a splitting algorithm for the propositional satisfiability problem (SAT) is DLL-type algorithm [DLL62].

Many NP-hard problems can be formulated in terms of Boolean formulas (for example, [GHNR00] shows how the NP-hard graph problem MAXCUT can be formulated in terms of Boolean formulas). Due to this fact we consider splitting algorithms for problems dealing with formulas in CNF (e.g., SAT, MAXSAT). Let us briefly describe the idea of a splitting algorithm for such problems. At each step it firstly simplifies an input formula using *simplification rules*. Then, if the obtained formula F is not trivial, it makes two recursive calls for formulas resulting from F by setting $l = true$ and $l = false$, where l is a literal of the formula F (these formulas are usually denoted $F[l]$ and $F[\bar{l}]$). Finally, it returns the answer according to the answers returned by the recursive calls. For example,

R. Downey, M. Fellows, and F. Dehne (Eds.): IWPEC 2004, LNCS 3162, pp. 248–259, 2004.
© Springer-Verlag Berlin Heidelberg 2004

in case of the SAT problem it must return the answer "Satisfiable" if and only if at least one of the recursive calls returns this answer. In general case, a splitting algorithm may call itself on a bigger number of formulas. The only condition on these formulas is the possibility to construct the solution for initial formula from the solutions for these formulas (in polynomial time).

There is a lot of literature where splitting algorithms for different problems are described. Moreover, for a number of problems there is a "hierarchy" of articles containing such algorithms, where each paper improves the result of a previous one by more involved case analysis or by introducing a new simplification rule. For example, there exist splitting algorithms for Satisfiability (see [DHIV01] for survey), Maximum Satisfiability ([BR99], [CK02], [GHNR00]), Exact Satisfiability ([BMS03], [K02]), Maximum Cut ([FK02], [GHNR00]), and many other problems. Usually the size of a case analysis in this hierarchy increases with the improvement of a result. But this is not always true: sometimes a new simplification rule allows to prove a better result by considering fewer cases.

In this paper we show how it is possible to write a program that given simplification rules would automatically generate a proof of an upper bound on the running time of a splitting algorithm that uses these rules. We present several automatically generated proofs of upper bounds for the SAT, MAXSAT and $(n, 3)$-MAXSAT problems.

Organization of the Paper. Our paper is organized as follows: In Sect. 2 we give basic definitions. Section 3 contains the detailed description of our program. In Sect. 4 we discuss related work. Section 5 contains further directions (in particular there we indicate several bottlenecks of all currently known frameworks for automated proofs of upper bounds for NP-hard problems).

2   General Setting

2.1   Basic Definitions

Let V be a set of Boolean variables. The negation of a variable v is denoted by $\bar{v}$, by $\bar{V}$ we denote the set $\{\bar{v} \mid v \in V\}$. *Literals* are the members of the set $W = V \cup \bar{V}$. If w denotes a literal $\bar{l}$, then $\bar{w}$ denotes a literal l. A *clause* is the disjunction of a finite set of literals that does not contain simultaneously any variable together with its negation. The empty clause is interpreted as *False*. A *formula in CNF* is the conjunction of a finite set of clauses. The empty formula is interpreted as *True*. The *length of a clause* is the number of its literals. The *length of a formula* is the sum of lengths of all its clauses. We say that *a literal l occurs* in a clause or in a formula, if this clause or this formula contains the literal l. However, we say that *a variable v occurs* in a clause or in a formula, if this clause or this formula contains the literal v, or it contains the literal $\bar{v}$. An (i, j)-*literal* is a literal that occurs in a formula i times positively and j times negatively.

An *assignment* is a finite subset of W that does not contain any variable together with its negation. Informally speaking, if an assignment I contains a literal l, it means that l has the value $True$ in I. To obtain a formula $F[I]$ from a formula F and an assignment $I = \{l_1, \ldots, l_s\}$, we first remove all clauses containing the literals l_i from F, and then delete all occurrences of the literals $\bar{l_i}$ from the other clauses.

2.2   Estimation of the Size of a Splitting Tree

Kullmann and Luckhardt introduced in [KL98] a notion of a splitting tree. One can consider an execution of a splitting algorithm as a tree whose nodes are labeled with CNF-formulas such that if a node is labeled with a CNF-formula F, then its sons are labeled with simplified formulas $F[I_1], F[I_2], \ldots, F[I_k]$ for assignments $I_1, I_2, \ldots, I_k$. To each formula F of this tree we attach a non-negative integer $\mu(F)$, which denotes the *complexity of F*. We use the following measures of complexity in our program:

1. $\mu(F) = N(F)$ is the number of variables in F;
2. $\mu(F) = K(F)$ is the number of clauses in F;
3. $\mu(F) = L(F)$ is the length of F.

The tree is a *splitting tree* if, for each node, the complexity of the formula labelling this node is strictly greater than the complexity of each of the formulas labelling its sons.

Let us consider a node in our tree labeled with a formula F_0. Suppose its sons are labeled with formulas $F_1, F_2, \ldots, F_m$. The *splitting vector* of this node is the m-tuple $(t_1, t_2, \ldots, t_m)$, where t_i are positive numbers not exceeding $\mu(F_0) - \mu(F_i)$. The *characteristic polynomial* of this splitting vector is defined as follows: $h(x) = 1 - \sum_{i=1}^m x^{-t_i}$. The only positive root of this polynomial is called the *splitting number* and denoted by $\tau(t_1, t_2, \ldots, t_m)$. The *splitting number of a tree* T is the largest of the splitting numbers of its nodes. We denote it by $\tau_{max,T}$.

Now we have provided all the necessary definitions to present a lemma proved by Kullmann and Luckhardt in [KL98] that allows to estimate the number of leaves in a splitting tree using its splitting number.

Lemma 1. *Let T be a splitting tree, and let its root be labeled with a formula F_0. Then the number of leaves in T does not exceed $(\tau_{max,T})^{\mu(F_0)}$.*

Let us show how this lemma helps to estimate the running time of a splitting algorithm. As we already mentioned a splitting algorithm at first simplifies the input formula and then makes several recursive calls of itself for formulas with smaller complexity. Clearly, the total running time of such an algorithm is the total running time of all recursive calls plus the time spent to make these calls (note that we consider only splitting algorithms that at each step make a number of recursive calls bounded by a constant). Therefore, the running time is within a polynomial factor of the number of nodes (or leaves) of the recursion tree.

2.3   SAT, MAXSAT, and $(n, 3)$-MAXSAT Problems

In this subsection we give definitions of NP-hard problems on which we tested our program.

The propositional satisfiability problem (SAT) is: given a formula in CNF, check whether there exists an assignment of Boolean values to all variables of this formula that satisfies all its clauses. In the maximum satisfiability problem (MAXSAT) the question is to find the maximum possible number of simultaneously satisfied clauses in a given CNF formula. The $(n, 3)$-MAXSAT problem is the version of MAXSAT, where each variable appears in the input formula at most three times.

Currently best upper bounds for these problems are the following:

- SAT: $O(1.238823^K)$, $O(1.073997^L)$ ([H00]),
- MAXSAT: $O(1.341294^K)$ ([CK02]), $O(1.105729^L)$ ([BR99]),
- $(n, 3)$-MAXSAT : $O(1.324719^N)$ ([BR99]).

All these bounds are obtained by using the splitting method. Note that there are no known upper bounds of the form $O(c^N)$, where $c < 2$ is a constant, for SAT and MAXSAT problems. We do not know any upper bounds for $(n, 3)$-MAXSAT w.r.t. number of clauses and length of the formula that improve the corresponding bounds for MAXSAT.

3   The Algorithm

We start with giving several non-standard definitions that are convenient in the context of this paper and then present the algorithm on which our program is based.

A typical splitting algorithm first simplifies an input formula and then splits the resulting formula. Usually it makes splittings according to some properties of a formula. That is, in a code of a splitting algorithm the application of simplification rules to the input formula is followed by several cases of the form as: if a formula contains such clauses or literals . . . , then split as . . . . To prove the correctness of such an algorithm one has to prove that each simplified formula (we say that a formula is simplified if no simplification rule is applicable to it) satisfies the condition of at least one of these cases. Note that a set of formulas satisfying the condition of a case corresponds to each such case, and the algorithm is correct if the union of all these sets contains the set of simplified formulas. In the following subsection we introduce a notion of a class of formulas which is convenient for describing splitting algorithms.

3.1   Class of Formulas

Let C be a clause consisting of literals $l_1, l_2, \ldots, l_k$. We define a *clause with unfixed length* as a set of clauses containing all these literals and probably some more literals and denote it by $(l_1 \vee l_2 \vee \ldots \vee l_k \ldots)$. The literals $l_1, l_2, \ldots, l_k$ we call the *basis* of this clause. For example, $(l \ldots)$ is a set of all clauses containing the literal l. In the following we use the word "clause" to refer both to clauses in its standard definition and to clauses with unfixed length.

Similarly we define a *class of formulas*. Let $C'_1, \ldots, C'_k$ be clauses (some of them may have unfixed lengths). Then a *class of formulas* is a set of formulas represented as $C_1 \wedge \ldots \wedge C_m$, such that the following conditions hold:

1. $m \geq k$,
2. for $1 \leq i \leq k$, $C'_i \subseteq C_i$ (as sets of literals), if C'_i is a clause with unfixed length, and $C_i = C'_i$ otherwise,
3. for $1 \leq i \leq k$, $k + 1 \leq j \leq m$, C_j does not contain any variable from the basis of C'_i.

We denote this set by $C'_1 \wedge \ldots \wedge C'_k \ldots$ and call the clauses $C'_1, \ldots, C'_k$ the *basis* of this class of formulas. We say that variables from clauses $C'_1, \ldots, C'_k$ are *known* for this class. When we say that a class of formulas contains some literal we mean that its basis contains this literal.

Informally speaking, one can obtain a class of formulas by replacing all occurrences of some variables in a formula by "$\ldots$". For example, the formula $(x \vee y \vee z) \wedge (\bar{x}) \wedge (\bar{y} \vee z \vee u) \wedge (\bar{u} \vee x) \wedge (\bar{z})$ is a member of the following class of formulas: $(z \ldots) \wedge (z \vee u \ldots) \wedge (\bar{u} \ldots) \wedge (\bar{z}) \ldots$.

Note that in most situations we can work with a class of formulas in the same way as with a CNF formula. For example, if we eliminate from the basis of a class of formulas all clauses containing literal x and all occurrences of literal $\bar{x}$ from the other clauses, we obtain the class of formulas resulting from all formulas of the initial class by setting the value of x to $True$. Also it is easy to see that if after assigning a Boolean value to a variable of a class of formulas or applying a (considered in this paper) simplification rule to it, its complexity measure decreases by Δ, then the complexity measure of each formula of this class decreases *at least* by Δ.

For a class of formulas $\mathcal{F}$ and a clause C we define a class of formulas $\mathcal{F} + \{C\}$ as a class resulting from $\mathcal{F}$ by adding clause C to its basis. Similarly we define the clause with unfixed length $C + \{l\}$, where C is a clause with unfixed length and l is a literal. By *fixing the length of the clause* $(l_1 \vee \ldots \vee l_k \ldots)$ we mean replacing it by the clause $(l_1 \vee l_2 \vee \ldots \vee l_k)$.

We say that a simplification rule is applicable to a class of formulas, if this rule is applicable to every formula of this class. For example, each formula of the class $(x) \wedge (x \vee \bar{y} \ldots) \wedge (y \ldots) \ldots$ contains pure literal (i.e., a literal that does not occur negated in the formula). Similarly, we say that a class of formulas has a splitting if each formula of this class has this splitting. For example, every formula of the class $(x \vee y) \wedge (\bar{x} \vee y \vee \bar{t} \ldots) \wedge (\bar{y} \vee u) \wedge (u \vee \bar{z}) \wedge (\bar{x} \vee t) \wedge (\bar{u} \vee \bar{y} \vee z) \ldots$ always has the $(1, 2)$-splitting (on variable x) w.r.t. the number of variables (t is the other variable, which is eliminated), the $(2, 2)$-splitting (on y) w.r.t. the number of clauses, and $(4, 3)$-splitting (on z) w.r.t. the length of the formula. However, when we deal with specific problem we can find even better splittings (i.e. with smaller splitting numbers) since simplification rules can reduce the formula after splitting.

3.2   Simplification Rules

We say that simplification rule is applicable to a formula F if it can replace F by a formula F' in polynomial time, so that both following conditions hold:

- the complexity of F' is smaller than the complexity of F,
- the solution for F can be constructed from the solution for F' (in polynomial time).

We say that simplification rule is applicable to a class of formulas if it is applicable to every formula of this class.

We use the following simplification rules in our program ($\mathcal{F}$ denotes a class of formulas):

1. *Pure literal.* If $\mathcal{F}$ contains a $(k, 0)$-literal l, replace $\mathcal{F}$ with $\mathcal{F}[l]$. Used for all considered problems.
2. *Unit clause.* If $(l) \in \mathcal{F}$, replace $\mathcal{F}$ with $\mathcal{F}[l]$. Used for SAT only.
3. *Resolution.* By elimination from $\mathcal{F}$ a variable x by resolution we mean adding all resolvents by x to $\mathcal{F}$ and elimination of all clauses containing x or $\bar{x}$ from $\mathcal{F}$ (for definition of the resolvent see, for example, [H00]). This rule eliminates by resolution from $\mathcal{F}$ all $(1,1)$-variables in case of MAXSAT and $(n,3)$-MAXSAT , and all $(1,k)$- and $(2,2)$-variables in case of SAT.
4. *Dominating unit clause.* If $\mathcal{F}$ contains a literal l, such that the literal $\bar{l}$ occurs in $\mathcal{F}$ not more times than the literal l occurs as a unit clause, then replace $\mathcal{F}$ with $\mathcal{F}[l]$. Used for MAXSAT and $(n,3)$-MAXSAT .
5. *Closed subformula.* If $\mathcal{F}$ contains a closed subformula F', replace $\mathcal{F}$ with $\mathcal{F} - F'$. Used for all considered problems.
6. *Almost dominating unit clause.* If $\mathcal{F}$ contains a literal l, such that l occurs in $\mathcal{F}$ as a unit clause k times and $\bar{l}$ occurs in $k+1$ clauses, such that two of them contain a pair of complementary literals (it means that at least one of these two clauses is *always* satisfied), replace $\mathcal{F}$ with $\mathcal{F}[l]$. Used for MAXSAT and $(n,3)$-MAXSAT .
7. *Almost common clauses.* If $\mathcal{F}$ contains two clauses C_1, C_2 with fixed lengths such that for a literal l $C_1 - \{l\} = C_2 - \{\bar{l}\}$, replace them with a clause $C_1 - \{l\}$. Used for MAXSAT and $(n,3)$-MAXSAT .
8. *Satisfying assignment.* If there is an assignment for all known variables of $\mathcal{F}$ satisfying all clauses of the basis of $\mathcal{F}$, then reduce $\mathcal{F}$ w.r.t. this assignment (i.e., assign the value *True* to all literals of assignment). Used for all considered problems.

All these rules except for the last one can be found in [BR99], [GHNR00], and [H00], the correctness of the last rule is trivial.

3.3   Implementation and Data Structures

In this subsection we briefly describe the implementation of the main procedures and data structures of our program.

A clause is represented as a list of literals and a flag `IsLengthFixed`. When this flag is set, no literal can be added to a clause. This flag is not set by default, and a clause is interpreted as a clause with unfixed length. The operations on a clause are fixing its length and adding/deleting a literal.

A simplification rule is implemented as a procedure that takes a class of formulas as an input. It checks whether its own condition holds for the input class of formulas, and if it does, it makes corresponding changes in it. For example, the pure literal rule checks whether a class of formulas contains $(k, 0)$-literals and assigns the value *True* to all such literals.

A class of formulas is represented simply as a list of clauses. The main operation on a class of formulas is assignment of a value to a literal. To assign the value *True* to a literal l we remove from a class of formulas all clauses containing the literal l and delete all occurrences of the literal $\bar{l}$.

Another important operation on a class of formulas is finding its possible splittings. Our program considers the following splittings ($\mathcal{F}$ is a class of formulas, x, y are literals):

1. $\mathcal{F}[x], \mathcal{F}[\bar{x}]$;
2. $\mathcal{F}[x, y], \mathcal{F}[x, \bar{y}], \mathcal{F}[\bar{x}]$;
3. $\mathcal{F}[x, y], \mathcal{F}[x, \bar{y}], \mathcal{F}[\bar{x}, y], \mathcal{F}[\bar{x}, \bar{y}]$.

After the construction of the classes of formulas in one of the cases above our program applies all simplification rules (given to it as an input) to all these classes as long as at least one of them is applicable. And then it calculates the resulting splitting number according to the given measure.

3.4   Algorithm

The main goal of our program is to prove an upper bound on the running time of a splitting algorithm of a form as given in Fig. 1.

Given a set of simplification rules $\mathcal{S}$, a measure of complexity μ and a number *SplitNum*, our program tries to prove that such an algorithm can always make a splitting with a splitting number not exceeding *SplitNum*. Thus, it tries to prove an upper bound $O(SplitNum^{\mu})$ for this algorithm.

The main part of our program is implemented in the procedure *AnalyzeCase*, which is given in Fig. 2. Informally speaking, this procedure tries to find a good splitting for the input class of formulas. If such a splitting is found, it returns, otherwise it constructs several new classes of formulas such that their union contains the input class of formulas and recursively calls for them. It is easy to see that if this procedure terminates on a given class of formulas, then all formulas of this class have a good splitting.

At the first step the procedure *AnalyzeCase* checks whether at least one simplification rule from the set $\mathcal{S}$ is applicable to $\mathcal{F}$ and returns if such a simplification rule exists. This is motivated by the fact that a splitting algorithm (of the form described above) makes splittings only on already simplified formulas. At step 2 the procedure tries to find a good splitting (i.e., a splitting with a

Procedure SplittingAlgorithm

Input:
formula F, set of simplification rules $\mathcal{S}$, measure of complexity μ.
Method

1. apply all simplification rules from the set $\mathcal{S}$ to F as long as at least one of them is applicable
2. if the solution for F can be constructed in polynomial time (in particular, if F is empty), return the answer
3. for each pair of variables x and y consider the following splittings:
   (a) $F[x], F[\bar{x}]$
   (b) $F[x, y], F[x, \bar{y}], F[\bar{x}]$
   (c) $F[x, y], F[x, \bar{y}], F[\bar{x}, y], F[\bar{x}, \bar{y}]$
   choose the one that provides the best splitting number (w.r.t. μ)
4. construct the formulas corresponding to the splitting found at the previous step and call SplittingAlgorithm for each of them
5. return the answer according to the answers returned by the recursive calls

Fig. 1. Form of splitting algorithms.

splitting number not exceeding $SplitNum$) for the input class of formulas and returns if such exists.

At step 3 $AnalyzeCase$ simply selects a clause with unfixed length. Note that the situation when there is no such a clause is impossible, since otherwise all known variables of $\mathcal{F}$ form a closed subformula and hence the corresponding rule is applicable.

At steps 4 and 5 the procedure considers two cases: when the selected clause contains at least one more literal and when it does not. In some sense, it splits a class of formulas for which it cannot find a good splitting into several classes and recursively calls for them. The goal of these two steps is to construct the classes of formulas such that their union contains $\mathcal{F}$.

We use the relation α at step 5 to define the set of pairs (i, j), such that an (i, j)-literal does not give immediately a good (i.e., not worse than $SplitNum$) splitting. Clearly, for any problem this relation may be defined as follows:

$$\alpha(K, SplitNum) = \{(i, j) : \tau(i, j) > SplitNum\},$$

$$\alpha(L, SplitNum) = \{(i, j) : \tau(i + j, i + j) > SplitNum\},$$

where K is the number of clauses and L is the length of the input formula. These sets are finite due to the properties of splitting numbers [KL98]. Unfortunately, we cannot define such a finite set when we deal with the number of variables as a complexity measure. But in case of the $(n, 3)$-MAXSAT problem we can write the following:

$$\alpha(N, SplitNum) = \{(i, j) : i + j \leq 3\},$$

where N is the number of variables.

Note that at step 3 the procedure can select *any* clause with unfixed length, but, in practice, the size of the resulting case analysis (when we write down all the classes of formulas processed by this procedure) depends on the selected clause. Current version of our program selects this clause in the following way: for each clause with unfixed length it constructs classes of formulas as described at steps 4 and 5 and counts the number of bad classes among these constructed classes (here we say that a class is bad if it does not satisfy the conditions of the steps 1 and 2). And finally, the clause providing a minimal number of bad classes is selected. In most situations this heuristic allowed to produce smaller proofs than any other one considered by us. However, this heuristic is obviously not the best one: for example, it is possible to count bad classes out of bad classes. But when we tried this heuristic the running time of our program increased greatly. So, we are going to think about smart heuristics.

Procedure AnalyzeCase

Input:
class of formulas $\mathcal{F}$, set of simplification rules $\mathcal{S}$, number $SplitNum > 1$, measure of complexity μ.

Method

1. if any simplification rule from the set $\mathcal{S}$ is applicable to $\mathcal{F}$, then return
2. if there is a splitting w.r.t. μ in $\mathcal{F}$ providing splitting number not exceeding $SplitNum$, then return
3. select from $\mathcal{F}$ according to some heuristic a clause with unfixed length; call this clause C
4. (a) $C_1 = C$
   (b) fix length of C_1
   (c) call AnalyzeCase($\mathcal{F} + \{C_1\} - \{C\}$) (all other input parameters of the procedure AnalyzeCase are the same here)
5. (a) introduce new variable to $\mathcal{F}$; call it u
   (b) $C = C + \{u\}$
   (c) for all $i, j > 0$, such that $(i, j) \in \alpha(\mu, SplitNum)$
       i. construct the set of all possible classes of formulas resulting from $\mathcal{F}$ by adding occurrences of the variable u for u being an (i, j)-variable
       ii. call AnalyzeCase on each constructed class of formulas

---

Fig. 2. Procedure AnalyzeCase.

3.5   Automatically Proved Upper Bounds

Our program proved the following upper bounds:

- SAT: $O(1.272021^K)$, $O(1.101027^L)$,
- MAXSAT: $O(1.37226^K)$, $O(1.135889^L)$,
- $(n, 3)$-MAXSAT : $O(1.2852^N)$, $O(1.236507^K)$, $O(1.098267^L)$.

The output files proving these bounds can be found at
http://logic.pdmi.ras.ru/~kulikov (also there are several proofs of simpler
bounds given). The structure of these files allows to verify each step of our
program. Note that a big number of cases in all presented proofs does not mean
that the corresponding algorithms are sophisticated. (Actually the form of all
these algorithms is given in Fig. 1.)

4   Related Work

Independently from us, another two frameworks for automated proofs of upper
bounds were implemented. First of them, by Nikolenko and Sirotkin [NS03], was
developed to prove upper bounds for the SAT problem and it uses only one
simplification rule, namely pure literal rule. It was proved by this framework
that SAT can be solved in $O(1.56639^K)$. Another framework was developed
by Gramm et al. [GGHN03]. In this framework several new bounds for hard
graph modification problems were proved. In particular, an upper bound for the
Cluster Editing problem was improved to $O(1.92^k)$ (where k is the number of
edge modifications allowed). The program of Gramm et al. allows to work with
different simplification rules.

   In general, all contemporary frameworks (including ours) are based on the
following simple observation: it is possible to prove an upper bound for a problem
just by considering all possible subformulas of the size bounded by a constant.
The framework of Gramm et al. is, in some sense, a straightforward implementation of this fact: their program given a number s constructs a set S of graphs
each having s vertices such that any possible graph has at least one element of
the set S as a subgraph. For each graph from the set S all possible splittings are
enumerated and the one providing the best splitting number is selected. And as
a result, an upper bound corresponding to the worst of these selected splittings
is proved for a problem. Clearly, the resulting upper bound is as better, as the
input number s is larger. But with the increasing of s the number of all possible
splittings for a graph on s vertices increases greatly (as well as the size of S).

   The differences of our approach from the one by Gramm et al. are the following:

1. Our program is given a bound first, and then it tries to to prove this bound,
   while the program by Gramm et al. given a number s tries to prove as better
   bound as possible by considering graphs on s vertices.
2. The set of all possible small subformulas in out program is constructed in the
   process of proof search , while the construction of similar set is the first stage
   of the program by Gramm et al. Thus, we never know whether our program
   will terminate on a given bound. And also, it is impossible to say in advance
   for which number s (or whether such number s exists) the framework by
   Gramm et al. can prove a given bound.
3. Our framework do not consider all possible splittings for each class of formulas (actually it considers only splittings on assignments having not more
   than two literals), as framework by Gramm et al. does.

5   Further Directions

The first direction is thinking about heuristics of quick finding of good split-tings for small formulas (or graphs). For example, authors of [GGHN03] indicate that in most cases almost all the time (of their program execution) is spent for concatenation of splitting vectors.

Another direction is to think about construction of the set of all possible small subformulas (or subgraphs). Clearly, such a set is not unique (for example, different sets may be constructed by selecting different clauses at step 3 of *AnalyzeCase*) and the resulting upper bound may depend on this set.

All known frameworks do not provide a tool for converting their output files into splitting algorithms. The advantage of such an algorithm is in fact that it does not have to spend time for finding good splittings and simplifications, since all this information may be given in the output file. Thus, it seems like such an algorithm could work better than a simple splitting algorithm.

The currently best algorithm for SAT [H00] uses the following simplification rule: If each clause of a formula that contains a $(2, 3^+)$-literal contains also a $(3^+, 2)$-literal, then all $(3^+, 2)$-literals can be assigned the value *True*. Clearly, this rule cannot be added to any known framework, since these frameworks consider only formulas (graphs) of the size bounded by a constant. So, it will be interesting to somehow introduce this rule into existing frameworks.

Another non-standard trick is used in currently best algorithm for MAXSAT [CK02]: at each iteration it preserves a certain invariant on the input formula. We are going to add this feature into our framework.

Acknowledgements

The authors are very grateful to Edward A. Hirsch for bringing the problem to their attention.

References

[BMS03] J. M. Byskov, B. A. Madsen, and B. Skjernaa New Algorithms for Exact Satisfiability. *Theoretical Computer Science Preprint*, 2003.

[BR99]   N. Bansal and V. Raman. Upper Bounds for MaxSat: Further Improved. *Proceedings of ISAAC'99*, 247–258, 1999.

[CK02]   J. Chen and I. Kanj. Improved exact algorithms for MAX-SAT. *Proceedings of 5th LATIN*, number 2286 in LNCS, 341-355, 2002.

[DHIV01] E. Dantsin, E. A. Hirsch, S. Ivanov, and M. Vsemirnov, Algorithms for SAT and Upper Bounds on Their Complexity. ECCC Technical Report 01-012, 2001. Electronic address: ftp://ftp.eccc.uni-trier.de/pub/eccc/reports/2001/TR01-012/index.html. A Russian version appears in *Zapiski Nauchnykh Seminarov POMI*, 277: 14–46, 2001.

[DLL62] M. Davis, G. Logemann, and D. Loveland. A machine program for theorem-proving. *Comm. ACM*, 5:394–397, 1962.

[FK02]   S. S. Fedin, A. S. Kulikov, A $2^{|E|/4}$-time Algorithm for MAX-CUT. *Zapiski nauchnykh seminarov POMI*, 293: 129–138, 2002. English translation is to appear in Journal of Mathematical Sciences.

[GGHN03] J. Gramm, J. Guo, F. Hüffner, and R. Niedermeier. Automated Generation of Search Tree Algorithms for Hard Graph Modification Problems. *Proceedings of 11th Annual European Symposium on Algorithms*, 642–653, 2003.

[GHNR00] J. Gramm, E. A. Hirsch, R. Niedermeier, and P. Rossmanith. New worst-case upper bounds for MAX-2-SAT with application to MAX-CUT. *Discrete Applied Mathematics*, 130(2):139–155, 2003.

[H00]   E. A. Hirsch New worst-case upper bounds for SAT. *Journal of Automated Reasoning*, 24(4):397–420, 2000.

[K02]   A. S. Kulikov, An upper bound $O(2^{0.16254n})$ for Exact 3-Satisfiability: A simpler proof. *Zapiski nauchnykh seminarov POMI*, 293: 118–128, 2002. English translation is to appear in Journal of Mathematical Sciences.

[KL98]   O. Kullmann and H. Luckhardt. Algorithms for SAT/TAUT decision based on various measures. *Informatics and Computation*, 1998.

[NS03]   S. I. Nikolenko and A. V. Sirotkin. Worst-case upper bounds for SAT: automated proof. *Proceedings of the Eight ESSLI Student Session*, 225–232, 2003. Available from http://logic.pdmi.ras.ru/~sergey.

[RRR98]   V. Raman, B. Ravikumar, and S. Srinivasa Rao. A Simplified NP-complete MAXSAT Problem. *Information Processing Letters*, 65:1–6, 1998.

Improved Parameterized Algorithms for Feedback Set Problems in Weighted Tournaments

Venkatesh Raman[1] and Saket Saurabh[2]

[1] The Institute of Mathematical Sciences, Chennai 600 113.
vraman@imsc.res.in
[2] Chennai Mathematical Institute, 92, G. N. Chetty Road, Chennai-600 017.
saurabh@cmi.ac.in

Abstract. As our main result, we give an $O((2.415)^k n^\omega)$ algorithm for finding a feedback arc set of weight at most k in a tournament on n vertices with each vertex having weight at least 1. This improves the previously known bound of $O(\sqrt{k}^k n^\omega \log n)$ for the problem in unweighted tournaments. Here ω is the exponent of the best matrix multiplication algorithm.

We also investigate the fixed parameter complexity of weighted feedback vertex set problem in a tournament. We show that (a) when the weights are restricted to positive integers, the problem can be solved as fast as the unweighted feedback vertex set problem, (b) if the weights are arbitrary positive reals then the problem is not fixed parameter tractable unless $P = NP$ and (c) when the weights are restricted to be of at least 1, the problem can be solved in $O((2.4143)^k n^\omega)$ time.

1 Introduction and Motivation

Given a directed graph on n vertices and an integer parameter k, the feedback vertex (arc) set problem asks whether the given graph has a set of k vertices (arcs) whose removal results in an acyclic directed graph. In the weighted version of the problem we are given non negative weights on vertices (arcs) and the problem asks whether the graph has a set of vertices (arcs) of weight at most k, whose removal makes the graph acyclic. While these problems in undirected graphs are known to be fixed parameter tractable [10] (in fact the edge version in undirected graphs can be trivially solved), the parameterized complexity of these problems in directed graphs is a long standing open problem in the area. In fact, there are problems on sequences and trees in computational biology, that are related to the directed feedback vertex set problem [6].

In this paper, we consider these problems in the special class of directed graphs, tournaments. A tournament $T = (V, E)$ is a directed graph in which there is exactly one directed arc between every pair of vertices. We give efficient fixed parameter tractable algorithms for the feedback arc set and weighted feedback vertex set problem in tournaments under the framework introduced by

R. Downey, M. Fellows, and F. Dehne (Eds.): IWPEC 2004, LNCS 3162, pp. 260–270, 2004.

Downey and Fellows[4]. Weighted feedback arc set problem in tournaments finds application in rank aggregation methods. Dwork et.al [5] have shown that the problem of computing the so called Kemeny optimal permutation for k full lists, where k is an odd integer, is reducible to the problem of computing a minimum feedback arc set problem on a weighted tournament with weights between 1 and $k - 2$.

The feedback vertex set problem in a tournament is a set of vertices that hits all the triangles in the tournament. But for the feedback arc set problem (FAS) this is not the case. In [11], an $O(\sqrt{k}^k n^\omega \log n)$ time algorithm, where ω is the exponent of the best matrix multiplication algorithm, was developed for FAS using the fact that a directed graph with at most k arcs away from a tournament has a cycle of length at most $\sqrt{k}$. In this paper, we use a different characterization of a minimal feedback arc set in directed graphs to develop new algorithms for the feedback arc set problem in a tournament. In Section 2, we first show that if a subset F of arcs forms a minimal feedback arc set in a directed graph then the graph formed after reversing these arcs is acyclic. Such a characterization helps us to maintain the tournament structure (since in every recursive step we reverse but not delete arcs). We first apply this characterization to develop an algorithm for FAS in tournaments taking $O(3^k n^\omega)$ time. We then improve this by using a simple branching technique along with the new characterization to obtain an $O((2.415)^k n^\omega)$ time algorithm. We also observe that the algorithm, and hence the bound, applies for the FAS problem in weighted tournaments as well, where weights on the arcs are at least 1.

In section 3, we consider the weighted version of the feedback vertex set problem in tournaments that is known to be NP-complete[12].

As observed before, the feedback vertex set problem in an unweighted tournament can be written as a 3-hitting set problem and hence we can apply the algorithm of [9] to get an $O((2.27)^k + n^3)$ algorithm as mentioned in [11]. However, this algorithm uses some preprocessing rules which don't naturally generalize for the weighted hitting set problem. We consider the following variants of the weighted feedback vertex set (WFVS) problem:

1. Integer-WFVS, where the weights are arbitrary positive integers,
2. Real-WFVS, where the weights are real numbers ≥ 1, and
3. General-WFVS, where the weights are positive real numbers.

We show that the Integer-WFVS and Real-WFVS are fixed parameter tractable but General-WFVS is not fixed parameter tractable (FPT) unless $P = NP$. More specifically, we show that the Integer-WFVS can be solved as fast as the feedback vertex set problem in an unweighted tournament, which currently has running time of $O((2.27)^k + n^3)$ [9], and that Real-WFVS can be solved in $O((2.4143)^k n^\omega)$ time.

In section 4, we conclude with some remarks and open problems. Throughout this paper, by $\log n$ and ω, we mean, respectively, the logarithm to the base 2 of n and the exponent of the running time of the best matrix multiplication algorithm. By $rev(x)$, where $x = (u, v)$ is an arc of a directed graph, we mean the arc (v, u).

2   Feedback Arc Set Problem

Feedback arc set problem is known to be NP-complete for weighted tournaments [5] but is still open for unweighted tournaments. It is believed to be NP-complete. In this section, we give two fixed parameter tractable algorithm for the feedback arc set problem in a tournament. Both of these algorithms are based on the fact that a subset of arcs forms a minimal feedback arc set only if the original graph becomes acyclic when the direction of these arcs are reversed. The first algorithm is a direct application of this and achieves the time bound of $O(3^k n^\omega)$. We further improve this to $O((2.415)^k n^\omega)$ using a new branching technique along with the reversal characterization of minimal feedback arc set. We conclude this section by observing that some of these algorithms are applicable for weighted tournaments as well as for dense directed graphs.

2.1   Main Algorithm

Though some authors use the characterization below as the definition, at least for the minimum feedback arc set [8], we could not find this precise statement (of the lemma) in the literature. We give a proof here for the sake of completeness.

Lemma 1. Reversal Lemma
Let $G = (V, E)$ be a directed graph and F be a minimal feedback arc set (FAS) of G. Let G' be the graph formed from G by reversing the arcs of F in G, then G' is acyclic.

Proof. Assume to the contrary that G' has a cycle C. Then C can not contain all the arcs of $E - F$, as that will contradict the fact that F is a FAS. Define the set $rev(F) = \{(u, v) \mid (v, u) \in F\}$. Let $C \cap rev(F) = \{f_1, f_2, \cdots, f_k\}$ and $e_i = rev(f_i)$. Then the set $\{e_1, e_2, \cdots, e_k\}$ is a set of arcs of G which are reversed and are part of C. Now since each $e_i \in F$, and F is minimal, there exists a cycle C_i such that $F \cap C_i = \{e_i\}$. Now consider the directed graph L induced by the arcs of $\{C, C_1, \cdots, C_k\} - F - rev(F)..$ It is clear that L is a directed closed walk with all the arcs in the original graph G. In fact, if $\forall i$, $C_i \cap C = \emptyset$, then L is a simple cycle in G, such that $L \cap F = \emptyset$, contradicting the fact that F is a FAS. If L is not a simple cycle then we can extract a simple directed cycle from it not having any arcs of F, violating the definition of F and that completes the proof.

$\square$

Now we use Lemma 1 to give a simple algorithm for the feedback arc set problem in a tournament.

Algorithm TFAS(T,k, F)(*T is a tournament, $k \geq 0$, and F is a set of arcs.*) (Returns a minimal feedback arc set of size at most k, if exists and returns 'NO' otherwise. F is a local variable containing arcs of a partial feedback arc set that are reversed from the original T. Initially the algorithm is called by TFAS($T, k, \emptyset$))

Step 0: If T does not have a directed triangle and $k \geq 0$, then answer YES and return F.

Step 1: If $k = 0$ and T has a triangle, then answer NO and EXIT.

Step 2: Find a triangle in T and let $\{a, b, c\}$ be the arcs of the triangle.

    Step 2a: If $rev(a), rev(b)$ and $rev(c)$ are in F, then answer NO and EXIT.

    Step 2b: If $TFAS(T\backslash\{x\} \cup rev\{x\}, k - 1, F \cup \{x\})$ is true for any arc x of the triangle such that $rev(x)$ is not in F, then answer YES and return F and exit. Otherwise return NO and exit.

We will need the following easy observations to prove the correctness and the runtime of the algorithm.

Lemma 2. *[1] A tournament $T = (V, E)$ has a directed cycle if and only if it has a directed triangle.*

Let M be the adjacency matrix of the tournament T. Then T has a directed triangle if and only if for some i, j such that $1 \leq i < j \leq n$, $M^2[i, j] \geq 1$ and $M[j, i] = 1$. This can be determined in $O(n^\omega)$ time. If such a pair (i, j) exists, then there exists a k such that $M[i, k] = M[k, j] = 1$ which can also be determined in $O(n)$ time. Such a triple $\{i, j, k\}$ forms a triangle. Further, T has a directed cycle of length 4 if and only there exists a pair (i, j) such that $1 \leq i < j \leq n$, $M^2[i, j] \geq 1$ and $M^2[j, i] \geq 1$. If such a pair exists, then as before, the witness 4-cycle can also be found in $O(n)$ time. So we have

Lemma 3. *Let T be a tournament on n vertices. Then we can find a directed triangle or a directed cycle of length 4 in T, if it exists, in $O(n^\omega)$ time.*

Theorem 1. *Given a tournament $T = (V, E)$ on n vertices, we can determine whether it has a feedback arc set of size at most k in $O(3^k n^\omega)$ time.*

Proof. First we will show that the algorithm $TFAS$ finds a minimal feedback arc set of size at most k if exists. Since T has a FAS of size at most k if and only if it has a minimal FAS of size at most k, then the theorem will follow. Correctness of Step 0 and Step 1 follow from Lemma 2. Step 2a answers correctly as by Reversal Lemma, the current F can not be extended to a minimal feedback arc set of G. In Step 2b, we branch on each arc x of the triangle such that $rev(x) \notin F$, because if none of these arcs is picked in the feedback arc set of G, then this triangle will survive in G', obtained by reversing the arcs of F. But then by Reversal Lemma, this F is not minimal. So this proves the correctness of the algorithm.

The claimed time bound can easily be seen by observing the fact that k decreases at every recursive Step 2b by 1. So the recursion depth is at most k. The branching factor at every recursion step is at most 3 and hence by Lemma 3, we have the desired time bound for the algorithm. $\square$

This already improves the $O(\sqrt{k}^k n^\omega \log n)$ time algorithm of [11] for the problem. Now we further improve the bound by a new branching technique, Lemma 1, and the following lemma.

Lemma 4. *Let $T = (V, A)$ be a tournament that contains no induced subgraph isomorphic to F_1 (see figure 1). Then the minimum feedback vertex and arc set problems are solvable in T in $O(n^\omega)$ time.*

Proof. It is easy to see that a tournament has a subgraph isomorphic to F_1 if and only if it has a directed 4-cycle. It is also an easy exercise to see that if a tournament has no subgraph isomorphic to F_1, then no pair of directed triangles in the tournament has a vertex in common. Hence the minimum feedback vertex or arc set is obtained by finding all triangles, and picking a vertex/arc from each of them.

Finding all triangles in such a tournament can be done in $O(n^\omega)$ time as follows. First compute M^2, the square of the adjacency matrix of the tournament. Since the tournament can have at most $n/3$ triangles, there can be at most $n/3$ pairs (i, j) such that $1 \le i < j \le n$ and $M^2[i, j] \ge 1$ and $M[j, i] = 1$. For each such pair, the corresponding witness triangle can be found in $O(n)$ time.

□

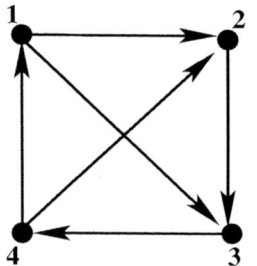

Fig. 1. F_1

Algorithm BTFAS(T,k, F)(*T is a tournament, $k \ge 0$, F is a set of arcs*) (Returns a minimal feedback arc set of size at most k, if exists and returns 'NO' otherwise. F is a local variable containing arcs of a partial feedback arc set that are reversed from the original T. Initially the algorithm is called by BTFAS($T, k, \emptyset$))

Step 0: If T does not have a directed triangle, then answer YES and return F.
Step 1: If $k = 0$ and T has a triangle, then answer NO and EXIT.
Step 2: Find an induced subgraph on 4 vertices isomorphic to F_1, if exists, in T. Such a subgraph is simply a tournament on 4 vertices having at least two directed triangles. Let the vertex set of such an F_1 be $\{1, 2, 3, 4\}$ and the adjacencies be as in figure 1 (in particular $(1, 2)$ is the only arc not part of any directed triangle). If no such subgraph exists in T, then go to Step 6.
Step 3: Let $\{a, b, c\}$ be the arcs of a triangle in F_1, such that there exists an arc $x \in \{a, b, c\}$ for which $rev(x) \in F$. If there is no such triangle in F_1, then go to Step 4.

Step 3a: If $rev(a), rev(b)$ and $rev(c)$ are in F, then answer NO and EXIT.

Step 3b: If $BTFAS(T \setminus \{x\} \cup rev(x), k - 1, F \cup \{x\})$ is true for any arc x of the triangle such that $rev(x)$ is not in F, then answer YES and exit after returning F; else answer NO and exit.

Step 4: If $rev((1, 2)) \notin F$ then if any of the following recursive calls returns true, then answer YES and return the corresponding F and exit, and answer NO and exit otherwise.

In the following, T' is obtained from T by reversing the 'newly included' arcs of F.

1. $BTFAS(T', k - 1, F \cup \{(3, 4)\})$,
2. $BTFAS(T', k - 2, F \cup \{(4, 1), (4, 2)\})$,
3. $BTFAS(T', k - 2, F \cup \{(4, 1), (2, 3)\})$,
4. $BTFAS(T', k - 2, F \cup \{(1, 3), (2, 3)\})$,
5. $BTFASF(T', k - 3, F \cup \{(1, 2), (1, 3), (4, 2)\})$

Step 5: If $rev((1, 2)) \in F$, then branch only on the first 4 cases enumerated in Step 4.

Step 6: Solve the problem in polynomial time for the resultant tournament using Lemma 4.

Note that all the induced subgraphs on 4 vertices with at least two directed triangles are isomorphic to F_1. In the above algorithm at every step, we first find a graph isomorphic to F_1, and then if there exists a directed triangle in F_1 with all its arcs included in the partial feedback arc set (F) obtained so far, then we apply Lemma 1 and answer NO. Otherwise we branch on all the arcs x of the triangle such that $rev(x) \notin F$ as by Lemma 1 at least one such arc must be part of F.

If none of the arcs of F_1 is part of F, then we branch on all possible minimal feedback arc sets of F_1. The only remaining case is when all the arcs x appearing in some triangle in F_1 are not in F but $rev((1, 2)) \in F$. In this case, Lemma 1 implies that item 5 of *Step 2b* is not applicable (because the set $\{(1, 3), (4, 2)\}$ is not a minimal FAS of F_1). So when we reach *Step 6* of the above algorithm, all the induced subgraphs on 4 vertices have at most one triangle. And the problem now can be solved in polynomial time by Lemma 4 .

The following lemma follows from Lemma 3 and the fact that a tournament has a subgraph isomorphic to F_1 if and only if it has a 4 cycle.

Lemma 5. *Given a tournament, a subgraph isomorphic to F_1 (see figure 1) can be found in $O(n^\omega)$ time.*

Thus, we get the following recurrence for the time complexity of the algorithm:

$$T(n, k) \leq \max \begin{cases} 2T(n, k - 1) + O(n^\omega) \text{ or} \\ T(n, k - 1) + 3T(n, k - 2) + T(n, k - 3) + O(n^\omega) \end{cases}$$

The above recurrences solve to $O((2.415)^k n^\omega)$. So we get the following theorem.

Theorem 2. *Given a tournament $T = (V, E)$, we can determine whether it has a feedback arc set of size at most k in $O((2.415)^k n^\omega)$ time.*

We remark that the above algorithm can also be applied for weighted feedback arc set problem in a tournament where the weight of every arc is at least 1. The only modification is in the last step application of Lemma 4 where we choose the minimum weight arc from each triangle. Hence we have

Theorem 3. *Given a tournament $T = (V, E)$, and a weight function $\pi : E \to \Re^+$, such that $\pi(e)$ is at least 1 for every $e \in E$, we can determine whether T has a feedback arc set of weight at most k in $O((2.415)^k n^\omega)$ time.*

2.2   Feedback Arc Set Problem in Dense Directed Graphs

In this section, we show that the feedback arc set problem is fixed parameter tractable for directed graphs which are at most $n^{1+o(1)}$ arcs away from a tournament. We need following lemma to show the desired result.

Lemma 6. *[2] Let $G = (V, E)$ be a strong directed graph with n vertices, m arcs and let $l \geq 2$. Then if $m \geq \frac{n^2 + (3 - 2l)n + (l^2 - l)}{2}$, then the girth of the graph $(g(G))$ is bounded by l.*

Theorem 4. *Let G be a directed graph with n vertices and $m \geq \binom{n}{2} - n^{1+o(1)}$ arcs. Then the feedback arc set (FAS) problem is fixed parameter tractable for G.*

Proof. For the feedback arc set problem, we can assume without loss of generality, that the given directed graph is a strongly connected directed graph. (Otherwise, try values up to k in each strongly connected subgraph and take the minimum.)

Then Lemma 6 implies that if a strong directed graph is at most $n(g-2)/2$, ie. $O(ng)$, arcs away from a tournament, then its girth is bounded by g, as

$$\binom{n}{2} - \frac{n^2 + (3 - 2g)n + (g^2 - g)}{2} = \frac{2n(g-2) + g - g^2}{2}$$

and

$$\frac{2n(g-2) + g - g^2}{2} \geq n(g-2) - \frac{g^2}{2} \geq \frac{n(g-2)}{2}$$

for $n \geq k^2/(k-2)$ and $3 \leq g \leq n/2$. Hence if the strong directed graph is at most $n^{1+o(1)}$ arcs away from a tournament, then its girth is bounded by $n^{o(1)}$. So we find the shortest cycle in G and then by applying Lemma 1, we branch on each arc by reversing the arc. This way we don't delete any arc and hence at every recursive step Lemma 6 ensures a cycle of length at most $n^{o(1)}$. So we have an algorithm for feedback arc set problem in G which takes $O((n^{o(1)})^k n^{O(1)})$. Cai and Judes [3] have observed that $O((n^{o(1)})^k)$ is bounded by $f(k)n^{O(1)}$, where f is some function of k, for every fixed n and k. Hence it follows that the feedback arc set problem is fixed parameter tractable in G.   □

3   Feedback Vertex Set in Weighted Tournaments

In this section, we show that the Integer-WFVS and Real-FVS are in FPT, while General-FVS is not in FPT unless $P = NP$. In fact we show that Integer-WFVS can be solved as fast as feedback vertex set problem in an unweighted tournament, which currently has a running time of $O((2.27)^k + n^3)$. Then we show that the Real-WFVS can be solved in $O((2.4143)^k n^3)$ using a simple branching technique.

3.1   Integer-WFVS

Now we will show that Integer-WFVS can be reduced to unweighted feedback vertex set problem in a tournament via a simple parameterized many-one reduction.

Theorem 5. Integer $-$ WFVS *can be solved in the same time as the unweighted feedback vertex set in a tournament, up to an additive term polynomial in n and k, ie., $O((2.27)^k + (kn)^3)$.*

Proof. We can safely assume that all the weights on the vertices are at most $k + 1$ as any vertex having weight strictly more than k can not be a part of any minimal feedback vertex set of weight at most k. So given the weight function if some vertex v has $\pi(v) > k$ then we make $\pi(v) = k + 1$. It is easy to see that T has feedback vertex set of weight at most k if and only if it has feedback vertex set of weight at most k with the modified weight function.

We will construct a new tournament T' from T as follows: replace each vertex v having weight $\pi(v) = w > 1$ with a cluster v' consisting of w vertices. If there is an arc (u, v) in the original tournament T then we add an arc from every vertex of the cluster u' to every vertex in v'. Now we add intra-cluster arcs so that each cluster is transitive.

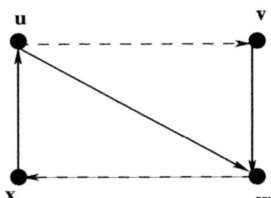

Fig. 2. A Witness Cycle

Let $\{v_1, v_2, \cdots, v_l\}$ be the feedback vertex set (FVS) of weight at most k in T. Then the vertices of the corresponding clusters $\{v'_1, v'_2, \cdots, v'_l\}$ form a FVS of size at most k in T'. The other direction follows from the observation that every minimal feedback vertex set (F) of size at most k in T' is such that either all the vertices of any cluster are there or none of them is. To see this, assume that there is a cluster v' such that there is a vertex $v \in v'$ in F and a $u \in v'$ not in F. Since $v \in F$, there exist a witness cycle C such that $F \cap C = \{v\}$. Now if $u \notin C$ then we get a cycle C' in T' by replacing v with u in C such that $C' \cap F = \emptyset$ contradicting the definition of F. Otherwise if u is part of this cycle then the length of the cycle is at least 4 and we can obtain a smaller cycle C' containing u such that $C' \cap F = \emptyset$ as shown in Figure 2. Here C is $\{u, \cdots, v, w, \cdots, x\}$ and we can take C' as $\{u, w, \cdots, x\}$. So this proves the observation.

Number of vertices in the new instance of the graph is bounded by $(k+1)n$, and hence by applying the hitting set algorithm on this instance, we solve the problem in time $O((2.270)^k + (kn)^3)$.

□

3.2   Real- and General-WFVS

If the weights are arbitrary reals, but at least 1, then the algorithm for un-weighted tournament cannot be directly applied. Here we give an algorithm which attains $O((2.4143)^k n^\omega)$ bound.

Algorithm TFVS(T,k,π, F)(*T is a tournament, $k \geq 0$, π is a weight function on V, F is a set of vertices*)
(Returns a minimal feedback vertex set of weight at most k, if exists and returns 'NO' otherwise. F is a local variable containing vertices of a partial feedback vertex set that are deleted from the original T. Initially the algorithm is called by TFVS($T, k, \emptyset$))

Step 0: If T does not have a directed triangle and $k \geq 0$, then answer YES and return F.
Step 1: If $k = 0$ and T has a triangle, then answer NO and EXIT.
Step 2: Find an induced subgraph on 4 vertices isomorphic to F_1 (as in figure 1) with vertex set, say $\{1, 2, 3, 4\}$ and the adjacencies be as in figure 1. If no such subgraph exists, then go to Step 4.
Step 3: If any of the following recursive calls result in true, then return YES and exit with the corresponding F, and return NO and exit otherwise. In the following, T' is obtained by deleting the 'newly included' vertices in F.
  1. $TFVS(T', k - \pi(3), \pi, F \cup \{3\})$,
  2. $TFVS(T', k - \pi(4), \pi, F \cup \{4\})$,
  3. $TFVS(T', k - \pi(1) - \pi(2), \pi, F \cup \{1, 2\})$
Step 4: Solve the minimum weighted feedback vertex problem in polynomial time for the resultant tournament using Lemma 4.

Correctness of Step 0 and 1 follow from Lemma 2. In Step 3, we branch on all possible minimal solutions of F_1. Step 4 follows from Lemma 4 with the modification that in each of the triangles we pick the minimum weight vertex.

The time taken by the algorithm is bounded by the following recurrence:

$$T(n, k) \leq 2T(n - 1, k - 1) + T(n - 2, k - 2) + O(n^\omega)$$

which solves to $O((2.4143)^k n^\omega)$. So we have following theorem:

Theorem 6. *Given a tournament $T = (V, E)$, and a weight function $\pi : V \to \Re^+$, such that $\pi(v)$ is at least 1 for every $v \in V$, then we can determine whether T has a feedback vertex set of weight at most k in $O((2.4143)^k n^\omega)$ time.*

General-WFVS problem, where weights can be arbitrary positive reals, is not fixed parameter tractable unless $P = NP$. We show this by proving that it is NP-complete for some fixed constant k (in fact, for $k = 1$). Our reduction is from the NP-complete unweighted feedback vertex set problem [12] in tournaments by simply defining the weight function π to be $\pi(v) = 1/k$ for all $v \in V$. Then the original tournament has a FVS of size k if and only if the resulting weighted tournament has a FVS of weight 1. This implies that there cannot be a $f(k)n^{O(1)}$ or even $n^{O(k)}$ time algorithm for General-WFVS problem unless $P = NP$.

Theorem 7. *General-WFVS problem is not fixed parameter tractable unless $P = NP$.*

4   Conclusions

In this paper, we have obtained improved algorithms for parameterized feedback arc and vertex set problem on weighted tournaments. We used a characterization for minimal feedback arc set which maintains the tournament structure at every recursive step. For the feedback arc set problem, the complexity of the algorithms in unweighted and weighted (with weights at least 1) are the same while this is not the case for the feedback vertex set problem.

It would be interesting to see whether the unweighted feedback vertex set problem on tournaments has some special structure that can be utilized to develop an algorithm better than that of the 3-hitting set problem. Similarly it would be interesting to see whether the best known algorithm for the 3-hitting set problem can be used for the feedback arc set problem in unweighted tournaments to achieve the same time bound.

References

1. J. Bang-Jensen and G. Gutin, 'Digraphs Theory, Algorithms and Applications', Springer Verlag 2001.
2. J. C. Bermond, A. Germa, M. C.Heydemann and D. Sotteau, 'Girth in Digraphs', *Journal of Graph theory*, **4** (3) (1980) 337-341.
3. L. Cai and D. Juedes, 'On the Existence of Subexponential Parameterized Algorithms', *Journal of Computer and System Sciences* **67** (4) (2003) 789-807.
4. R. Downey and M. R. Fellows, 'Parameterized Complexity', Springer Verlag 1998.
5. C. Dwork, R. Kumar, M. Naor and D. Sivakumar, 'Rank Aggregation Revisited', WWW10; selected as Web Search Area highlight, 2001.
6. M. Fellows, M. Hallett, C. Korostensky, U. Stege, 'Analogs and Duals of the MAST Problem for Sequences and Trees', *in Proceedings of 6th Annual European Symposium on Algorithms (ESA '98)*, Venice, Lecture Notes in Computer Science **1461** (1998) 103-114.

7. A. Itai and M. Rodeh, 'Finding a Minimum Circuit in a Graph', *Siam Journal of Computing* **7** (4) (1978) 413-423.
8. D. A. Narayan, 'The Reversing Number of a Digraph; A Disjoint Union of Directed Stars', *Congressus Numerantium* **145**, (2000) 53-64.
9. R. Niedermeier and P. Rossmanith, 'An efficient Fixed Parameter Algorithm for 3-Hitting Set', *Journal of Discrete Algorithms* **2** (1) (2001).
10. V. Raman, S. Saurabh and C. R. Subramanian, 'Faster Fixed Parameter Tractable Algorithms for Undirected Feedback Vertex Set', *in the Proceedings of the 13th International Symposium on Algorithms and Computation (ISAAC 2002)*, Lecture Notes in Computer Science **2518** (2002) 241-248.
11. V. Raman and S. Saurabh, 'Parameterized Complexity of Directed Feedback Set Problems in Tournaments', *in the Proceedings of Workshop on Data Structure and Algorithms (WADS 2003)*, Lecture Notes in Computer Science **2748** (2003) 484-492.
12. E. Speckenmeyer, 'On Feedback Problems in Digraphs', *in Proceedings of the 15th International Workshop WG'89*, Lecture Notes in Computer Science **411** (1989) 218-231.

Greedy Localization, Iterative Compression, and Modeled Crown Reductions: New FPT Techniques, an Improved Algorithm for Set Splitting, and a Novel $2k$ Kernelization for Vertex Cover

Frank Dehne[1], Mike Fellows[2], Frances Rosamond[2], and Peter Shaw[2]

[1] Griffith University, Brisbane QLD 4111, Australia
frank@dehne.net
[2] University of Newcastle, Callaghan NSW 2308, Australia
{mfellows,fran,pshaw}@cs.newcastle.edu.au

Abstract. The two objectives of this paper are: (1) to articulate three new general techniques for designing FPT algorithms, and (2) to apply these to obtain new FPT algorithms for SET SPLITTING and VERTEX COVER. In the case of SET SPLITTING, we improve the best previous $\mathcal{O}^*(72^k)$ FPT algorithm due to Dehne, Fellows and Rosamond [DFR03], to $\mathcal{O}^*(8^k)$ by an approach based on *greedy localization* in conjunction with *modeled crown reduction*. In the case of VERTEX COVER, we describe a new approach to $2k$ kernelization based on *iterative compression* and *crown reduction*, providing a potentially useful alternative to the Nemhauser-Trotter $2k$ kernelization.

1 Introduction

This paper has a dual focus on: (1) the exposition of some new general FPT algorithm design techniques, and (2) the description of two concrete applications of these techniques to the VERTEX COVER and SET SPLITTING problems. The latter is defined:

SET SPLITTING
Instance: A collection $\mathcal{F}$ of subsets of a finite set X, and a positive integer k.
Parameter: k
Question: Is there a subfamily $\mathcal{F}' \subseteq \mathcal{F}$ and a partition of X into disjoint subsets X_0 and X_1 such that for every $S \in \mathcal{F}', S \cap X_0 \neq \varnothing$ and $S \cap X_1 \neq \varnothing$, with $|\mathcal{F}'| \geq k$?

The SET SPLITTING problem is NP-complete [GJ79] and APX-complete [Pe94]. Andersson and Engebretsen [AE97], and Zhang and Ling [ZL01] presented approximation algorithms that provide solutions within a factor of 0.7240

R. Downey, M. Fellows, and F. Dehne (Eds.): IWPEC 2004, LNCS 3162, pp. 271–280, 2004.
© Springer-Verlag Berlin Heidelberg 2004

and 0.7499, respectively. A 1/2 approximation algorithm for the version of the SET SPLITTING problem where the size of X_0 is specified has been described by Ageev and Sviridenko [AS00].

It is a straightforward exercise to show that SET SPLITTING is fixed parameter tractable by the method of color-coding [AYZ95]. One of the techniques we will discuss, which we refer to here as *greedy localization* was first used by Chen, Friesen, Jia and Kanj [CFJK01] (see also Jia, Zhang and Chen [JZC03]). This approach can potentially be employed in designing FPT algorithms for many different *maximization* problems. In the case of SET SPLITTING we present an example of the deployment of this approach that yields a significant improvement over the best previous $\mathcal{O}^*(72^k)$ FPT algorithm for this problem, due to Dehne, Fellows and Rosamond [DFR03]. Here we combine this technique with *crown reduction* (where the reduction rule is guided by a crown decomposition of an associated graph that models the situation) and obtain an $\mathcal{O}^*(8^k)$ FPT algorithm for SET SPLITTING.

The method of *iterative compression* could be viewed as in some sense "dual" to greedy localization, since it seems to be potentially applicable to a wide range of *minimization* problems. Both of these techniques are in the way of "opening moves" that can be used to develop some initial structure to work with. Neither is technically deep, but still they can be pointed to as potentially of broad utility in FPT algorithm design. A simple application of iterative compression to VERTEX COVER yields a new $2k$ Turing kernelization that may offer practical advantages over the Nemhauser-Trotter $2k$ many:1 kernelization algorithm.

We assume that the reader has a basic familiarity with the fundamental concepts and techniques in the FPT toolkit, as exposited in [DF99, Nie02] (and also with the definition of basic combinatorial problems such as VERTEX COVER). We also assume that the reader is familiar with the research program in "worst-case exponential complexity" articulated in the survey paper by Woeginger [Woe03]. In particular, we employ the handy $\mathcal{O}^*$ notation introduced there for FPT results, that suppresses the polynomial time contribution of the overall input size and focuses attention on the exponential time-complexity contribution of the declared parameter. An FPT algorithm that runs in time $\mathcal{O}^*(8^k)$ thus runs in time $\mathcal{O}(8^k n^c)$ for some constant c independent of the parameter k.

2   The New Techniques

There are three new FPT design techniques to which we wish to draw attention:
- greedy localization
- iterative compression
- modeled crown reductions

The first two are only relatively simple opening moves, but nevertheless these deserve wider recognition in the context of FPT algorithm design.

2.1   Greedy Localization

This is an approach that can often be applied to maximization problems. The idea is to start off with an attempted greedy solution. For example, in the case of the SET PACKING algorithm due to Jia, Zhang and Chen, the first step is to greedily compute a maximal collection of pairwise disjoint sets. If k are found, then of course we are done. Otherwise, we can make the observation that if there is any solution (k pairwise disjoint sets) then every set in the solution must intersect our (small) maximal collection. Thus we have gained some initial information that narrows our search, "localizes" our efforts to this initial structure.

As SET SPLITTING is a maximization problem, we will similarly employ here an opening move that attempts to find a greedy solution, which similarly either succeeds and we are done, or provides us with some initial structure to work with. Greedy localization has been employed in a few other recent FPT algorithms [FHRST04, PS04, MPS04, FKN04].

2.2   Iterative Compression

This "opening move" to develop initial structure seems first to have been used in an FPT algorithm recently described by Reed, Smith and Vetta for the problem of determining for a graph G whether k vertices can be deleted to obtain a bipartite graph G' (an important breakthrough as this problem has been open for some time) [RSV03]. Their approach can be briefly described as follows.

First, the problem is respecified constructively: we aim for an FPT algorithm that either outputs NO, or constructively produces the set of k vertices whose removal will make the graph bipartite.

Second, we attempt a recursive solution (which we will see has a simple iterative interpretation). Choose a vertex v, and call the algorithm on $G - v$. This either returns NO, and we can therefore return NO for G, or it returns a solution set of size k. By adding v to this set, we obtain a solution of size $k + 1$ for G, and what remains to be done is to address the following (constructive) *compression* form of the problem:

Input: G and a solution S of size $k + 1$
Output: Either NO, or a solution of size k, if one exists.

The iterative interpretation is that we are building the graph up, vertex by vertex, and at each step we have a small solution (of size $k + 1$) and attempt to compress it. This interpretation makes clear that our overall running time will be $\mathcal{O}(n \cdot f(n, k))$ where $f(n, k)$ is the running time of our FPT algorithm for the compression form of the problem. Of course, all the real work lies there, but this overall algorithmic approach, simple as it is, gives us some initial structure to work with. The approach is clearly of potential utility for many different minimization problems. (For another recent application see [Ma04].)

2.3   Modeled Crown Reductions

Both of our concrete applications, to the SET SPLITTING and to the VERTEX COVER problems, also use the recently developed techniques of crown decompositions and crown reduction rules. This technique was first introduced by Chor, Fellows and Juedes [CFJ04] (a brief exposition can also be found in the recent survey [F03]). In [CFJ04] the technique is applied to the problems GRAPH COLORING WITH $(n - k)$ COLORS and to VERTEX COVER. Crown reduction has turned out to be effective for a surprisingly wide range of parameterized problems; see also [PS03, FHRST04, PS04, MPS04]. Here we show that crown reductions can even be employed on problems that are not about graphs. Our $\mathcal{O}^*(8^k)$ FPT algorithm for SET SPLITTING employs a kernelization rule that is based on a crown decomposition in an associated auxiliary graph that models some of the combinatorics of the SET SPLITTING problem.

The machinery from [CFJ04] that we employ here is next described.

Definition 1. *A crown decomposition of a graph $G = (V, E)$ is a partition of the vertices of G into three disjoint sets H, C and J with the following properties:*

1. *C is an independent set in G.*
2. *H separates C from J, that is, there are no edges between C and J.*
3. *H is matched into C, that is, there is an injective assignment $m : H \to C$ such that $\forall h \in H$, h is adjacent to $m(h)$.*

The Crown Rule for VERTEX COVER transforms (G, k) into (G', k'), where $G' = G - C - H$, and $k' = k - |H|$. The Crown Rule for the GRAPH COLORING WITH $(n - k)$ COLORS problem is (surprisingly) the same rule applied to $\bar{G}$ [CFJ04].

We will use the following lemma from [CFJ04].

Lemma 1. *If a graph $G = (V, E)$ has an independent set $I \subseteq V(G)$ such that $|N(I)| < |I|$ then a nontrivial crown decomposition (C, H, J) with $C \subseteq I$ for G can be found in time $\mathcal{O}(|V| + |E|)$.*

3   An $\mathcal{O}^*(8^k)$ FTP Algorithm for Set Splitting

The input to the SET SPLITTING problem consists of a family $\mathcal{F} \subseteq 2^X$ of subsets of a base set X, and a positive integer k. We can trivially assume that every set $S \in \mathcal{F}$ consists of at least two elements of X.

The first step of our algorithm is a greedy computation of what we will term a *witness structure* for the instance. The witness structure consists of a collection of sets $\mathcal{F}' \subseteq \mathcal{F}$, and for each of the sets $S_i \in \mathcal{F}'$ a choice of two distinct elements $b_i \in S_i$ and $w_i \in S_i$. It is allowed that these chosen elements may coincide, that is, for $S_i \neq S_j$ possibly $b_i = b_j$ or $w_i = w_j$ (or both). What is also required of the witness structure is that the sets $B = \{b_1, b_2, ..., b_r\}$ of *black* witness elements, and $W = \{w_1, w_2, ..., w_r\}$ of *white* witness elements are disjoint. It is clear that if we succeed in greedily computing a witness structure with $|\mathcal{F}'| = r$, then any extension of the disjoint subsets B and W of X to a bipartition of X will split the r sets in $\mathcal{F}'$.

The first step (greedy localization) is to compute a *maximal* witness structure by the obvious greedy algorithm of repeatedly adding sets to $\mathcal{F}'$ so long as this is possible. If $r \geq k$ then we are done.

At the end of the greedy step, if we are not done, then the following structural claims hold.

Claim 1. Every set S not in the maximal witness structure collection $\mathcal{F}'$ consists entirely of black or entirely of white elements, that is, either $S \subseteq B$ or $S \subseteq W$.

Proof. We have assumed that every set contains at least two elements. Consider a set $S \in \mathcal{F}$ that is not in the maximal witness structure family $\mathcal{F}'$. If $S \subseteq B \cup W$, then clearly either $S \subseteq B$ or $S \subseteq W$ else S is split and could be added to $\mathcal{F}'$. Hence suppose that there is an element $x \in S$, where $x \notin B \cup W$. If S contains an element of B (or W) then x could be assigned to W (or B) and $\mathcal{F}'$ augmented by S, contradicting our assumption that the witness structure is maximal. Since S has at least two elements, the only remaining case is that S contains two distinct elements that do not belong to $B \cup W$. But then, one could be assigned to B and one to W and $\mathcal{F}'$ could again be augmented, contradicting our assumption.

The following claim is obvious (but crucial):

Claim 2. $|B| \leq k - 1$ and $|W| \leq k - 1$.

Our algorithm is described as follows:

Step (1): Greedily compute a maximal witness structure. If k sets are split, then report YES and STOP. (If not, then $|\mathcal{F}'| \leq k - 1$.)

Step (2): Branch on all ways of "recoloring" the (at most) $2(k - 1)$ elements that were colored (placed in either B or W) in the witness structure.

Subproblem
For each recoloring (bipartition) of $B \cup W$ into B' and W'

Step (3): Determine the number of sets that have been split. If k sets are split then report YES and STOP.

Otherwise
Step (4): Generate an auxiliary graph G describing the sets that remain unsplit and the elements of $X - (B \cup W)$ contained in them.

Step (5): Repeatedly apply the Crown Reduction Rule (described below) to the subproblem represented by this graph until a subproblem kernel consisting of at most $(k - 1)$ elements not in $B' \cup W'$ remains.

Step (6): After we have exhausted the ability to reduce the subproblem instance using the Crown Reduction Rule, there can be at most $k - 1$ vertices still remaining to be assigned a color. Try all 2^{k-1} ways to color these elements.

3.1   The Subproblem

After re-coloring (partitioning) $B \cup W$ into B' and W', some number t of sets in $\mathcal{F}$ will have been split (that is, have nonempty intersection with both B' and W'). If $t \geq k$ then of course we will be done (Step 3). Let $\mathcal{G}$ denote the subfamily of $\mathcal{F}$ that is not split by B' and W'. The subproblem is whether the disjoint sets B' and W' can be extended to a bipartition of X that splits k sets. In other words, the subproblem is to determine if the remaining (yet uncolored) elements, those in $X - (B' \cup W')$ can be colored (black and white, extending B' and W') in such a way that at least $k' = k - t$ further sets in $\mathcal{G}$ are split. Note that the fate of any set that is a subset of $B \cup W$ is completely determined by the recoloring into B' and W': it is either among those split, or no extension can split it. Thus in the subproblem, we can restrict our attention (by Claim 1) to the sets in $\mathcal{G}' = \mathcal{G} - 2^{B \cup W} \subseteq \mathcal{F}'$. That is, the only candidates for further splitting belong to our greedy collection $\mathcal{F}'$ (!) and there are at most $k - 1$ of these. We can therefore observe the following claims concerning the subproblem:

Claim 3. Every set in $\mathcal{G}'$ contains either two distinct elements of B' (denote these sets $\mathcal{B}$) or two distinct elements of W' (denote these sets $\mathcal{W}$). Furthermore, every set in $\mathcal{G}'$ contains at least one element of $X - (B' \cup W')$.

Claim 4. $|\mathcal{B} \cup \mathcal{W}| \leq k - 1$.

3.2   Crown Reduction for the Subproblem

The subproblem is modeled by a bipartite graph with vertex sets $V_{\mathcal{B}} \cup V_{\mathcal{W}}$, and $V_{\mathcal{U}}$. The vertices v_S of $V_{\mathcal{B}} \cup V_{\mathcal{W}}$ correspond, respectively, to the unsplit sets S in $\mathcal{B}$ and $\mathcal{W}$. The vertices u_x of $V_{\mathcal{U}}$ correspond to the uncolored elements in $\mathcal{U} = X - (B' \cup W')$. There is an edge between v_S and u_x if and only if $x \in S$. See Figure 2.

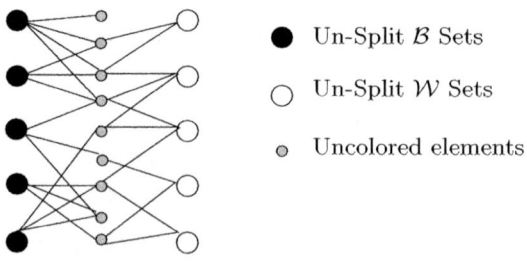

Fig. 1. Auxiliary Graph

The graph model of the subproblem may now be used to locate a crown decomposition that can be used to kernelize the subproblem instance.

By Lemma 1, if $|\mathcal{U}| \geq k$ then we can efficiently compute a nontrivial crown decomposition (C, H, J) with $C \subseteq V_{\mathcal{U}}$. Interpreting what this means for the subproblem instance, we have identified a nonempty subset $\mathcal{H}$ of the unsplit sets in $\mathcal{B} \cup \mathcal{W}$ (the head) that is matched into a subset C of the uncolored elements $\mathcal{U}$, the crown. Furthermore, by the properties of a crown decomposition, the elements of C do not belong to any other unsplit sets in $\mathcal{B} \cup \mathcal{W}$.

We can kernelize the subproblem instance according to the following rule:

Crown Reduction Rule: In the situation described above, we can reduce the subproblem instance by using the matched elements in C to split the sets in $\mathcal{H}$, augmenting B' and W' accordingly. Thus the reduced subproblem instance is modeled by the graph obtained by deleting the vertices that correspond to C and $\mathcal{H}$ and recalculating k'.

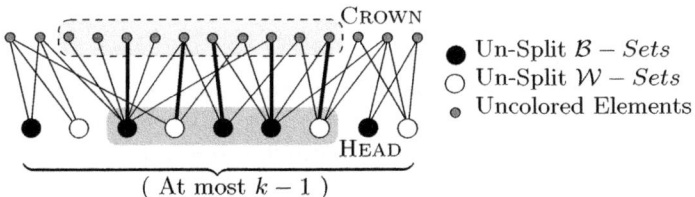

Fig. 2. Crown Decomposition

Lemma 2. *The crown rule can be used to reduce the number of elements not assigned a color to be less than $k - 1$ in polynomial time.*

Proof. Lemma 1 states that if a graph $G = (V, E)$ has an independent set $I \subseteq V(G)$ such that $|N(I)| < |I|$ then G admits a nontrivial crown decomposition where the crown set is a subset of I. As long as the number of elements not assigned a color is greater than $|\mathcal{W} \cup \mathcal{B}|$ we can find a crown in polynomial time. Thus as $|\mathcal{W} \cup \mathcal{B}| \leq k - 1$, by continually applying the crown rule we can reduce the number of elements that still need to be assigned a color to $k - 1$.

3.3   Complexity Analysis of the Algorithm

Theorem 1 *The* SET SPLITTING *problem for parameter k can be solved in $\mathcal{O}^*(8^{k-1})$ time.*

Proof. Finding a maximal witness structure can be performed in $\mathcal{O}(n)$ time. By Lemma 3, $|B \cup W| \leq 2(k - 1)$. The algorithm will branch into at most 4^{k-1} subproblems. Each branch is a completely contained subproblem with the partial splitting of the base set depending on which branch we are on. The Crown Rule kernelization results in a subproblem kernel having at most $k - 1$ uncolored elements. Thus there are at most 2^{k-1} colorings of these to be explored. With at most 4^{k-1} subproblems, each yielding (after kernelization) at most 2^{k-1} branches to explore, we get an $\mathcal{O}^*(8^{k-1})$ FPT algorithm.

4   A New $2k$ Kernelization for Vertex Cover

If we apply the iterative compression technique to the VERTEX COVER problem, then we are concerned with the following *solution compression* form of the problem, which is specified constructively, rather than as a decision problem:

Input: A graph $G = (V, E)$ and a $(k + 1)$-element vertex cover $V' \subseteq V$.
Parameter: k
Output: Either a k-element vertex cover, or NO if none exists.

Lemma 1 guarantees a nontrivial crown decomposition if the number of vertices in $V - V'$ exceeds $k + 1$. Thus we immediately obtain a problem kernel having at most $2(k + 1)$ vertices. This improves the $3k$ kernel based on crown reduction described in [CFJ04].

Note. The astute reader will note that this is not a "kernel" in the usual sense of the word (which is generally taken in the sense of many:1 polynomial-time reductions). Here the $2k$ kernel that we achieve is actually a series of n kernels, which can be formalized as a Turing form of parameterized problem kernelization.

5   Conclusions and Open Problems

We have described some new approaches in the design of FPT algorithms, and have applied these to two concrete problems. The more substantial of these applications is an $\mathcal{O}^*(8^k)$ FPT algorithm for SET SPLITTING, significantly improving the best previous $\mathcal{O}^*(72^k)$ algorithm. While our contribution in the case of VERTEX COVER is really little more than a small observation, it is still somewhat surprising that after so much attention to this problem there is *anything* new to be said about it. Whether $2k$ kernelization via iterative compression and crown reduction has any practical advantages over Nemhauser-Trotter kernelization is an interesting question for further research along the lines of [ACFLSS04], where it is demonstrated that crown reduction is indeed useful in a practical sense. In general, it seems that the articulation of the basic toolkit for FPT algorithm design is still, surprisingly, in its infancy.

Acknowledgement. We thank Daniel Marx for helpful discussions and suggestions, particularly about the iterative compression technique, for which he suggested the name.

References

[ACFLSS04] F. N. Abu-Khzam, R. L. Collins, M. R. Fellows, M. A. Langston, W. H. Suters, and C. T. Symons. Kernelization algorithms for the Vertex Cover problem: theory and experiments, in *Proceedings ALENEX'04*, ACM/SIAM, 2004.

[AS00] A. A. Ageev and M.I. Sviridenko. An approximation algorithm for hypergraph max k-cut with given sizes of parts, in *Proceedings of the European Symposium on Algorithms (ESA) 2000*, Springer-Verlag, *Lecture Notes in Computer Science* 1879 (2000), 32–41.

[AYZ95] N. Alon, R. Yuster, and U. Zwick. Color-Coding, in *Journal of the ACM*, 42 (1995), 844-856.

[AE97] G. Andersson and L. Engebretsen. Better approximation algorithms for set splitting and Not-All-Equal SAT, in *Information Processing Letters*, Vol. 65, pp. 305-311, 1998.

[CFJ04] B. Chor, M. Fellows, and D. Juedes. Linear Kernels in Linear Time, or How to Save k Colors in $O(n^2)$ Steps). *Proceedings WG 2004 - 30th Workshop on Graph Theoretic Concepts in Computer Science*, Springer-Verlag, *Lecture Notes in Computer Science*, 2004 (to appear).

[CFJK01] J. Chen, D. K. Friesen, W. Jia and I. A. Kanj. Using Nondeterminism to Design Efficient Deterministic Algorithms, in *Proceedings 21st Annual Conference on Foundations of Software Technology and Theoretical Computer Science*, Springer-Verlag, *Lecture Notes in Computer Science* 2245 (2001), 120–131. (Journal version to appear in *Algorithmica*.)

[DFR03] F. Dehne, M. Fellows, and F. Rosamond. An *FPT* Algorithm for Set Splitting, in *Proceedings WG 2003 - 29th Workshop on Graph Theoretic Concepts in Computer Science*, Springer-Verlag, *Lecture Notes in Computer Science* 2880 (2003), 180-191.

[DF99] R. G. Downey and M. R. Fellows. *Parameterized Complexity*. Springer-Verlag, 1999.

[F03] M. Fellows. Blow-ups, Win/Win's and Crown Rules: Some New Directions in *FPT*, in *Proceedings WG 2003 - 29th Workshop on Graph Theoretic Concepts in Computer Science*, Springer-Verlag, *Lecture Notes in Computer Science* 2880 (2003), 1-12.

[FKN04] M. Fellows, C. Knauer, N. Nishimura, P. Ragde, F. Rosamond, U. Stege, D. Thilikos and S. Whitesides. Faster Fixed-Parameter Tractable Algorithms for Matching and Packing Problems. *Proceedings of the European Symposium on Algorithms (ESA) 2004*, Springer-Verlag, *Lecture Notes in Computer Science*, 2004 (to appear).

[FHRST04] M. Fellows, P. Heggernes, F. Rosamond, C. Sloper, and J.A. Telle. Exact Algorithms for Finding k Disjoint Triangles in an Arbitrary Graph, to appear in *Proceedings WG 2004 - 30th Workshop on Graph Theoretic Concepts in Computer Science*, Springer-Verlag, *Lecture Notes in Computer Science*, 2004 (to appear).

[GJ79] M. R. Garey and D. S. Johnson. *Computers and Intractability: A Guide to the Theory of NP-Completeness*. W.H. Freeman, 1979.

[JZC03] W. Jia, C. Zhang, and J. Chen. An Efficient Parameterized Algorithm for Set Packing. Manuscript, 2003, to appear in *Journal of Algorithms*.

[Ma04] D. Marx. Chordal Deletion is Fixed-Parameter Tractable. Manuscript, 2004.

[MPS04] L. Mathieson, E. Prieto, and P. Shaw. Packing Edge Disjoint Triangles: A Parameterized View. *Proceedings of the International Workshop on Parameterized and Exact Computation*, Springer-Verlag, *Lecture Notes in Computer Science* (this volume), 2004.

[Nie02] R. Niedermeier. *Invitation to Fixed-Parameter Algorithms,* Habilitationschrift, University of Tubingen, 2002.

[Pe94] E. Petrank. The hardness of approximation: Gap location. *Computational Complexity*, 4 (1994), 133–157.

[PS03] E. Prieto and C. Sloper. Either/Or: Using Vertex Cover Structure in Designing *FPT* Algorithms–the Case of *k*-Internal Spanning Tree. *Proceeding of the Workshop on Algorithms and Data Structures (WADS) 2003*, Springer-Verlag, *Lecture Notes in Computer Science* 2748 (2003), 474-483.

[PS04] E. Prieto and C. Sloper. Looking at the Stars. *Proceedings of the International Workshop on Parameterized and Exact Computation*, Springer-Verlag, *Lecture Notes in Computer Science* (this volume), 2004.

[RSV03] B. Reed, K. Smith, and A. Vetta. Finding Odd Cycle Transversals. *Operations Research Letters* 32 (2004), 299–301.

[Woe03] G. J. Woeginger. Exact Algorithms for NP-Hard Problems: A Survey. *Proceedings of 5th International Workshop on Combinatorial Optimization-Eureka, You Shrink! Papers dedicated to Jack Edmonds,* M. Junger, G. Reinelt, and G. Rinaldi (Festschrift Eds.) Springer-Verlag, *Lecture Notes in Computer Science* 2570 (2003), 184-207.

[ZL01] H. Zhang and C.X. Ling. An Improved Learning Algorithm for Augmented Naive Bayes. *Proceedings of the Pacific-Asia Conference on Knowledge Discovery and Data Mining (PAKDD)*, Springer-Verlag, *Lecture Notes in Computer Science* 2035 (2001), 581–586.

Space and Time Complexity of Exact Algorithms: Some Open Problems

Gerhard J. Woeginger

TU Eindhoven, The Netherlands
gwoegi@win.tex.nl

Abstract. We discuss open questions around worst case time and space bounds for NP-hard problems. We are interested in exponential time solutions for these problems with a relatively good worst case behavior.

1 Introduction

Every problem in NP can be solved in exponential time by exhaustive search: Recall that a decision problem is in NP, if and only if there exists a polynomial time decidable relation $R(x, y)$ and a polynomial $m(|x|)$ such that for every YES-instance x, there exists a YES-certificate y with $|y| \leq m(x)$ and $R(x, y)$. A trivial exact algorithm for solving instance x enumerates all possible strings y with lengths up to $m(|x|)$, and checks whether any of them yields a YES-certificate. Up to polynomial factors that depend on the evaluation time of $R(x, y)$, this yields an exponential running time of $2^{m(x)}$.

A natural question is: Can we do better than this trivial enumerative algorithm? Interestingly, for many combinatorial optimization problems the answer is YES. Early examples include an $O^*(1.4422^n)$ algorithm for deciding 3-colorability of an n-vertex graph by Lawler [21]; an $O^*(1.2599^n)$ algorithm for finding a maximum independent set in an n-vertex graph by Tarjan & Trojanowski [24]; an $O^*(1.4142^n)$ algorithm for the SUBSET-SUM problems with n integers by Horowitz & Sahni [18]. (The notation $O^*(f(n))$ is explained at the end of this section.) Woeginger [26] surveys many results in this area.

For some optimization problems, we can reach an improved time complexity, but it seems that we have to pay for this with an *exponential* space complexity. Note that algorithms with exponential space complexities are absolutely useless for real life applications. In this paper, we discuss a number of results around fast exponential time algorithms that come with exponential space complexities. We present approaches, tricks, related polynomially solvable problems, and related open questions.

Notation. Throughout this paper, we will use a modified big-Oh notation that suppresses polynomially bounded terms. For a positive real constant c, we write $O^*(c^n)$ for a time complexity of the form $O(c^n \cdot \text{poly}(n))$. The notations $\Omega^*(c^n)$ and $\Theta^*(c^n)$ are defined analogously.

R. Downey, M. Fellows, and F. Dehne (Eds.): IWPEC 2004, LNCS 3162, pp. 281–290, 2004.

2 Integers and Their Sums

We start this section with a couple of polynomially solvable problems: An input to the first problem "k-SUM" consists of m integers $a_1, \ldots, a_m$ and a goal sum S. The problem is to decide whether there are k of these integers that add up to S. An input to the second problem "Table-k-SUM" consists of a $k \times m$ table and a goal sum S; the entries in row i of the table are denoted by $R_i(1), \ldots, R_i(m)$. The problem is to decide whether one can choose k integers from this table, exactly one from each row, that add up to S. In both problems, the number k is a fixed integer that is not part of the input. Both problems are closely related, and they can be reduced to each other in linear time (Erickson [12]). Both problems are trivially solvable in polynomial time $O(m^k)$.

Here is how to get a better time complexity for Table-2-SUM: Sort the entries in the first row. Then for $j = 1, \ldots, m$ perform a binary search for the value $S - R_2(j)$ in this sorted first row. If the search succeeds at $R_1(i)$, then $R_1(i) = S - R_2(j)$ and the answer is YES. If all searches fail, then the answer is NO.

Fact. *Table-2-SUM can be solved in $O(m \log m)$ time and $O(m)$ space.*

The same approach also yields fast algorithms for Table-k-SUM for all $k \geq 3$: Compute the sum of every $\lceil k/2 \rceil$-tuple of integers that has one entry in each of the first $\lceil k/2 \rceil$ rows; these sums form the first row in a new table. Compute the sum of every $\lfloor k/2 \rfloor$-tuple of integers that has one entry in each of the last $\lfloor k/2 \rfloor$ rows; these sums form the second row in the new table. Apply the above algorithm to this new instance of Table-2-SUM.

Fact. *Table-k-SUM can be solved in $O(m^{\lceil k/2 \rceil} \log m)$ time and $O(m^{\lceil k/2 \rceil})$ space.*

For odd k, the time complexity can be slightly improved to $O(m^{\lceil k/2 \rceil})$; see for instance Erickson [11]. In particular, the 3-SUM problem can be solved in $O(m^2)$ time. We will not go into details, since in this paper we really do not care about logarithmic factors. The main drawback of all these algorithms is their horrible space complexity.

Schroeppel & Shamir [23] improve the space complexity for Table-4-SUM by using a data structure that enumerates the m^2 sums $R_1(i) + R_2(j)$ with $1 \leq i, j \leq m$ in non-decreasing order. This data structure uses only $O(m)$ space. Every time we kick it, it starts working for $O(\log m)$ time steps, and then spits out the next larger sum $R_1(i) + R_2(j)$. The data structure is based on a balanced search tree that supports deletions, insertions, and extracting the minimum with logarithmic work per operation. It is built as follows: In a preprocessing step, we bring the entries in the second row into non-decreasing order. As a consequence, we have for every fixed index i that

$$R_1(i) + R_2(1) \ \leq \ R_1(i) + R_2(2) \ \leq \ \cdots \ \leq \ R_1(i) + R_2(m).$$

For every index i ($1 \leq i \leq m$), the data structure stores the pair (i, j) that corresponds to the first unvisited sum $R_1(i) + R_2(j)$ in this ordering. Whenever the data structure is kicked, it extracts and deletes the pair (i, j) with minimum sum, and inserts the pair $(i, j + 1)$ instead. All in all, the enumeration of the m^2 sums costs $O(m^2 \log m)$ time.

Schroeppel & Shamir [23] use two such data structures; the first one generates the sums $x = R_1(i) + R_2(j)$ in non-decreasing order, whereas the second one generates the sums $y = R_3(s) + R_4(t)$ in non-increasing order. Whenever $x + y < S$ holds, the current value of x is too small for reaching the goal sum S; we replace it by the next larger sum $R_1(i) + R_2(j)$ from the first data structure. Whenever $x + y > S$ holds, the current value of y is too large for reaching the goal sum S; we replace it by the next smaller sum $R_3(s) + R_4(t)$ from the second data structure. These steps are repeated over and over again, until one data structure becomes empty (answer NO) or until we reach $x + y = S$ (answer YES).

Fact. *Table-4-SUM can be solved in $O(m^2 \log m)$ time and $O(m)$ space.*

Open problem 1

(a) Is there an $O(m^3 \log m)$ time and $O(m)$ space algorithm for Table-6-SUM?

(b) Is there an $O(m^{\lceil k/2 \rceil - \alpha})$ time algorithm for Table-k-SUM for some integer $k \geq 3$ and some real $\alpha > 0$?

Now let us turn to negative results around the k-SUM and the Table-k-SUM problem. The 3-SUM problem plays a notorious role in computational geometry. Gajentaan & Overmars [15] have put together a long list of geometric problems: All problems on this list can be solved in quadratic time, and for all of them nobody knows how to do better. All problems on this list contain 3-SUM as a special case (under linear time reductions), and for all of them this 3-SUM special case (intuitively) seems to be the main obstacle for breaking through the quadratic time barrier. One example problem on this list is: Given m (possibly overlapping) triangles in the Euclidean plane, compute the area of their union. Another one: Given m pairwise non-intersecting straight line segments in the Euclidean plane, is there a straight line that separates them into two non-empty subsets? And another one: Given m points in the Euclidean plane, are some three of them on a common line? For instance, the linear time reduction from 3-SUM to 3-POINTS-ON-A-COMMON-LINE is based on the following observation: The x-coordinates of the intersection points of the line $y = ax + b$ with the curve $y = f(x) = x^3 - Sx^2$ are the roots of $x^3 - Sx^2 - ax - b = 0$; for every line the sum of these roots equals S, the coefficient of the quadratic term. Consequently, the point set $(a_1, f(a_1)), (a_2, f(a_2)), \ldots, (a_m, f(a_m))$ contains three points $(a_x, f(a_x)), (a_y, f(a_y)), (a_z, f(a_z))$ on a common line, if and only if $a_x + a_y + a_z = S$. The bottom-line of this paragraph is that research on the 3-SUM problem is severely stuck at the threshold $O(m^2)$.

What about the general k-SUM problem with $k \geq 4$? Here we are stuck around the threshold $O(m^{\lceil k/2 \rceil})$. Erickson [11] proved an $\Omega(m^{\lceil k/2 \rceil})$ lower bound on k-SUM in a certain restricted variant of the linear decision tree model. The additional restriction in his model is that every decision step must be based on testing the sign of some affine linear combination of at most k elements of the input. At first sight, this model seems to be strange, and the lower bound result seems to be quite weak. However, given our general failure in proving reasonable lower bounds for algorithmic problems and given the lack of tools in this area, Erickson's lower bound result in fact is a major breakthrough.

Open problem 2 *Prove a non-trivial lower bound for the k-SUM problem in the algebraic decision tree model or in the algebraic computation tree model (see Ben-Or [4]).*

Downey & Fellows [7,8] have proved that the k-SUM problem with parameter k is W[1]-hard. All these negative results for k-SUM translate into analogous negative results for Table-k-SUM.

After this long polynomial time prelude, we will spend the rest of this section on NP-hard problems. In the NP-hard SUBSET-SUM problem, the input consists of n positive integers $b_1, \ldots, b_n$ and a goal sum B. The problem is to decide whether there exists some subset of the b_i that add up to B. The strongest known negative result for SUBSET-SUM is an $\Omega(n^2)$ lower bound in the algebraic computation tree model of computation [6,4].

On the positive side, Horowitz & Sahni [18] have come up with the following approach for SUBSET-SUM: They split the instance into two parts, one part with $b_1, \ldots, b_{\lfloor n/2 \rfloor}$ and another part with $b_{\lfloor n/2 \rfloor+1}, \ldots, b_n$. They construct a table with two rows, where the first row consists of all the subset sums for the first part, and where the second row consists of all the subset sums for the second part. The table can be computed in $O^*(2^{n/2})$ time. The SUBSET-SUM instance has answer YES, if and only if the constructed Table-2-SUM instance with $S = B$ has answer YES. Our above discussion of Table-2-SUM yields the following result.

Fact. *SUBSET-SUM can be solved in $O^*(2^{n/2})$ time and in $O^*(2^{n/2})$ space.*

Schroeppel & Shamir [23] follow essentially the same idea, but instead of splitting the SUBSET-SUM instance into two parts, they split it into four parts of size approximately $n/4$. This leads to a corresponding instance of Table-4-SUM, and to a substantially improved space complexity.

Fact. *SUBSET-SUM can be solved in $O^*(2^{n/2})$ time and in $O^*(2^{n/4})$ space.*

Generally, if we split the SUBSET-SUM instance into $k \geq 2$ parts, then we get a corresponding table with k rows and $O(2^{n/k})$ elements per row. Applying the fastest known algorithm to the corresponding instance of Table-k-SUM gives a time complexity of $O^*(2^{f(n,k)})$ with $f(n,k) = n \lceil k/2 \rceil / k \geq n/2$. Hence, this approach will not easily lead to an improvement over the time complexity $O^*(2^{n/2})$. Schroeppel & Shamir [23] also construct $t(n)$ time and $s(n)$ space algorithms for SUBSET-SUM for all $s(n)$ and $t(n)$ with $\Omega^*(2^{n/2}) \leq t(n) \leq O^*(2^n)$ and $s^2(n) \cdot t(n) = \Theta^*(2^n)$.

Open problem 3

(a) *Construct an $O^*(1.4^n)$ time algorithm for SUBSET-SUM.*
(b) *Construct an $O^*(1.99^n)$ time and polynomial space algorithm for SUBSET-SUM.*

(c) We have seen that positive results for Table-k-SUM yield positive results for SUBSET-SUM. Can we establish some reverse statement? Do fast (exponential time) algorithms for SUBSET-SUM yield fast (polynomial time) algorithms for Table-k-SUM?

Another NP-hard problem in this area is the EQUAL-SUBSET-SUM problem: Given n positive integers $b_1, \ldots, b_n$, do there exist two disjoint non-empty subsets of the b_i that both have the same sum. A translation of EQUAL-SUBSET-SUM into a corresponding Table-4-SUM instance leads to an $O^*(2^n)$ algorithm for EQUAL-SUBSET-SUM. It might be interesting to design faster algorithms for EQUAL-SUBSET-SUM, and to get some understanding of the relationship between fast algorithms for SUBSET-SUM and fast algorithms for EQUAL-SUBSET-SUM.

3   Graphs and Their Cliques and Cuts

We start this section with the polynomially solvable k-CLIQUE problem: An input consists of an undirected, simple, loopless p-vertex graph $G = (V, E)$. The problem is to decide whether G contains a clique on k vertices. We stress that k is not part of the input. The k-CLIQUE problem is easily solved in polynomial time $O(p^k)$.

Itai & Rodeh [19] observed that fast matrix multiplication can be used to improve this time complexity for 3-CLIQUE: Recall that the product of two $p \times p$ matrices can be computed in $O(p^\omega)$ time, where $\omega < 2.376$ denotes the so-called *matrix multiplication exponent*; see Coppersmith & Winograd [5]. Recall that in the ℓth power A^ℓ of the adjacency matrix A of graph G, the entry at the intersection of row i and column j counts the number of walks with $\ell + 1$ vertices in G that start in vertex i and end in vertex j. Furthermore, a 3-clique $\{x, y, z\}$ yields a walk $x - y - z - x$ with four vertices from x to x, and vice versa, every walk with four vertices from vertex x to vertex x corresponds to a 3-clique. Hence, G contains a 3-clique if and only if A^3 has a non-zero entry on its main-diagonal.

Fact. *The 3-CLIQUE problem for a p-vertex graph can be solved in $O(p^\omega)$ time (where $\omega < 2.376$ is the matrix multiplication exponent) and in $O(p^2)$ space.*

Nešetřil & Poljak [22] extend this idea to the $3k$-CLIQUE problem: For every k-clique C in G, create a corresponding vertex $v(C)$ in an auxiliary graph. Two vertices $v(C_1)$ and $v(C_2)$ are connected by an edge in the auxiliary graph, if and only if $C_1 \cup C_2$ forms a $2k$-clique in G. Note that the auxiliary graph has $O(p^k)$ vertices. Furthermore, graph G contains a $3k$-clique if and only if the auxiliary graph contains a 3-clique.

Fact. *The $3k$-CLIQUE problem for a p-vertex graph can be solved in $O(p^{\omega k})$ time and $O(p^{2k})$ space.*

This approach yields a time complexity of $O(p^{\omega k+1})$ for $(3k+1)$-CLIQUE, and a time complexity of $O(p^{\omega k+2})$ for $(3k+2)$-CLIQUE. Eisenbrand & Grandoni [9] slightly improve on these bounds for $(3k+2)$-CLIQUE (with $k \geq 2$) and for

$(3k+1)$-CLIQUE (with $1 \leq k \leq 5$). In particular, for 4-CLIQUE [9] gives a time complexity of $n^{3.334}$.

Open problem 4

(a) Design algorithms with better time and/or space complexities for k-CLIQUE!
(b) Is there an $O(p^{7.5})$ time algorithm for 10-CLIQUE?
(c) Is 3-CLIQUE as difficult as Boolean matrix multiplication?

On the negative side, we have that the k-CLIQUE problem with parameter k is W[1]-hard (Downey & Fellows [7,8]). For the variant where k is part of the input and $k \approx \log n$, Feige & Kilian [14] show that a polynomial time algorithm highly unlikely to exist.

Now let us turn to NP-hard problems. In the MAX-CUT problem, the input consists of an n-vertex graph $G = (V, E)$. The problem is to find a cut of maximum cardinality, that is, a subset $X \subseteq V$ of the vertices that maximizes the number of edges between X and $V - X$. The MAX-CUT problem can be solved easily in $O^*(2^n)$ time by enumerating all possible certificates X. Fedin & Kulikov [13] present an $O^*(2^{|E|/4})$ time algorithm for MAX-CUT; however, it seems a little bit strange to measure the time complexity for this problem in terms of $|E|$ and not in terms of $n = |V|$.

Williams [25] developed the following beautiful approach for MAX-CUT: We partition the vertex set V into three parts V_0, V_1, V_2 that are of roughly equal cardinality $n/3$. We introduce a complete tri-partite auxiliary graph that contains one vertex for every subset $X_0 \subseteq V_0$, one vertex for every subset $X_1 \subseteq V_1$, and one vertex for every subset $X_2 \subseteq V_2$. For every subset $X_i \subseteq V_i$ and every $X_j \subseteq V_j$ with $j = i+1 \pmod 3$, we introduce the directed edge from X_i to X_j. This edge receives a weight $w(X_i, X_j)$ that equals the number of edges in G between X_i and $V_i - X_i$ plus the number of edges between X_i and $V_j - X_j$ plus the number of edges between X_j and $V_i - X_i$. Note that for $X_i \subseteq V_i$ ($i = 0, 1, 2$) the cut $X_0 \cup X_1 \cup X_2$ cuts exactly $w(X_0, X_1) + w(X_1, X_2) + w(X_2, X_0)$ edges in G. Consequently, the following three statements are equivalent:

- The graph G contains a cut with z edges.
- The auxiliary graph contains a 3-clique with total edge weight z.
- There exist non-negative integers z_{01}, z_{12}, z_{20} with $z_{01} + z_{12} + z_{20} = z$, such that the auxiliary graph contains a 3-clique on three vertices $X_i \subseteq V_i$ ($i = 0, 1, 2$) with $w(X_0, X_1) = z_{01}$ and $w(X_1, X_2) = z_{12}$ and $w(X_2, X_0) = z_{20}$.

The condition in the third statement is easy to check: There are $O(|E|^3)$ possible triples (z_{01}, z_{12}, z_{20}) to consider. For each such triple, we compute a corresponding simplified version of the auxiliary graph that only contains the edges of weight z_{ij} between vertices $X_i \subseteq V_i$ and $X_j \subseteq V_j$. Then everything boils down to finding a 3-clique in the simplified auxiliary graph on $O(2^{n/3})$ vertices.

Fact. *The MAX-CUT problem can be solved in $O^*(2^{\omega n/3})$ time and $O^*(2^{\omega n/3})$ space. Note that $2^{\omega n/3} < 1.732^n$.*

Of course, William's algorithm could also be built around a partition of the vertex set V into four parts of roughly equal cardinality $n/4$, or around a partition of the vertex set V into k parts of roughly equal cardinality n/k. The problem then boils down to finding a k-clique in some simplified auxiliary graph on $O(2^{n/k})$ vertices. With the currently known k-CLIQUE algorithms, this will not give us an improved time complexity.

Open problem 5

(a) Design a faster exact algorithm for MAX-CUT.
(b) Construct an $O^*(1.99^n)$ time and polynomial space algorithm for MAX-CUT.

An input of the NP-hard BISECTION problem consists of an n-vertex graph $G = (V, E)$. The problem is to find a subset $X \subseteq V$ with $|X| = n/2$ that minimizes the number of edges between X and $V - X$. The approach of Williamson yields an $O^*(2^{\omega n/3})$ time algorithm for BISECTION. Can you do better?

4   Sets and Their Subsets

There is a number of exact algorithms in the literature that attack an NP-hard problem by running through all the subsets of an underlying n-element ground set, while generating and storing useful auxiliary information. Since an n-element ground set has 2^n subsets, the time complexities of these approaches are typically $\Omega^*(2^n)$. And also the space complexities of these approaches are typically $\Omega^*(2^n)$, since they store and remember auxiliary information for every subset.

A good example for this approach is the famous dynamic programming algorithm of Held & Karp [17] for the travelling salesman problem (TSP): A travelling salesman has to visit the cities 1 to n. He starts in city 1, runs through the cities $2, 3, \ldots, n-1$ in arbitrary order, and finally stops in city n. The distance $d(i, j)$ from city i to city j is specified as part of the input. The goal is to find a path that minimizes the total travel length of the salesman. The dynamic program of Held & Karp [17] introduces for every non-empty subset $S \subseteq \{2, \ldots, n-1\}$ of the cities and for every city $i \in S$ a corresponding state $[S; i]$. By LENGTH$[S; i]$ we denote the length of the shortest path that starts in city 1, then visits all cities in $S - \{i\}$ in arbitrary order, and finally stops in city i. Clearly, LENGTH$[\{i\}; i] = d(1, i)$ holds for every $i \in \{2, \ldots, n-1\}$. And for every $S \subseteq \{2, \ldots, n-1\}$ with $|S| \geq 2$ we have

$$\text{LENGTH}[S; i] \;=\; \min\left\{\text{LENGTH}[S - \{i\}; j] + d(j, i) : \; j \in S - \{i\}\right\}.$$

By processing the subsets S in order of increasing cardinality, we can compute the value LENGTH$[S; i]$ in time proportional to $|S|$. In the end, the optimal travel length is given as the minimum $\min_{2 \leq k \leq n-1}$ LENGTH$[\{2, \ldots, n-1\}; k] + d(k, n)$.

Fact. *The TSP can be solved in $O^*(2^n)$ time and $O^*(2^n)$ space.*

Open problem 6

(a) *Construct an exact algorithm for the n-city TSP with $O^*(1.99^n)$ time complexity!*

(b) *Construct an exact algorithm for the n-city TSP with $O^*(2^n)$ time complexity and polynomial space complexity!*

In the Hamiltonian path problem, we have to decide for a given graph $G = (V, E)$ with vertices $1, \ldots, n$ whether it contains a spanning path starting in vertex 1 and ending in vertex n. The Hamiltonian path problem forms a simpler special case of the TSP. Karp [20] (and independently Bax [1]) provided a cute solution for the restriction of Problem 6.(b) to this Hamiltonian special case. We use the following definitions. A *walk* in a graph is a sequence $v_1, \ldots, v_k$ of vertices such that every pair of consecutive vertices is connected by an edge; vertices and edges may show up repeatedly in a walk. For a subset $S \subseteq V$ we denote by WALK(S) the set of all walks with n vertices in G that start in vertex 1, end in vertex n, and avoid all the vertices in S. Let A be the adjacency matrix of $G - S$. Recall that in the kth power A^k of A, the entry at the intersection of row i and column j counts the number of walks with $k + 1$ vertices in $G - S$ that start in vertex i and end in vertex j. Therefore, the number of walks in WALK(S) can be read from matrix A^{n-1}:

Fact. *For every subset $S \subseteq V$, the cardinality $|\text{WALK}(S)|$ can be determined in polynomial time.*

If a walk through n vertices in G does not avoid any vertex k, then it must visit all the vertices, and hence must form a Hamiltonian path. Consequently, the number of Hamiltonian paths from 1 to n in G equals

$$|\text{WALK}(\emptyset)| \; - \; \left| \bigcup_{k=2}^{n-1} \text{WALK}(\{k\}) \right| \;\; = \;\; \sum_{S \subseteq V} (-1)^{|S|} \cdot |\text{WALK}(S)|.$$

Here we have used the inclusion-exclusion principle. The sum in the right hand side of the displayed equation is straightforward to compute by applying the fact discussed above. We only need to remember the partial sum of all the terms evaluated so far, and the space used for evaluating one term can be reused in evaluating the later terms. All in all, evaluating and adding up the values of $O(2^n)$ terms yields an $O^*(2^n)$ time and polynomial space algorithm for *counting* the number of Hamiltonian paths. The following fact is a trivial consequence of this:

Fact. *The Hamiltonian path problem in an n-vertex graph can be solved in $O^*(2^n)$ time and polynomial space.*

Eppstein [10] improves on this polynomial space result for Hamiltonian path in the special case of *cubic* graphs: He presents an algorithm that uses $O^*(1.297^n)$ time and linear space. Bax [2] and Bax & Franklin [3] have extended the inclusion-exclusion approach to a number of counting problems around paths and cycles in n-vertex graphs. For all these problems, the time complexity is $O^*(2^n)$ and the space complexity is polynomial.

Open problem 7 *Construct $O^*(1.99^n)$ time exact algorithms for the following counting problems in n-vertex graphs G:*

(a) Count the number of paths between a given pair of vertices in G.
(b) Count the number of cycles in G.
(c) Count the number of cycles through a given vertex in G.
(d) Count the number of cycles of a given length ℓ in G.

Now let us turn to some relatives of the n-city TSP. For a fixed Hamiltonian path from city 1 to city n and for a fixed city k, we denote by the *delay* of city k the length of the subpath between city 1 and city k. In the travelling repairman problem (TRP), the goal is to find a Hamiltonian path from city 1 to city n that minimizes the sum of delays over all cities. In the precedence constrained travelling repairman problem (prec-TRP), the input additionally specifies a partial order on the cities. A Hamiltonian path is feasible, if it obeys the partial order constraints.

Here is a related scheduling problem SCHED: There are n jobs $1, \ldots, n$ with processing times $p_1, \ldots, p_n$. The jobs are partially ordered (precedence constrained), and if job i precedes job j in the partial order, then i must be processed to completion before j can begin its processing. All jobs are available at time 0, and job preemption is not allowed. The goal is to schedule the jobs on a single machine such that all precedence constraints are obeyed and such that the total job completion time $\sum_{j=1}^{n} C_j$ is minimized; here C_j is the time at which job j completes in the given schedule. SCHED is the special case of prec-TRP where the distances between cities $i \neq j$ are given by $d(i, j) = p_j$. It is quite straightforward to design an $O^*(2^n)$ time and $O^*(2^n)$ space dynamic programming algorithm for prec-TRP (and for its special cases TRP and SCHED).

Open problem 8

(a) Construct an $O^*(1.99^n)$ time exact algorithm for TRP or for SCHED or for prec-TSP.
(b) Provide evidence in favor of or against the following claim: If there exists an $O^*(c^n)$ time exact algorithm with $c < 2$ for one of the four problems TSP, TRP, SCHED, prec-TSP, then there exist $O^*(c^n)$ time exact algorithms for all four problems.

References

1. E.T. BAX (1993). Inclusion and exclusion algorithm for the Hamiltonian path problem. *Information Processing Letters 47*, 203–207.
2. E.T. BAX (1994). Algorithms to count paths and cycles. *Information Processing Letters 52*, 249–252.
3. E.T. BAX AND J. FRANKLIN (1996). A finite-difference sieve to count paths and cycles by length. *Information Processing Letters 60*, 171–176.
4. M. BEN-OR (1983). Lower bounds for algebraic computation trees. In *Proceedings of the 15th Annual ACM Symposium on the Theory of Computing (STOC'1983)*, 80–86.

5. D. COPPERSMITH AND S. WINOGRAD (1990). Matrix multiplication via arithmetic progressions. *Journal of Symbolic Computation 9*, 251–280.

6. D. DOBKIN AND R.J. LIPTON (1978). A lower bound of $\frac{1}{2}n^2$ on linear search programs for the knapsack problem. *Journal of Computer and System Sciences 16*, 413–417.

7. R.G. DOWNEY AND M.R. FELLOWS (1995). Fixed-parameter tractability and completeness II: On completeness for W[1]. *Theoretical Computer Science 141*, 109–131.

8. R.G. DOWNEY AND M.R. FELLOWS (1999). *Parameterized complexity*. Springer Monographs in Computer Science.

9. F. EISENBRAND AND F. GRANDONI (2004). On the complexity of fixed parameter clique and dominating set. *Theoretical Computer Science*, to appear.

10. D. EPPSTEIN (2003). The traveling salesman problem for cubic graphs. In *Proceedings of the 8th International Workshop on Algorithms and Data Structures (WADS'2003)*, Springer-Verlag, LNCS 2748, 307–318.

11. J. ERICKSON (1999). Lower bounds for linear satisfiability problems. *Chicago Journal of Theoretical Computer Science 1999(8)*.

12. J. ERICKSON (2004). Private communication.

13. S.S. FEDIN AND A.S. KULIKOV (2002). Solution of the maximum cut problem in time $2^{|E|/4}$. (In Russian). *Zapiski Nauchnykh Seminarov Sankt-Peterburgskoe Otdeleniya Matematicheskiĭ Institut imeni V.A. Steklova 293*, 129–138.

14. U. FEIGE AND J. KILIAN (1997). On limited versus polynomial nondeterminism. *Chicago Journal of Theoretical Computer Science 1997*.

15. A. GAJENTAAN AND M.H. OVERMARS (1995). On a class of $O(n^2)$ problems in computational geometry. *Computational Geometry 5*, 165–185.

16. M.R. GAREY AND D.S. JOHNSON (1979). *Computers and Intractability: A Guide to the Theory of NP-Completeness*. Freeman, San Francisco.

17. M. HELD AND R.M. KARP (1962). A dynamic programming approach to sequencing problems. *Journal of SIAM 10*, 196–210.

18. E. HOROWITZ AND S. SAHNI (1974). Computing partitions with applications to the knapsack problem. *Journal of the ACM 21*, 277–292.

19. A. ITAI AND M. RODEH (1978). Finding a minimum circuit in a graph. *SIAM Journal on Computing 7*, 413–423.

20. R.M. KARP (1982). Dynamic programming meets the principle of inclusion and exclusion. *Operations Research Letters 1*, 49–51.

21. E.L. LAWLER (1976). A note on the complexity of the chromatic number problem. *Information Processing Letters 5*, 66–67.

22. J. NEŠETŘIL AND S. POLJAK (1985). On the complexity of the subgraph problem. *Commentationes Mathematicae Universitatis Carolinae 26*, 415–419.

23. R. SCHROEPPEL AND A. SHAMIR (1981). A $T = O(2^{n/2})$, $S = O(2^{n/4})$ algorithm for certain NP-complete problems. *SIAM Journal on Computing 10*, 456–464.

24. R.E. TARJAN AND A.E. TROJANOWSKI (1977). Finding a maximum independent set. *SIAM Journal on Computing 6*, 537–546.

25. R. WILLIAMS (2004). A new algorithm for optimal constraint satisfaction and its implications. In *Proceedings of the 31st International Colloquium on Automata, Languages and Programming (ICALP'2004)*, Springer Verlag, 2004.

26. G.J. WOEGINGER (2003). Exact algorithms for NP-hard problems: A survey. In *Combinatorial Combinatorial Optimization – Eureka, you shrink!"*, LNCS 2570, Springer Verlag, 185–207.

Practical FPT Implementations and Applications

Mike Langston

University of Tennessee and Oak Ridge National Laboratory

Abstract. When combined with high performance computational architectures, methods born of FPT can be used as a practical basis for launching systematic attacks on large-scale combinatorial problems of significance. Efficient sequential techniques for kernelization and highly parallel algorithms for branching will be discussed. The importance of maintaining a balanced decomposition of the search space turns out to be critical to achieving scalability. Applications abound, perhaps most notably in high-throughput computational biology. A toolchain will be described that transforms immense quantities of mRNA microarray data into instances of the clique problem, which are then solved via vertex cover to derive sets of putatively co-regulated genes. This makes it possible to narrow the search for cis and trans regulatory structures on scales that were previously unthinkable.

R. Downey, M. Fellows, and F. Dehne (Eds.): IWPEC 2004, LNCS 3162, pp. 291–291, 2004.
© Springer-Verlag Berlin Heidelberg 2004

Author Index

Lecture Notes in Computer Science

For information about Vols. 1–3116

please contact your bookseller or Springer